Study Guide

to accompany

Basic Engineering Circuit Analysis

Sixth Edition

J. David Irwin
Auburn University

Chwan-Hwa Wu
Auburn University

Prepared by
Bill Dillard
Auburn University

JOHN WILEY & SONS, INC.
NEW YORK • CHICHESTER • WEINHEIM • BRISBANE • SINGAPORE • TORONTO

To order books or for customer service call 1-800-CALL-WILEY (225-5945).

ISBN 0-471-36648-X

Printed in the United States of America

10 9 8 7 6 5 4 3 2 1

Printed and bound by Victor Graphics, Inc.

Preface

This study guide is an ancillary text to the sixth edition of *Basic Engineering Circuit Analysis*, hereafter referred to as BECA, by J. David Irwin and John Wu of Auburn University. The content and format here have been carefully selected with four goals in mind.

1. The format of the guide should follow that of the text. Accordingly, chapter titles in the guide and in BECA are the same.
2. Information in the guide should support the text, not repeat it.
3. The scope of the contents should be considerable. We have included solved example problems, demonstrative computer simulations and multimedia presentations.
4. Computer simulation techniques should treated in more depth. These include PSPICE, Electronics Workbench, MATLAB and EXCEL.

A CD-ROM is bundled with the guide, which includes most of the abovementioned simulations and demonstrations. Also on the CD ROM is a folder called Visual Tutors. The files in that folder are executable videos that demonstrate some of the key PSpice examples within this study guide. Since Visual Tutor files are executables, you don't need video software to view them. Simply run them as you would any other executable file. Finally, the CD contains a folder called BECA libraries which contains four PSpice library files. These libraries contain some custom parts we have created for your enjoyment. Procedures for adding the BECA libraries to PSpice are discussed in Chapter Four of this guide. Each of the major file types on the CD-ROM and the software version used to create them are listed in the table below. Compatibility with older or newer versions is not guaranteed.

FILE TYPE	CREATED WITH
PSPICE	VERSION 8.0
ELECTRONICS BENCH	VERSION 5.0
EXCEL	VERSION 7.0
WORD	VERSION 7.0
MATLAB	VERSION 5.1

Trademarks

PSPICE is a registered trademark of ORCAD Inc.

Electronics Workbench is a registered trademark of Interactive Image Technologies, Ltd.

EXCEL and Word are registered trademarks of Microsoft Corp.

MATLAB is a registered trademark of the MathWorks Inc.

CONTENTS

Chapter 1 BASIC CONCEPTS

THE PASSIVE SIGN CONVENTION

In chapter 1, we are learning a very fundamental law in circuit analysis, Ohm's Law, which states that the voltage-current relationship for a resistor is

$$V = IR \tag{1.1}$$

There is however an underlying requirement for Ohm's Law to be used properly, the *passive sign convention.* This convention, demonstrated in Figure 1.1, sets the relationship between the polarity of the voltage, *V,* and the direction of the current, *I.* In the passive sign convention, current is ASSUMED to enter the positive side of the voltage polarity. Each of the circuits in Figure 1.1 represent the exact same situation: 2 Amps of current flowing from node A to node B resulting in a potential of 20 Volts at node A with respect to node B. Applying Ohm's Law, we find the resistance value is 10 Ω in each case.

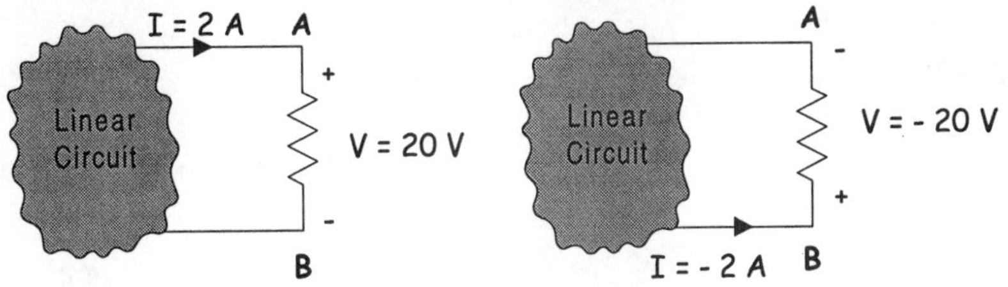

$$R = \frac{V}{I} = \frac{20}{2} = 10\ \Omega \qquad R = \frac{V}{I} = \frac{-20}{-2} = 10\ \Omega$$

Figure 1.1 Demonstrating the passive sign convention.

The passive sign convention is absolutely critical in determining power flow in circuit analyses. We know that the power consumed or generated by an element is

$$P = VI \tag{1.2}$$

But this equation tells us nothing about the power flow, generation or consumption? As the name implies, when using the passive sign convention definitions for current direction and voltage polarity, a positive result for *P* means power is being consumed - the element is passive. Conversely, if *P* is negative, the element is supplying power. Consider the circuit in Figure 1.2 where the passive sign convention has been applied to all three elements. Based on our calculations, the resistor and the 1-V source are using power while the 10-V source delivers power.

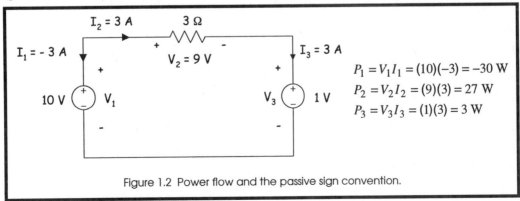

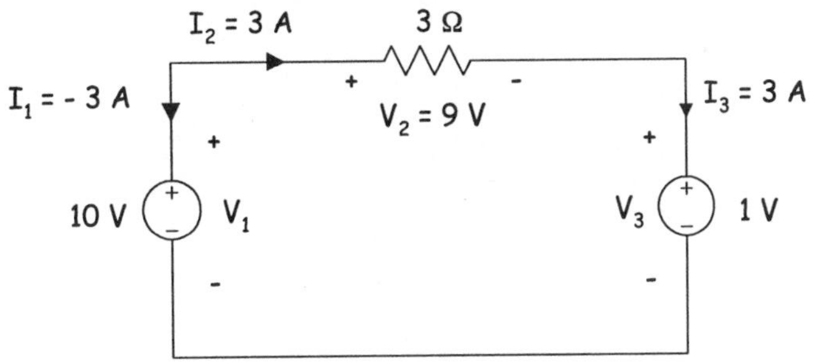

$$P_1 = V_1 I_1 = (10)(-3) = -30 \text{ W}$$
$$P_2 = V_2 I_2 = (9)(3) = 27 \text{ W}$$
$$P_3 = V_3 I_3 = (1)(3) = 3 \text{ W}$$

Figure 1.2 Power flow and the passive sign convention.

PROBLEM SOLVING EXAMPLES

1. The charge entering the circuit below is given by the expression $q(t) = 4t^2$. Determine the current entering the circuit at $t = 2$ s.

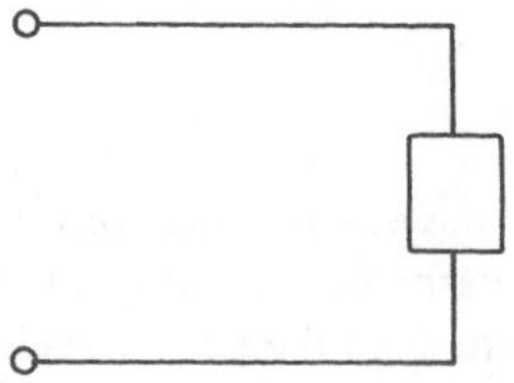

Solution: $i(t) = \frac{dq}{dt} = \frac{d}{dt}(4t^2) = 8t$

$i(t=2) = 16A$

2. Find the magnitude and direction of the voltage across the following elements.

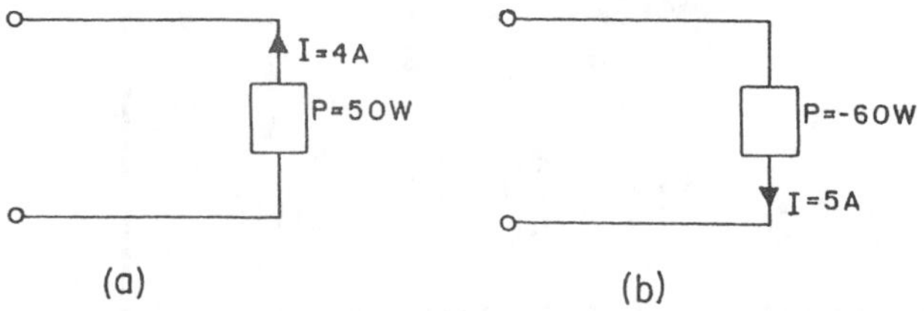

Solution: (a) $V = \frac{P}{I} = \frac{50}{4} = 12.5V$

bottom terminal is positive

(b) $V = \frac{P}{I} = \frac{60}{5} = 12V$

bottom terminal is positive

3. How much current does a 1500-W direct-current hair dryer draw when connected to a constant 120-V line?

Solution: $P = VI$, $1500 = 120I$ $\quad I = 12.5A$

4. Find the power absorbed or supplied by the elements in the following circuit.

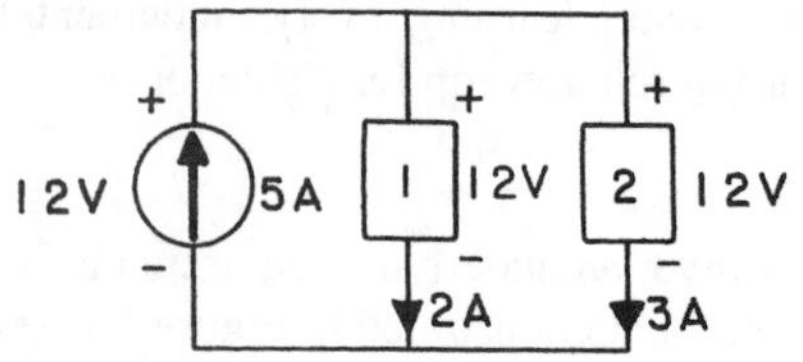

Solution: $P_S = (5)(12) = 60W$ Supplied

$P_1 = (2)(12) = 24W$ Absorbed

$P_2 = (3)(12) = 36W$ Absorbed

5. Determine the power absorbed or supplied by the elements in the network below.

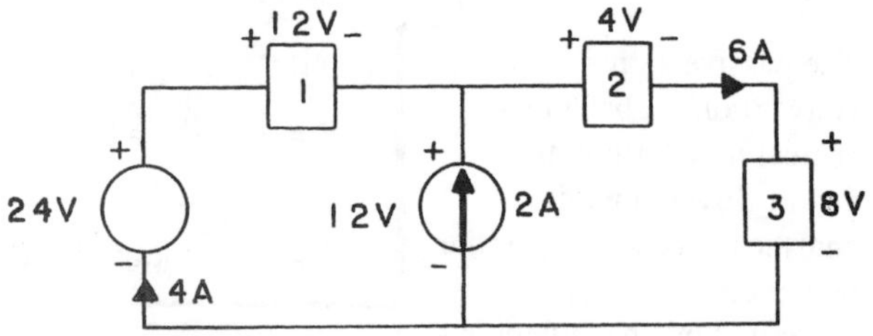

Solultion: $P_{24V} = (4)(24) = 96W$ Supplied

$P_{2A} = (2)(12) = 24W$ Supplied

$P_1 = (4)(12) = 48W$ Absorbed

$P_2 = (4)(6) = 24W$ Absorbed and

$P_3 = (6)(8) = 48W$ Absorbed

Chapter 2 RESISTIVE CIRCUITS

RESISTOR CONSTRUCTION

Resistors are the most basic and numerous circuit elements. Commercial resistors obey Ohm's Law quite well, are available in values from mΩ (1/1000) to GΩ (10^9), cost less than $0.05 each (in 1998) and are sold by the millions each month. Two manufacturing processes that account for the majority of the general purpose resistor market, film and wirewound technology.

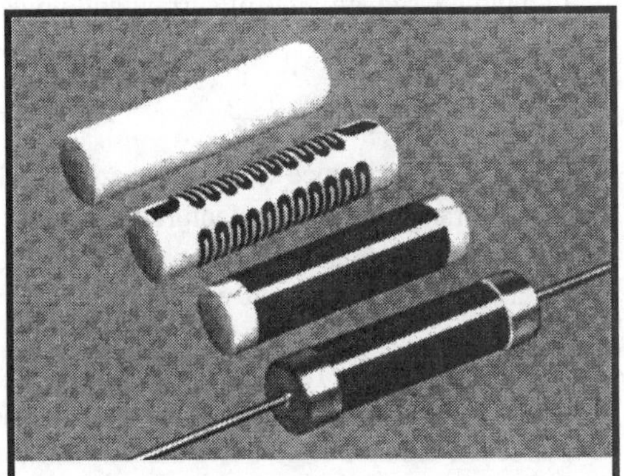

Figure 2.1 A film resistor in various stages of construction.

Figure 2.1 shows a film resistor in various stages of construction. The principle components are the ceramic bobbin, the film, the end caps and the encapsulation. The ceramic bobbin is an insulator, conducting no current. It serves only as a mechanical support for the film layer which is the true resistive material.

Film materials fall into two categories: metal films, sometimes called thin films, and carbon based films sometimes called thick films. In metal film resistors, a poorly conducting metal such as manganin™ or nichrome™ is deposited onto the bobbin in a very thin layer, about 1 millionth of a meter. External wires, called leads, are attached to the film via the metal end caps. Finally, the body of the resistor is coated in a plastic encapsulant for mechanical and environmental protection. In carbon film resistors, the film starts as a slurry containing carbon (insulator) and metal (conductor) particles. The bobbin is coated with the mixture, which is then it is cured in an oven. End caps and encapsulant are added to finish the process.

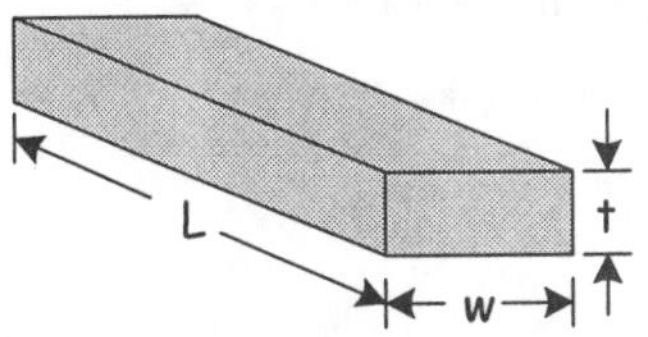

Figure 2.2 A block of resistive material.

How is the resistor's value controlled? First consider the resistive block in Figure 2.2. The resistance of any piece of material can be express as

$$R = \frac{\rho L}{tw} = \frac{\rho L}{A} \tag{2.1}$$

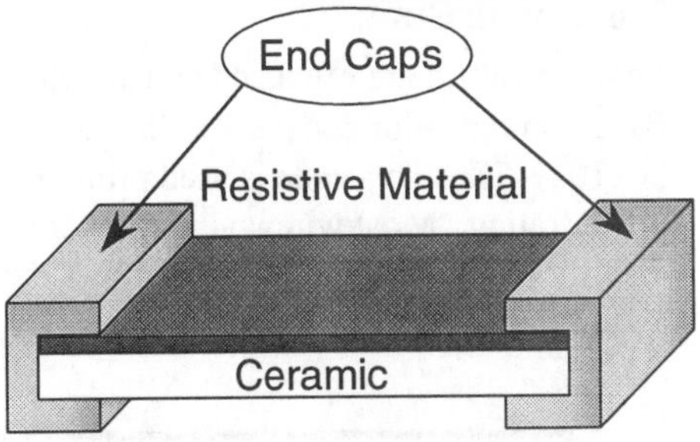

Figure 2.3 Internal diagram of a chip resistor.

where w is the width, t is the thickness, A is the cross-sectional area, L is the length and ρ is the resistivity of the material. The resisitivity of the metal in metal film resistors is determined by which metal is used. Carbon film resistivity is controlled by adjusting the ratio of carbon to metal in the slurry. After deposition, the film can be ablated using a laser or a miniature sandblaster to create the serpentine effect in Figure 2.1. This increases the resistors' value by simultaneously increasing L and decreasing w. Ablation also allows the manufacturer to trim the resistor's value to meet tolerance requirements.

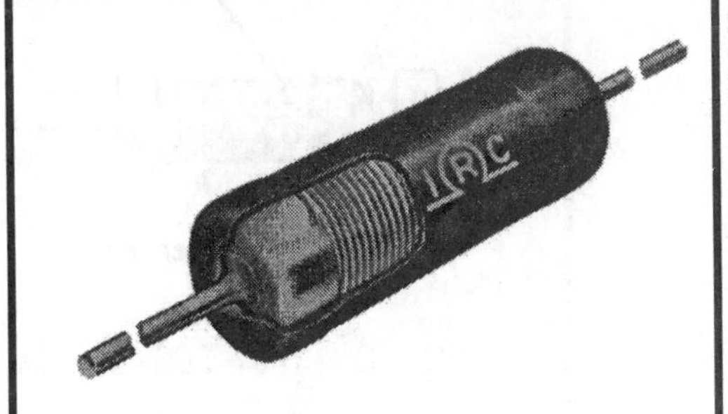

Figure 2.4 Internal view of an enamel coated wirewound resistor.

Another class of film resistor is the chip resistor, shown in Figure 2.3. It is manufactured using either thin metal films or thick ruthenium-based films on a ceramic base. Used extensively in printed circuit boards, chip resistors attach directly to the surface of the PC board. This eliminates the need for end cap wires making the part much smaller, lighter and less expensive.

A wirewound resistor, shown in cross-section in Figure 2.4, is little more than a long piece of wire that has been wound up into a coil. To keep the length of the wire reasonable, the wire is made of a poor conducting metal such as nichrome™ (ρ = 110 μΩ-cm). Wirewound resistors, often packaged in silicone or in hollow enamel tubes, are inherently bulkier and more expensive than carbon film resistors. They do however, have two advantages. First, they can be wound to very precise and custom values. Second, they can be physically sized to handle large amounts of current and power. Commercial wirewound resistors range in value from about 0.1 Ω to 1 MΩ.

POWER AND TOLERANCE

Besides the resistance value, there are two other important resistor specifications - tolerance and power rating. The tolerance is the manufacturer's guaranteed range for the resistance. For example, a 1000 Ω, 5% resistor will have an actual resistor between 950 and 1050 Ω. Standard tolerances are 5% and 2%. Tighter tolerances are available but are more expensive.

A resistor's power rating is the maximum continuous power the element can absorb. The problem is that as a resistor absorbs power (energy) it gets warmer. More power - more heat. Eventually, the heat causes something inside the resistor to give and the part fails. In extreme cases, the resistor catches on fire! Standard power ratings for carbon film resistors are $^1/_8$, ¼, ½ and 1 Watt. Wirewound resistor power ratings are typically between 1 and 200 Watts. As an example of how easy it can be to exceed a power rating, consider an ordinary ¼ W, 5-Ω resistor connected to a fresh 1.5-V, D cell battery. What an innocent scenario. The power absorbed by the resistor is

$$P = \frac{V^2}{R} = \frac{1.5^2}{5} = 0.45 \text{ W} \tag{2.2}$$

almost twice the power rating! So, always consider the expected power dissipation when purchasing resistors.

THE COLOR CODE

Resistor values are often coded using the color code described and demonstrated in Figure 2.5 and Table 1. Based on the color code, a 200-kΩ, 5% resistor will have the color code RED-BLACK-YELLOW-GOLD = $20\text{x}10^4 \pm 5\%$. This code is used primarily on carbon film resistors and has no information about the part's power rating. Most wirewound resistors have their specifications, resistance, power rating and tolerance, printed directly on the element. Table 2 shows the standard commercial resistor values for 5% and 10% tolerances with 10% values in boldface.

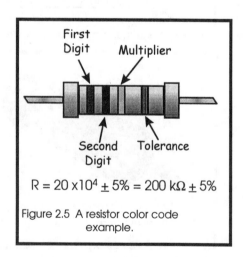

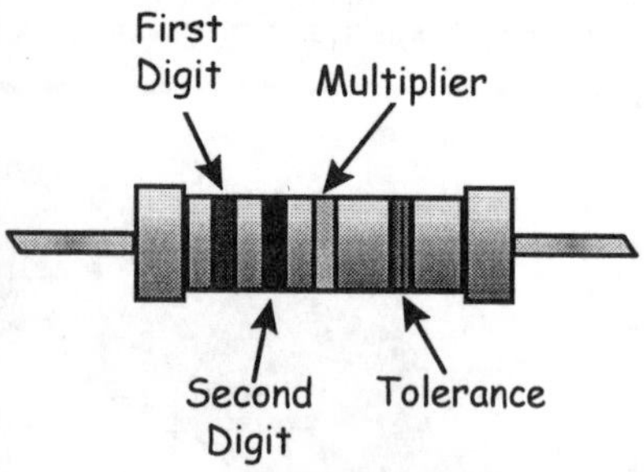

$R = 20 \text{ x}10^4 \pm 5\% = 200 \text{ k}\Omega \pm 5\%$

Figure 2.5 A resistor color code example.

TABLE 1. THE RESISTOR COLOR CODE

Color	Value	Tolerance
Silver		10%
Gold		5%
Black	0	
Brown	1	
Red	2	2%
Orange	3	
Yellow	4	
Green	5	
Blue	6	
Violet	7	
Gray	8	
White	9	

TABLE 2. STANDARD RESISTOR VALUES FOR 5% AND 10% TOLERANCES

(10% values shown in boldface)

1.0	10	100	1.0k	10k	100k	1.0M	10M
1.1	11	110	1.1k	11k	110k	1.1M	11M
1.2	**12**	**120**	**1.2k**	**12k**	**120k**	**1.2M**	**12M**
1.3	13	130	1.3k	13k	130k	1.3M	13M
1.5	**15**	**150**	**1.5k**	**15k**	**150k**	**1.5M**	**15M**
1.6	16	160	1.6k	16k	160k	1.6M	16M
1.8	**18**	**180**	**1.8k**	**18k**	**180k**	**1.8M**	**18M**
2.0	20	200	2.0k	20k	200k	2.0M	20M
2.2	**22**	**220**	**2.2k**	**22k**	**220k**	**2.2M**	**22M**
2.4	24	240	2.4K	24K	240K	2.4M	
2.7	**27**	**270**	**2.7k**	**27k**	**270k**	**2.7M**	
3.0	30	300	3.0k	30k	300k	3.0M	
3.3	**33**	**330**	**3.3k**	**33k**	**330k**	**3.3M**	
3.6	36	360	3.6k	36k	360k	3.6M	
3.9	**39**	**390**	**3.9k**	**39k**	**390k**	**3.9M**	
4.3	43	430	4.3k	43k	430k	4.3M	
4.7	**47**	**470**	**4.7k**	**47k**	**470k**	**4.7M**	
5.1	51	510	5.1k	51k	510k	5.1M	
5.6	**56**	**560**	**5.6k**	**56k**	**560k**	**5.6M**	
6.2	62	620	6.2k	62k	620k	6.2M	
6.8	**68**	**680**	**6.8k**	**68k**	**680k**	**6.8M**	
7.5	75	750	7.5k	75k	750k	7.5M	
8.2	**82**	**820**	**8.2k**	**82k**	**820k**	**8.2M**	
9.1	91	910	9.1k	91k	910k	9.1M	

NOTES

PROBLEM SOLVING EXAMPLES

1. Determine the number of nodes and branches in the circuit below.

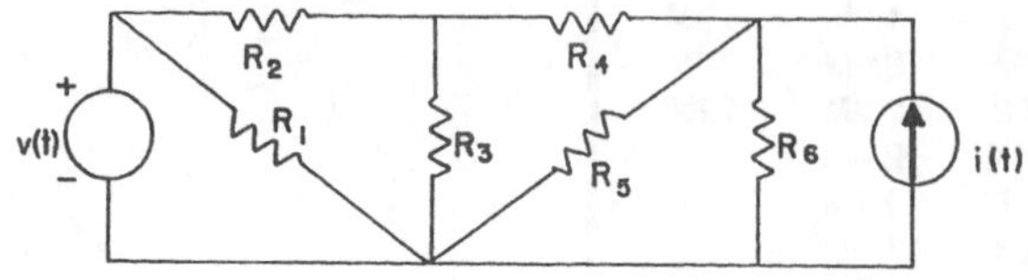

Solution: There are 3 non-reference nodes (4 nodes total) and there are 8 branches (one voltage source, one current source and 6 resistors).

2. Given the network below (a) find the current I, (b) compute the power dissipated in the resistor and (c) show that the power generated by the source is equal to that dissipated by the resistor.

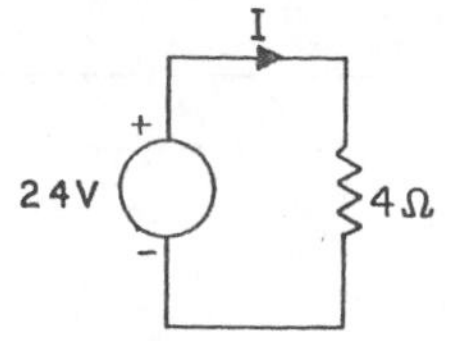

Solution: (a) $I = \frac{V}{R} = \frac{24}{4} = 6A$

(b) $P_R = I^2R = (6)^2(4) = 144W$

(c) $P_S = VI = (24)(6) = 144W$

3. Given the following circuit, find I and V_o.

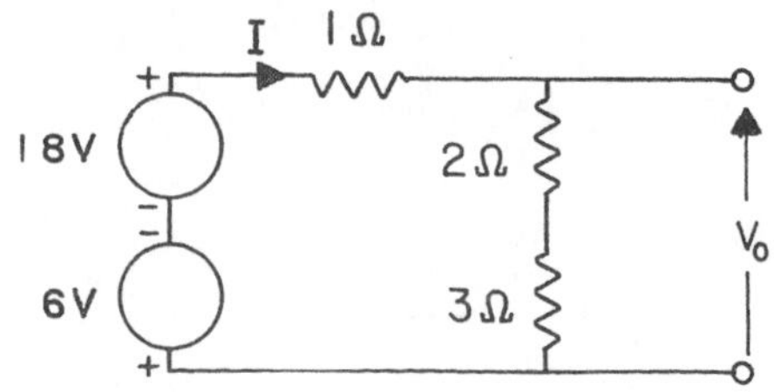

Solution: Using KVL $-6+18-I-2I-3I = 0$ therefore $I = 2A$

$V_o-2I-3I = 0$ hence $V_o = 10V$

4. Using Kirchhoff's current law, compute the unknown currents in the circuits below.

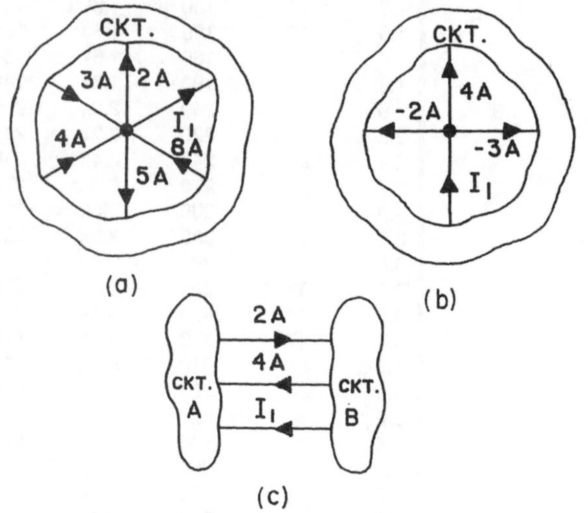

Solution: (a) using KCL $2-3-4+5-8+I_1 = 0$ therefore $I_1 = 8A$, (b) $4-2-I_1-3 = 0$ therefore $I_1 = -1A$ and (c) $2-4-I_1 = 0$ therefore $I_1 = -2A$.

5. If in the following circuit, V_{R_2} is known to be 6V, find V_{R_1} and V_A.

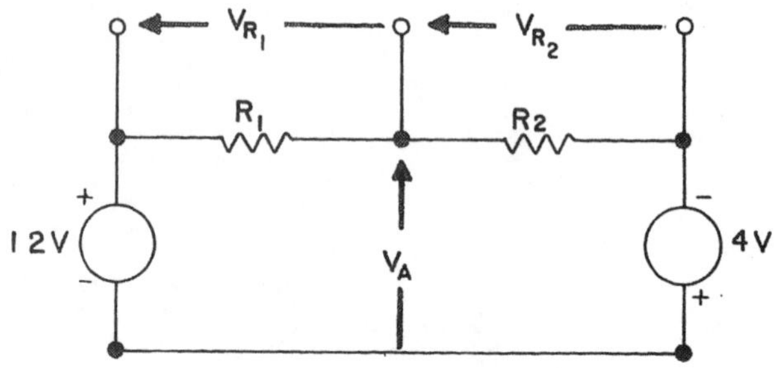

Solution: using KVL $12-V_{R_1}-V_{R_2}+4 = 0$ therefore $V_{R_1} = 10V$. Also $12-V_{R_1}-V_A = 0$, hence $V_A = 2V$ or $V_A-V_{R_2}+4 = 0$ which also yields $V_A = 2V$.

6. Find I and V_o in the network below.

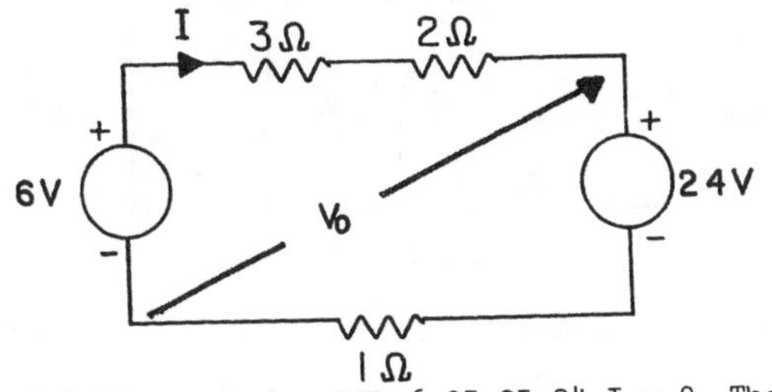

Solution: using KVL $6-3I-2I-24-I = 0$. Therefore $I = -3A$, then $V_o-24-I = 0$ and $V_o = 21V$.

7. If V_{R_1} is known to be 6V, find V_{R_3} in the following network.

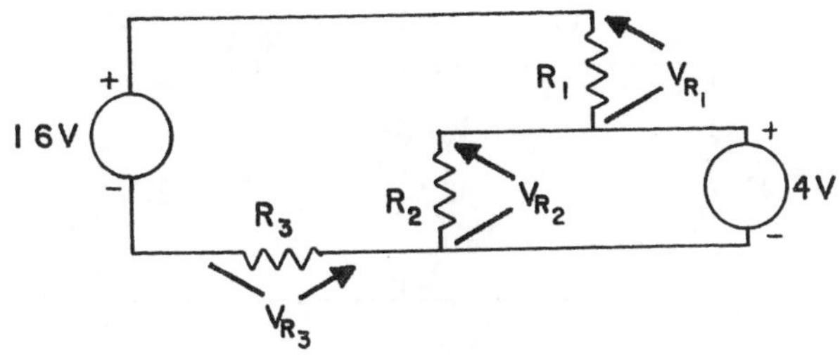

Solution: Note that $V_{R_2} = 4V$, hence using KVL $+16-6-4-V_{R_3} = 0$ and therefore $V_{R_3} = 6V$.

8. Use current division to determine the currents I_A, I_B, I_C in the networks below.

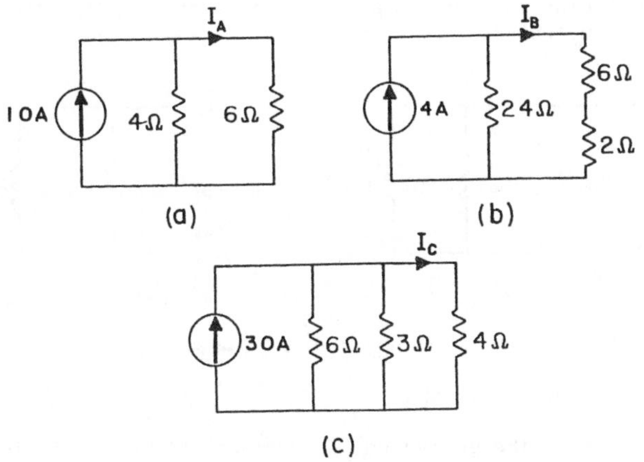

Solution: (a) $I_A = (10)(\frac{4}{6+4}) = 4A$, (b) $I_B = (4)(\frac{24}{24+8}) = 3A$ and (c) $I_C = 30(\frac{6||3}{6||3+4}) = 30(\frac{2}{2+4}) = 10A$.

9. Given the circuits below, find I_A, I_B, and I_C, using current division.

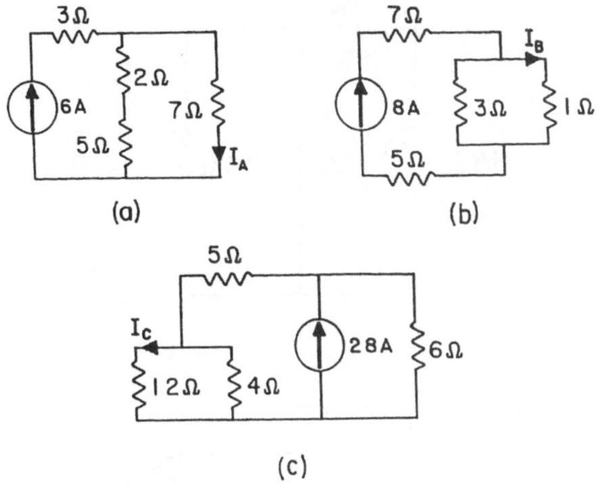

Solution: (a) $I_A = (6)(\frac{2+5}{2+5+7}) = 3A$, (b) $I_B = 8(\frac{3}{3+1}) = 6A$ and (c) $I_C = (28)(\frac{6}{6+5+12||4})(\frac{4}{4+12}) = 3A$.

10. Using voltage division, determine the voltages V_A, V_B, and V_C in the following networks.

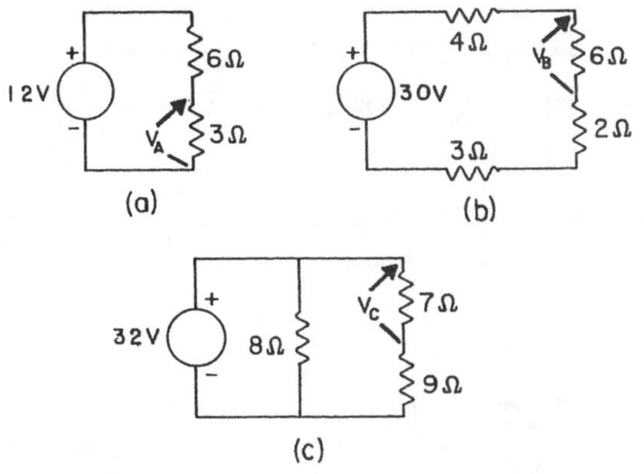

Solution: (a) $V_A = 12(\frac{3}{6+3}) = 4V$, (b) $V_B = 30(\frac{6}{4+6+3+2}) = 12V$ and (c) $V_C = 32(\frac{7}{7+9}) = 14V$.

11. Determine the voltages V_A, V_B, and V_C in the following networks using voltage division.

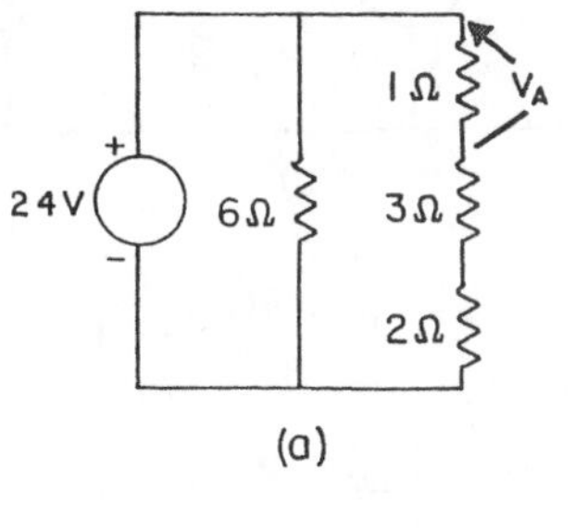

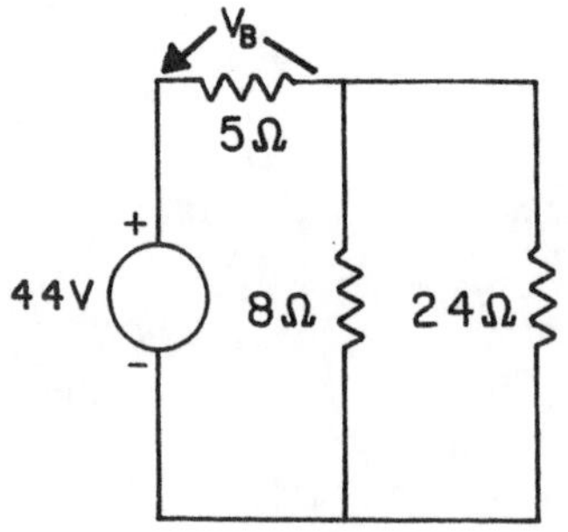

(b)

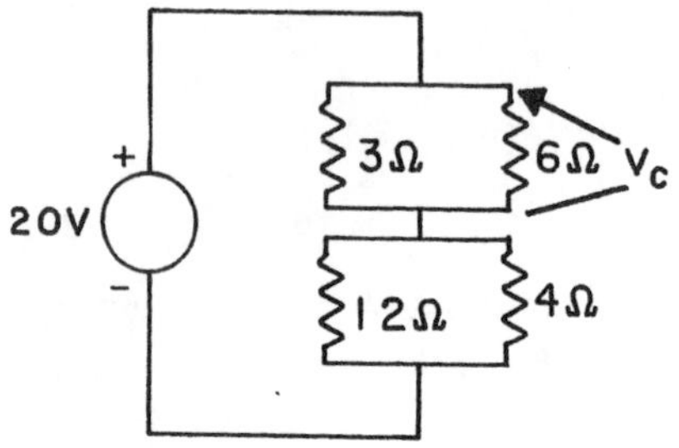

(c)

Solution: (a) $V_A = 24(\frac{1}{1+3+2}) = 4V$, (b) $V_B = 44(\frac{5}{5+8||24}) = 20V$ and (c) $V_C = 20(\frac{3||6}{3||6+4||12}) = 20(\frac{2}{2+3}) = 8V$.

12. Using voltage division, find V_A, V_B, and V_C in the networks shown below.

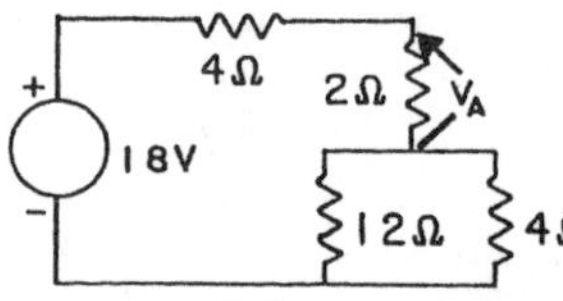

(a)

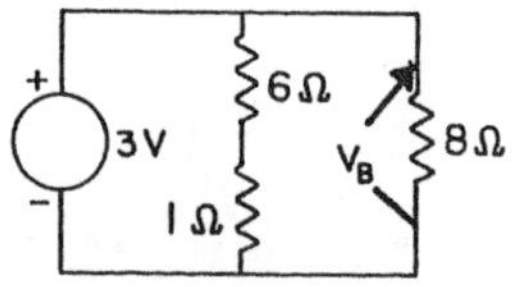

(b)

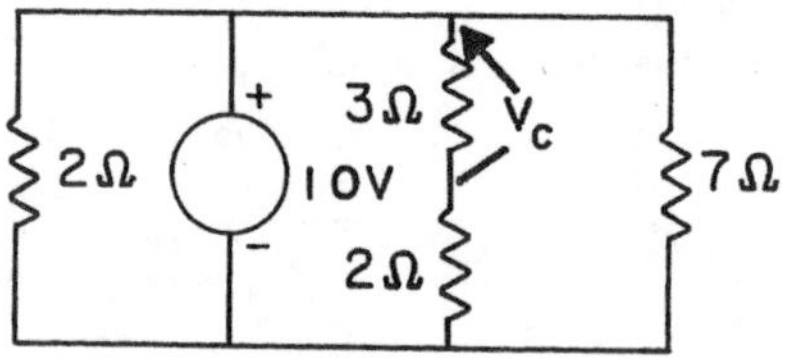

(c)

Solution: (a) $V_A = 18(\frac{2}{4+2+4||12}) = 4V$, (b) $V_B = 3V$ and (c) $V_C = 10(\frac{3}{3+2}) = 6V$.

13. Given the following network, find V and I_o.

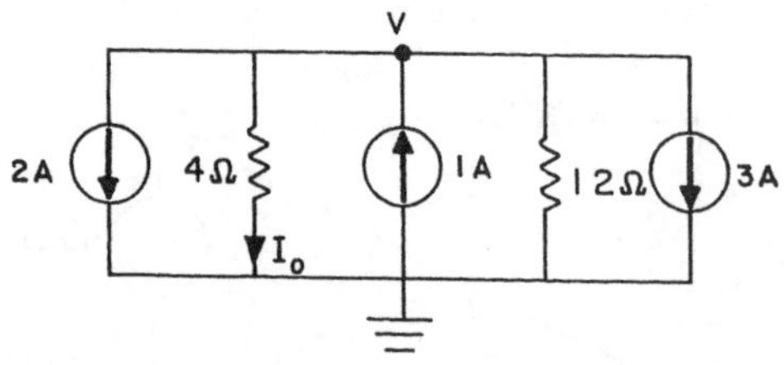

Solution: using KCL $2 + \frac{V}{4} - 1 + \frac{V}{12} + 3 = 0$. Therefore $V = -12V$ and hence $I_o = -\frac{12}{4} = -3A$.

14. Given $V_A = 8V$, find V_o in the circuit below.

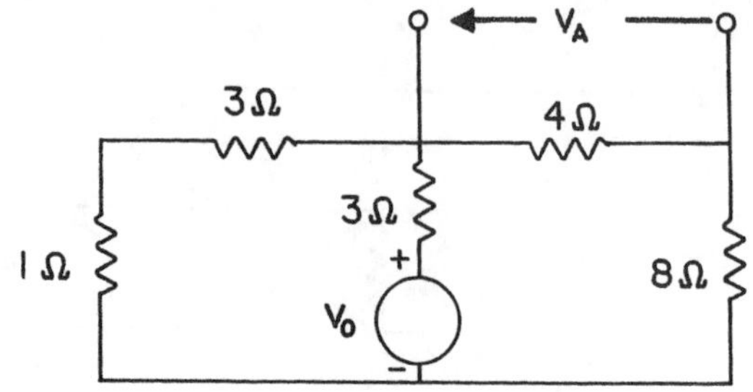

Solution: If $V_A = 8V$ then $I_{4\Omega} = \frac{8}{4} = 2A$ and $I_{8\Omega} = I_{4\Omega} = 2A$. Then $V_{8\Omega} = (2)(8) = 16V$. Hence $V_{4\Omega+8\Omega} = V_{3\Omega+1\Omega} = 24V$. Then $I_{3\Omega+1\Omega} = \frac{24}{3+1} = 6A$. Therefore using KCL $I_{V_o} = 2+6 = 8A$, and hence using KVL $V_o = 24+8(3) = 48V$.

15. Find the equivalent resistance of the following networks.

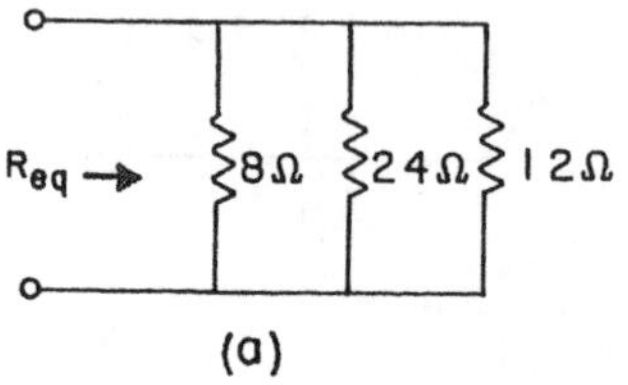

(a)

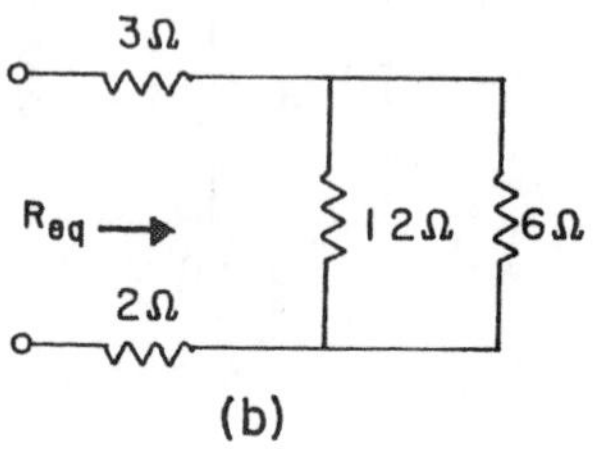

(b)

Solution: (a) $R_{eq} = 8||24||12 = 4\Omega$, (b) $R_{eq} = 3+6||12+2 = 9\Omega$.

16. Find the equivalent conductance of the following networks.

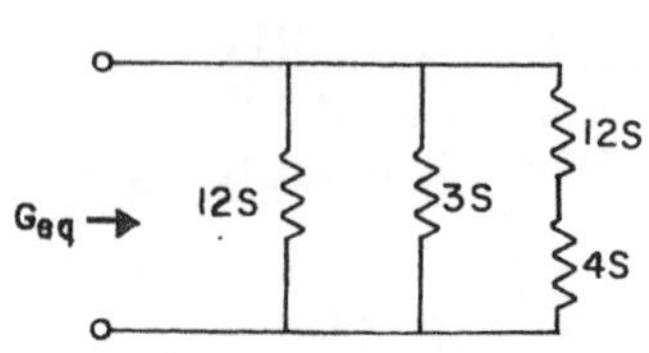

(a)

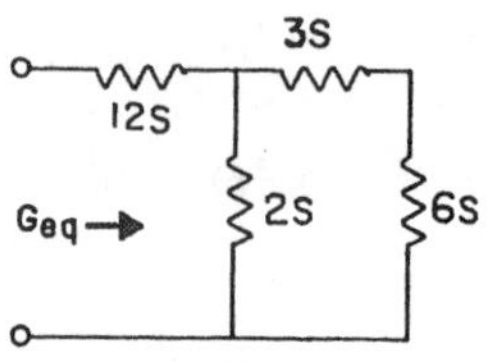

(b)

Solution: (a) $G_{eq} = 12||4+3+12 = 18S$

(b) $$G_{eq} = \frac{(\frac{(3)(6)}{(3+6)} + 2)(12)}{(\frac{(3)(6)}{(3+6)} + 2)+12} = 3S$$

17. Find R_{eq} in the circuit below.

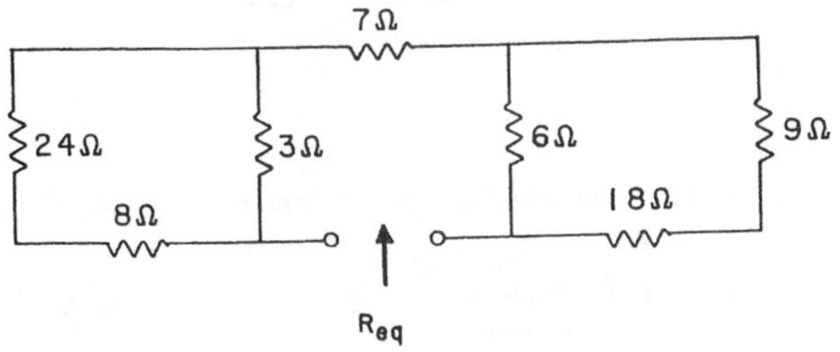

Solution: $R_{eq} = 32||3+7+6||27 = 14.65\Omega$

18. Find R_{eq} in the circuit shown below.

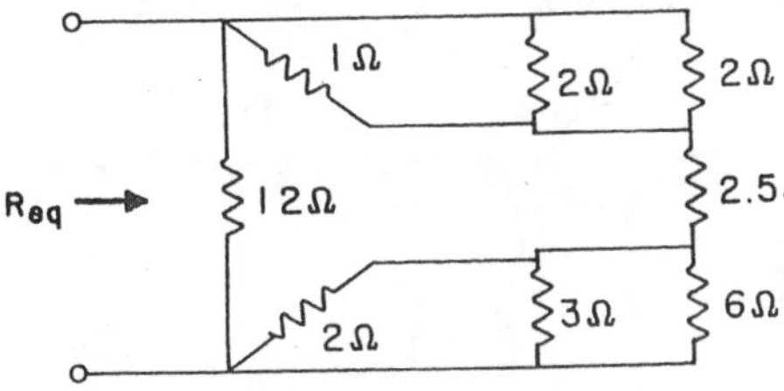

Solution: The circuit can be redrawn by combining resistances to yield the following network

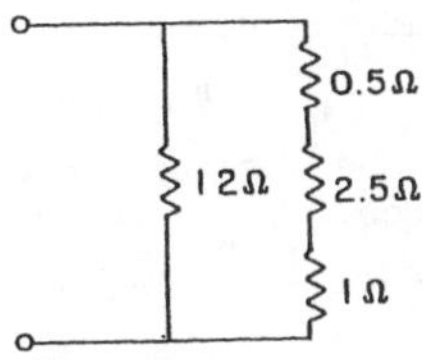

therefore $R_{eq} = 12||4 = 3\Omega$

19. Find the equivalent resistance of the following circuit.

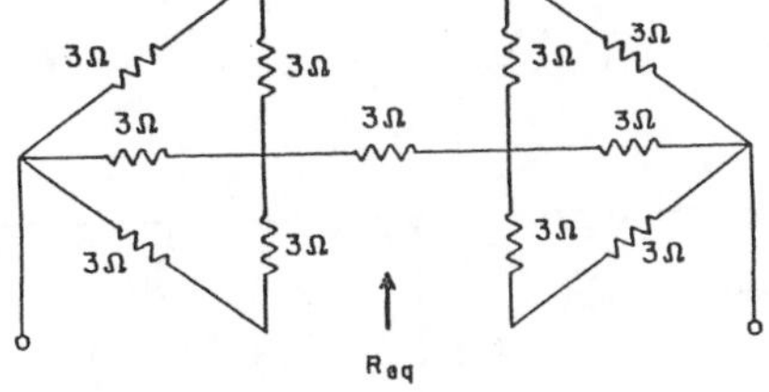

Solution: The network can be redrawn as follows by combining resistors

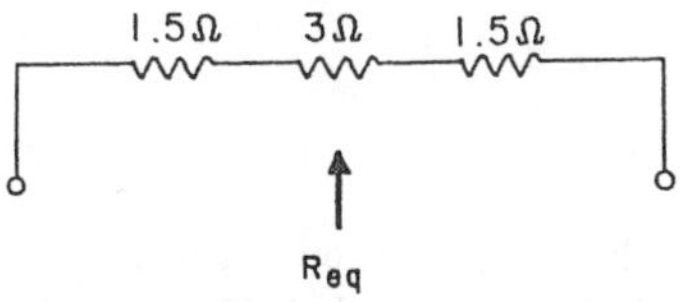

therefore $R_{eq} = 6\Omega$

20. Find the equivalent resistance of the following circuit.

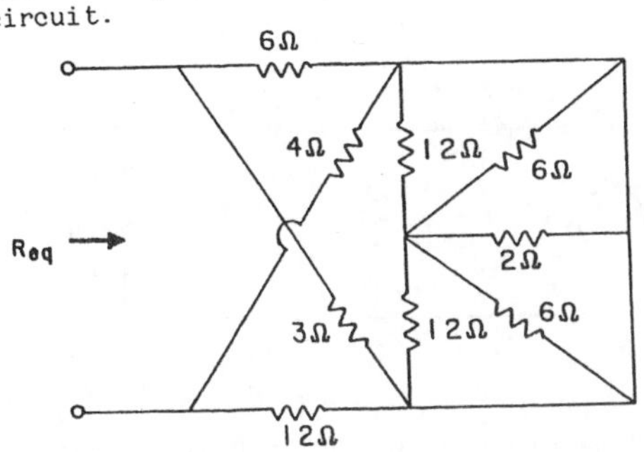

Solution: Note that the five resistors on the right-hand side of the network are short circuited and therefore the network can be redrawn as follows.

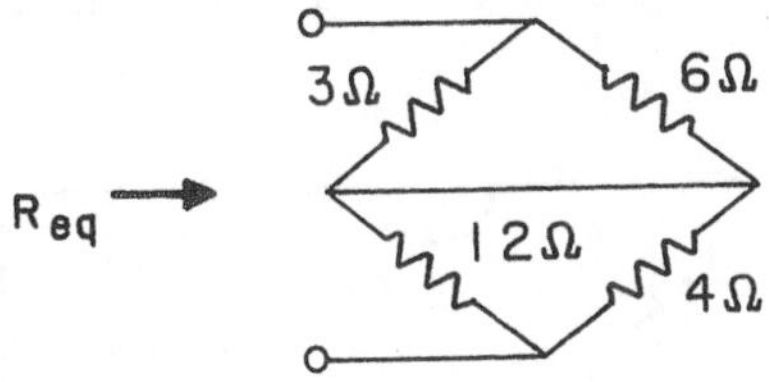

Therefore $R_{eq} = 3||6+12||4 = 5\Omega$

21. Find R_{eq} in the resistor network shown below if the 4Ω resistor is open circuited and the 1Ω resistor is shorted.

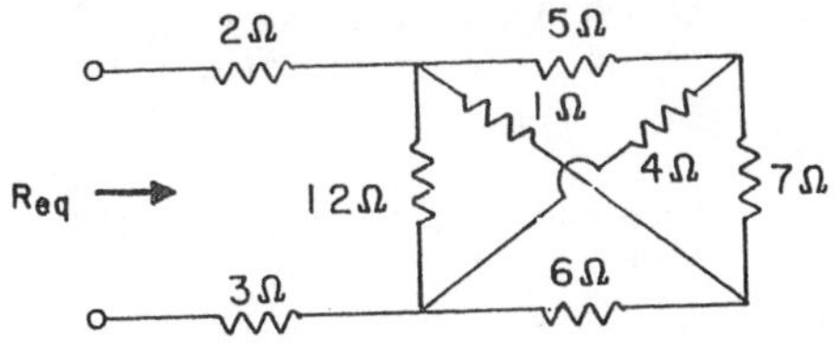

Solution: By open circuiting the 4Ω resistor and short circuiting the 1Ω resistor we short the 5 and 7 ohm resistors and place the 6 and 12 ohm resistors in parallel. Hence $R_{eq} = 2+4+3 = 9\Omega$.

22. Find I_o in the following network.

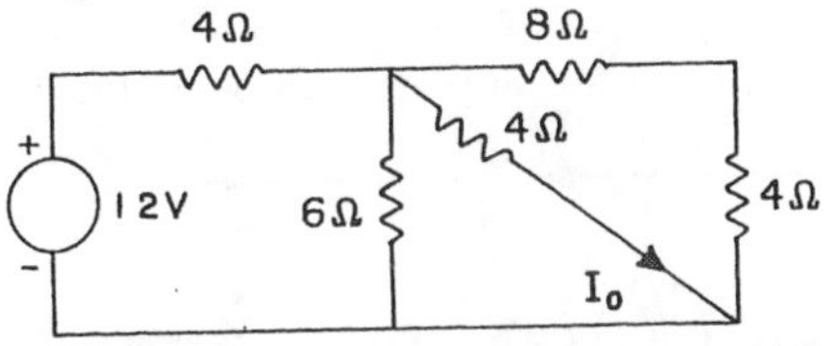

Solution: The total resistance seen by the source is $4+6||4||12 = 6\Omega$. Therefore $I_{12V} = \frac{12}{6} = 2A$. Using KVL the voltage across the parallel resistor combination is $12-(2)(4) = 4V$. Therefore $I_o = \frac{4}{4} = 1A$.

23. Find V in the network below, given that I = 4A.

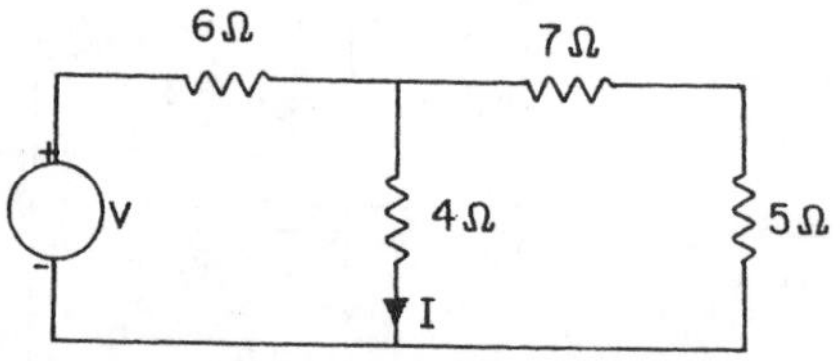

Solution: If I = 4A then $V_{4\Omega} = (4)(4) = 16V$. Then the current in the 5-and 7-ohm resistors is $I_{7\Omega} = \frac{16}{12} = \frac{4}{3}$ A. Hence $I_{6\Omega} = 4 + \frac{4}{3} = \frac{16}{3}$ A. Therefore $V = 6(\frac{16}{3}) + 16 = 48V$.

24. In the network below, if I = 3A, find V.

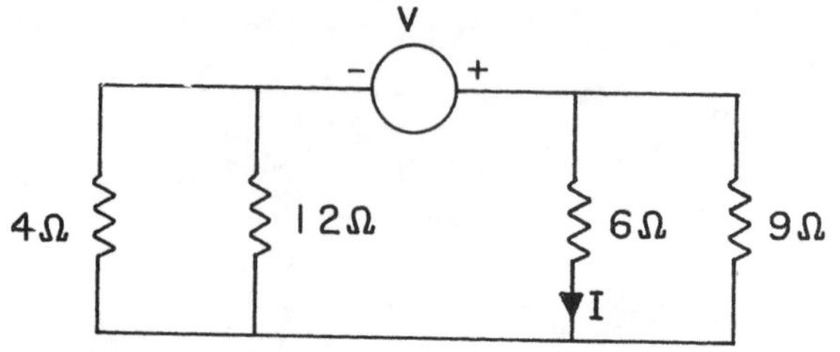

Solution: If $I_{6\Omega} = 3A$ then $V_{6\Omega} = 18V$ and therefore $I_{9\Omega} = \frac{18}{9} = 2A$. Using KCL $I_V = 2+3 = 5A$. The voltage across the 4-and 12-ohm resistors is $V_{12\Omega} = (4||12)(5) = 15V$. Hence $V = 15+18 = 33V$.

25. Find the power absorbed by the 6-ohm resistor in the network below.

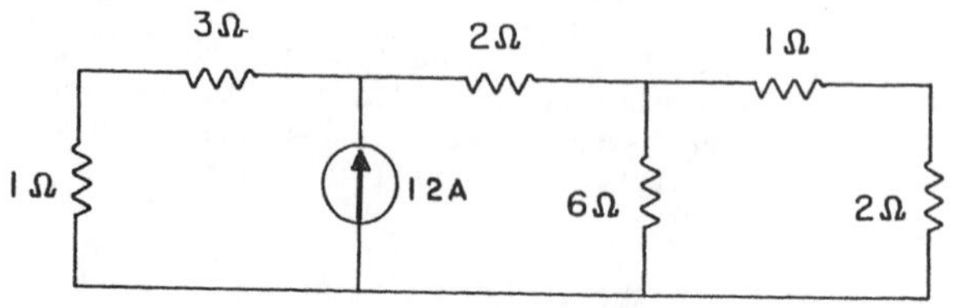

Solution: The network can be redrawn as follows.

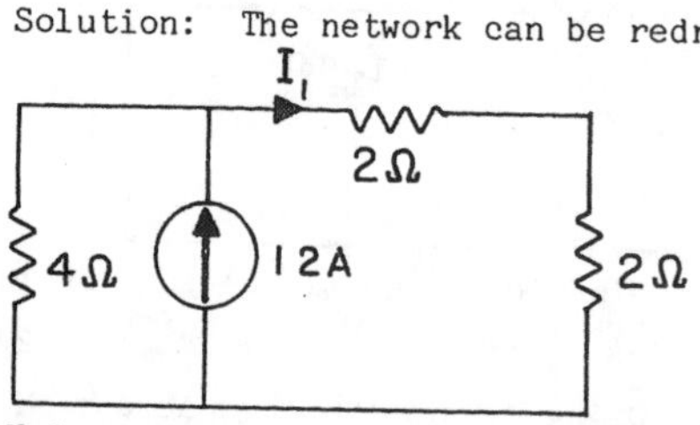

Using current division $I_1 = (12)(\frac{4}{4+4}) = 6A$.

Then applying current division again yields $I_{6\Omega} = I_1(\frac{2+1}{6+2+1}) = 2A$. Hence $P_{6\Omega} = (2)^2(6) = 24W$.

26. Given that the 1Ω resistor in the following network absorbs 16W of power, determine the value of the voltage source.

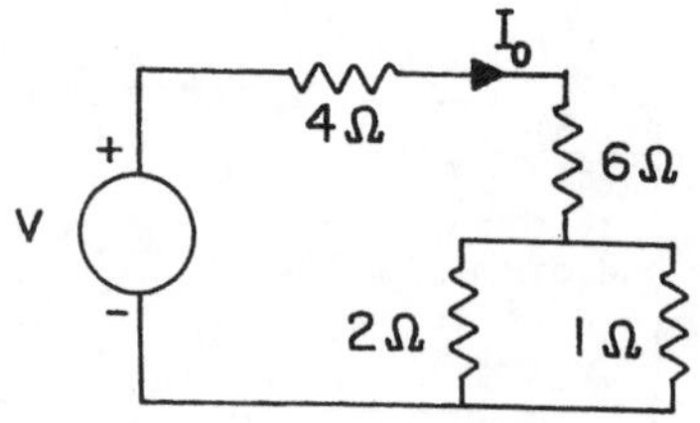

Solution: The magnitude of the current in the 1Ω resistor is $I_{1\Omega} = \sqrt{\frac{16}{1}} = 4A$. The current direction is downward because of the direction of the voltage source. $V_{1\Omega} = 4V$ and $V_{2\Omega} = V_{1\Omega}$. $I_{2\Omega} = \frac{4}{2} = 2A$ and hence $I_o = 2+4 = 6A$. Then $V = 6(4+6)+4 = 64V$.

27. The power absorbed by the 4Ω resistor in the following network is $P_{4\Omega} = 64W$. Find V.

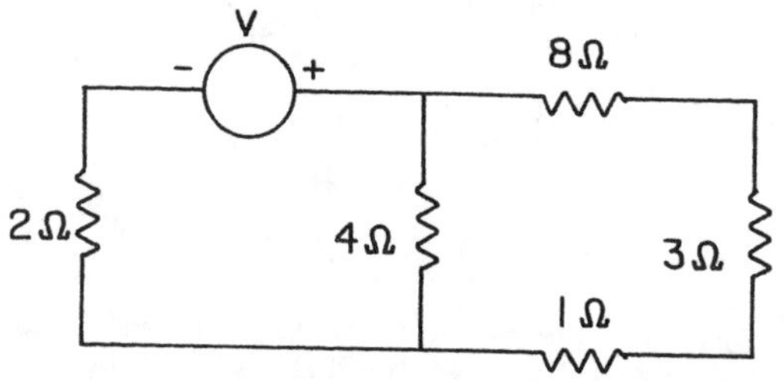

Solution: The magnitude of the current in the 4Ω resistor is $I_{4\Omega} = \sqrt{\frac{64}{4}} = 4A$. The current direction is downward because of the direction of the voltage source. $V_{4\Omega} = (4)(4) = 16V$ and hence $I_{8\Omega} = \frac{16}{8+3+1} = \frac{4}{3}$ A. The current in the voltage source is $I_V = 4 + \frac{4}{3} = \frac{16}{3}$A. Therefore $V = 16 + (\frac{16}{3})(2) = \frac{80}{3}$V.

28. The current I_A in the following network is known to be 4A. Determine the voltage source V_A.

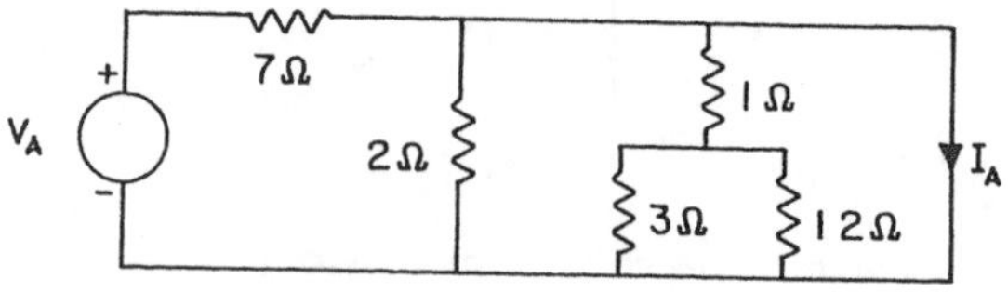

Solution: The short circuit at the right-hand side of the network shorts every resistor except the 7Ω resistor and hence $V_A = (4)(7) = 28V$.

29. Find V in the network shown below.

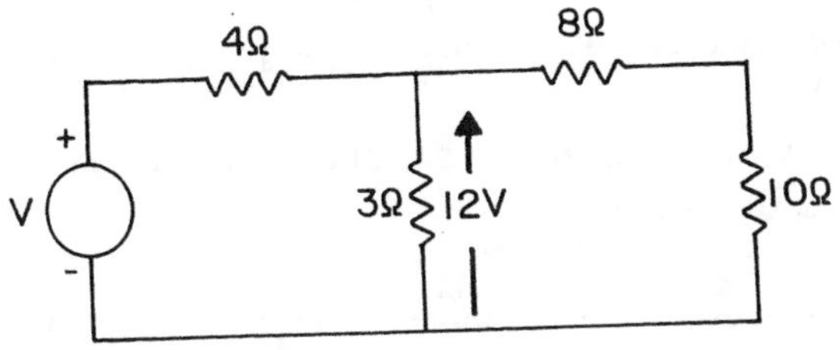

Solution: $I_{3\Omega} = \frac{12}{3} = 4A$ and $I_{8\Omega} = \frac{12}{18} = \frac{2}{3}$ A.

Therefore $I_V = 4 + \frac{2}{3} = \frac{14}{3}$A. Hence $V = (\frac{14}{3})(4) + 12 = \frac{92}{3}$ V.

30. Find V_o in the network below.

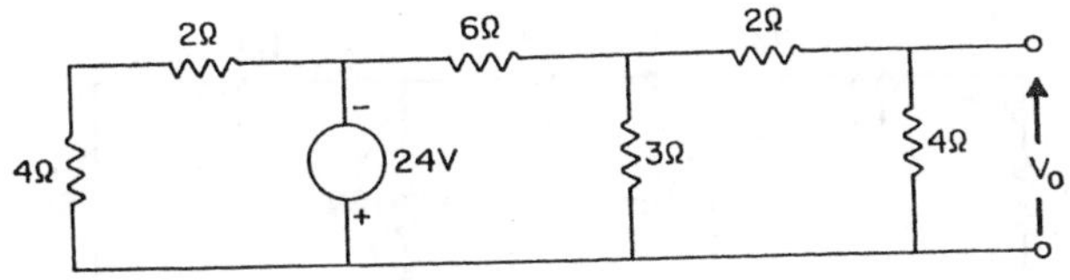

Solution: To determine V_o we will apply voltage division twice. $V_{3\Omega} = (-24)(\frac{6||3}{6||3+6}) = -6V$ and then $V_o = (-6)(\frac{4}{2+4}) = -4V$.

31. Find V_o in the following network.

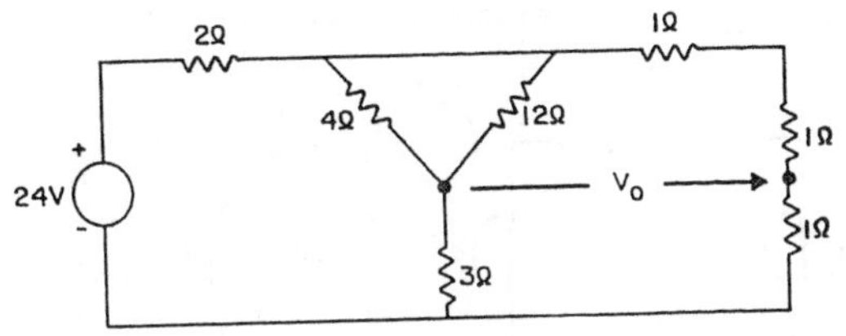

Solution: The network can be drawn as follows.

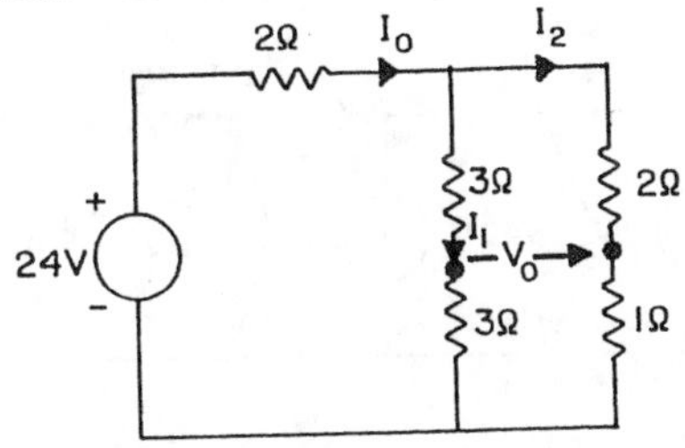

The current $I_o = \frac{24}{2+6||3} = 6A$. Then $I_1 = \frac{(6)(3)}{6+3} = 2A$ and $I_2 = \frac{(6)(6)}{6+3} = 4A$. Then using KVL $3I_1 - 2I_2 - V_o = 0$ and $V_o = -2V$.

32. Find I_o in the following network.

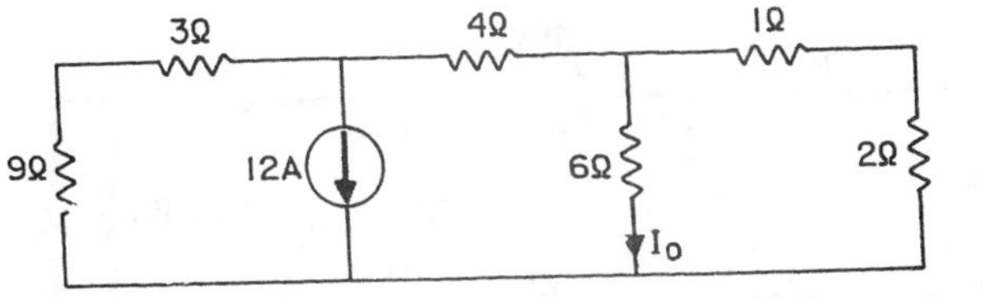

Solution: We can determine I_o by applying current division twice. $I_{4\Omega} = (-12)(\frac{12}{12+4+6||3}) = -8A$. Then $I_o = (-8)(\frac{3}{6+3}) = -\frac{8}{3}$ A.

33. Determine the current flowing through the 7Ω resistor in the network below.

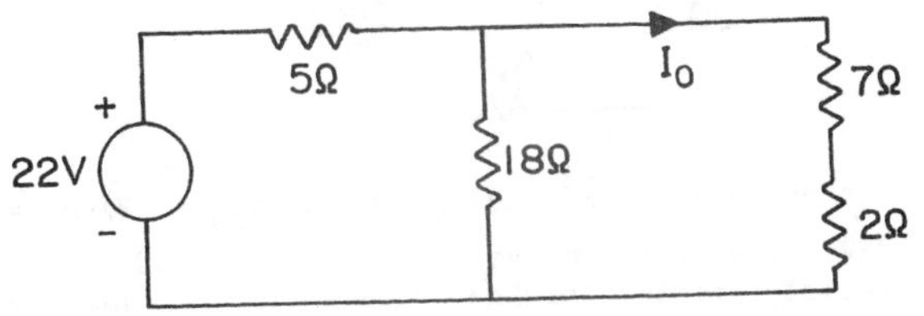

Solution: The voltage across the parallel combination of resistors is $V_{18\Omega} = (22)(\frac{18||9}{18||9+5}) = 12V$. Therefore $I_o = \frac{12}{7+2} = \frac{4}{3}$ A.

34. Given the following network, if $V_o = 12V$, what is I_o?

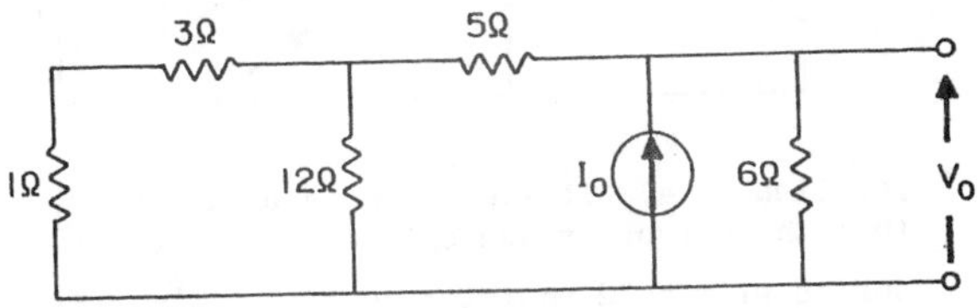

Solution: If $V_o = 12V$ then $I_{6\Omega} = \frac{12}{6} = 2A$. The current in the 5Ω resistor is $I_{5\Omega} = \frac{12}{5+12||4} = \frac{3}{2}$ A. Therefore $I_o = \frac{3}{2} + 2 = \frac{7}{2}$ A.

35. Find all the currents in the network shown below.

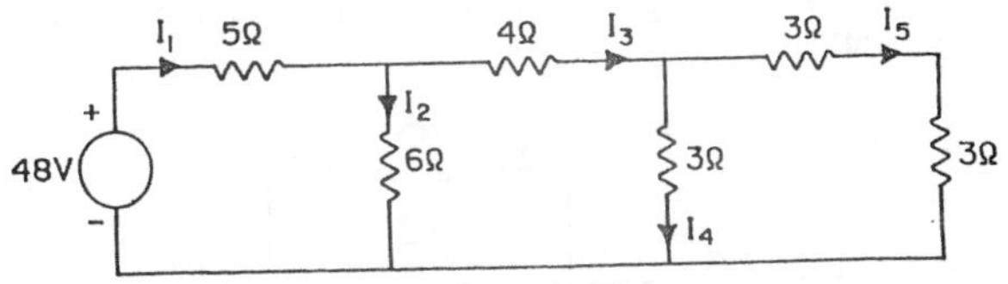

Solution: The total resistance to the right of the 6Ω resistor is 6Ω (i.e. 4+6||3). Then the total resistance seen by the 48V source is $5+6||6 = 8\Omega$. Hence $I_1 = \frac{48}{8} = 6A$. Then using current division $I_2 = I_3 = 3A$. Then $I_4 = (3)(\frac{6}{3+6}) = 2A$ and $I_5 = 1A$.

36. Find V_o in the following network.

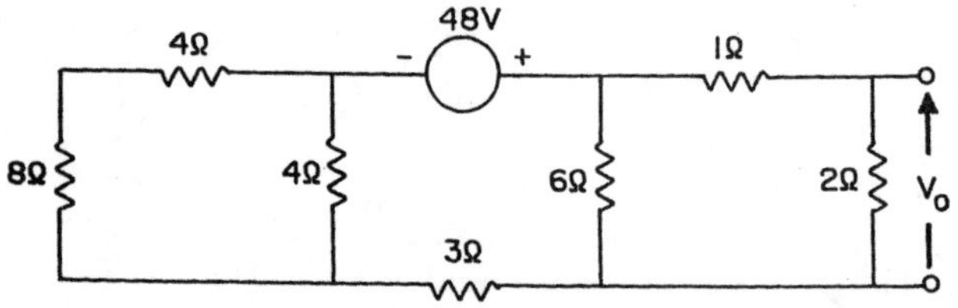

Solution: By combining resistors the network can be reduced to

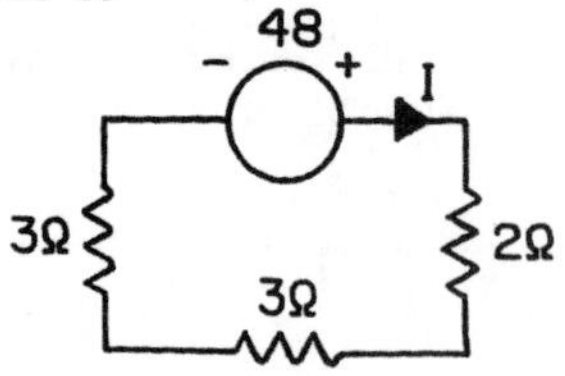

The current $I = \frac{48}{8} = 6A$ and hence $V_{2\Omega} = (2)(6) = 12V$. Therefore the voltage across the 6Ω resistor in the original network is 12V. Using a voltage divider $V_o = \frac{(12)(2)}{1+2} = 8V$.

37. Given the network below, (a) find the current I_o if the 6Ω resistor is short circuited and (b) find the voltage V_o if the 3Ω resistor is open circuited.

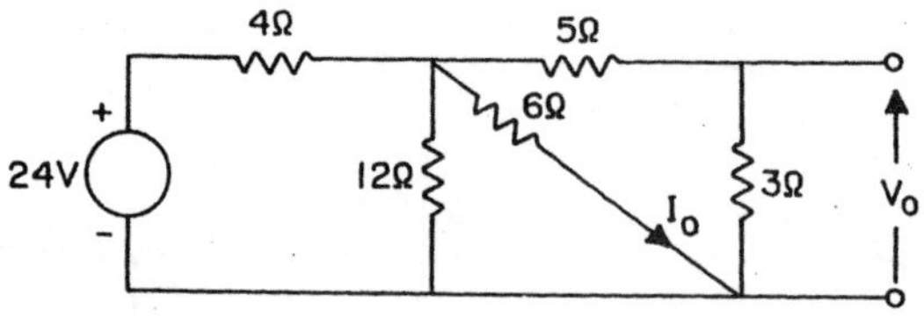

Solution: (a) If the 6Ω resistor is shorted, then the network reduces to a 24V source in series with a 4Ω resistor hence $I_o = \frac{24}{4} = 6A$.
(b) If the 3Ω resistor is open circuited, the voltage across the parallel combination of the 12-and 6-ohm resistors is $V_{12\Omega} = \left(\frac{24}{4+6||12}\right)(6||12) = 12V$. However, since there is no current in the 5Ω resistor $V_o = V_{12\Omega} = 12V$.

38. Given the circuit below, (a) find the current I_o if the 1Ω resistor is short circuited and (b) find the voltage V_o if the 1Ω resistor is open circuited.

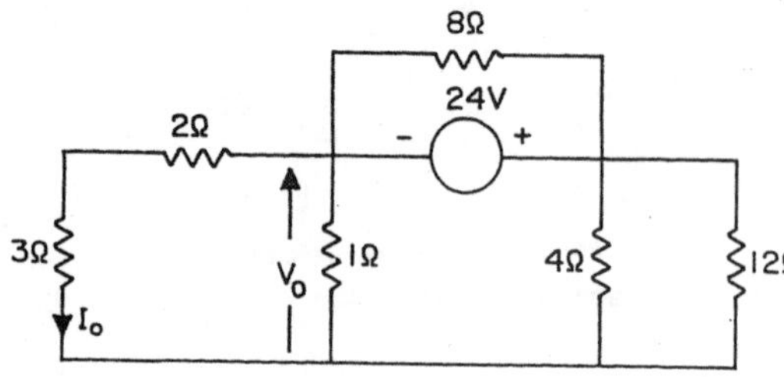

Solution: (a) If the 1Ω resistor is shorted, all the current will flow through the short circuit and therefore $I_o = 0$. (b) If the 1Ω resistor is open circuited $V_o = \frac{-24}{4||12+3+2}(3+2) = -15V$.

39. If V_1 is known to be 12V, find I_o in the following network.

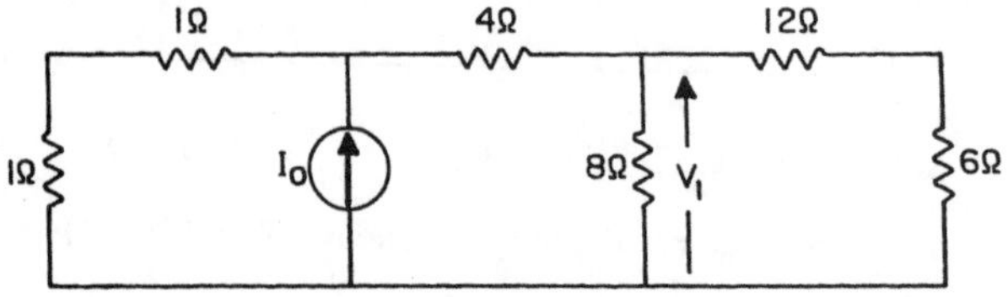

Solution: Consider the network labelled below.

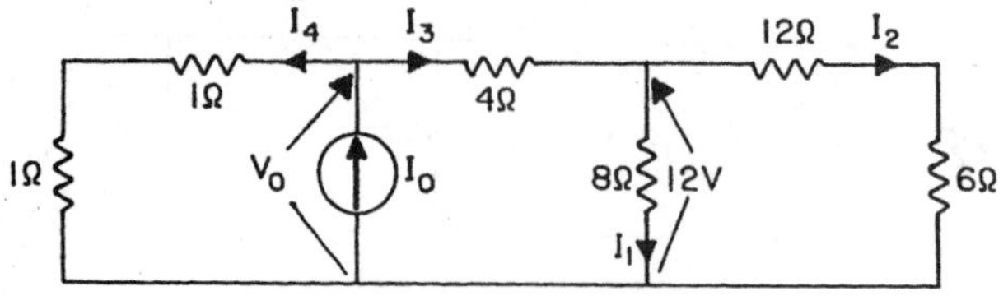

$I_1 = \frac{12}{8} = \frac{3}{2}$ A and $I_2 = \frac{12}{18} = \frac{2}{3}$ A therefore $I_3 = \frac{3}{2} + \frac{2}{3} = \frac{16}{3}$ A. Then $V_o = 12+4I_3 = \frac{62}{3}$ V. Hence $I_4 = \frac{\frac{62}{3}}{1+1} = \frac{62}{6}$ A. Therefore $I_o = \frac{16}{3} + \frac{62}{6} = \frac{25}{2}$ A.

40. Find I_o in the network below.

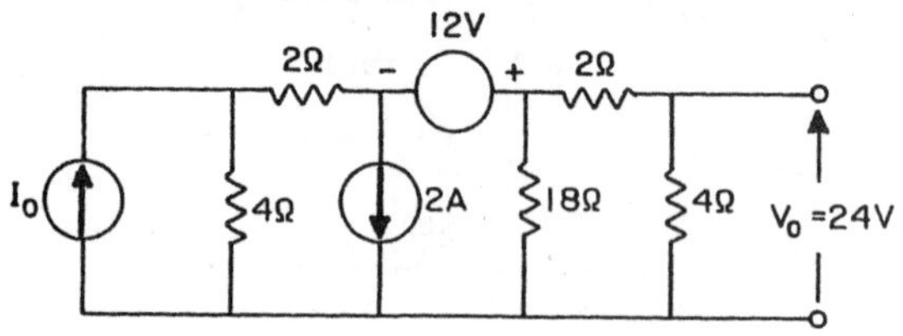

Solution: Consider the network labelled below.

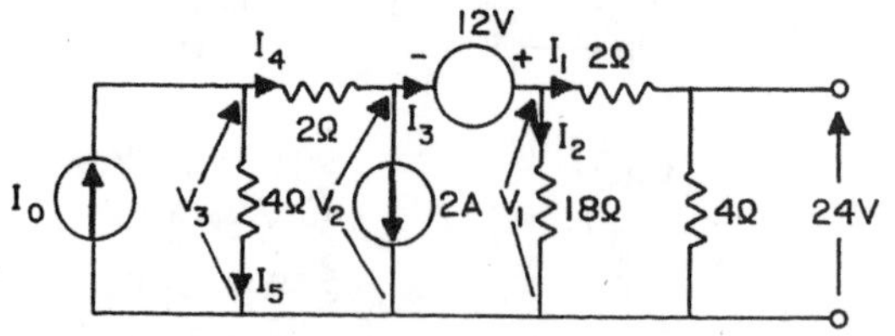

$I_1 = \frac{24}{4} = 6A$ and $V_1 = 2I_1+24 = 36V$. Therefore $I_2 = \frac{36}{18} = 2A$ and $I_3 = 6+2 = 8A$. $V_2 = -12+V_1 = 24V$ and $I_4 = I_3+2 = 10A$. Hence $V_3 = 2I_4+V_2 = 44V$. $I_5 = \frac{44}{4} = 11A$ and hence $I_o = 21A$.

41. Find V_A in the network below.

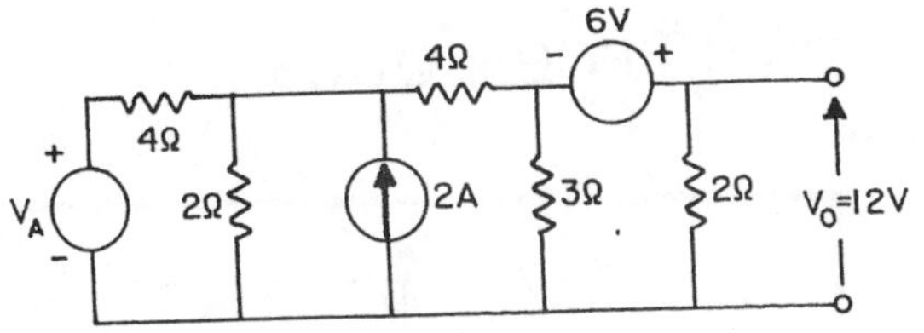

Solution: Consider the labelled network below.

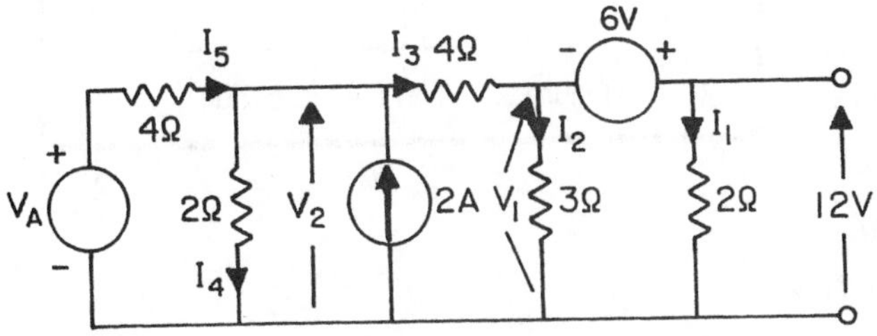

$I_1 = \frac{12}{2} = 6A$ and $V_1 = -6+12 = 6V$. Hence $I_2 = \frac{6}{3}$ $= 2A$ and therefore $I_3 = I_1+I_2 = 8A$. $V_2 = 4I_3+V_1 = 38V$ and $I_4 = \frac{38}{2} = 19A$. Then $I_5 = I_3-2+I_4 = 25A$. Hence $V_A = 4I_5+V_2 = 138V$.

42. Find the voltage V_o in the following network.

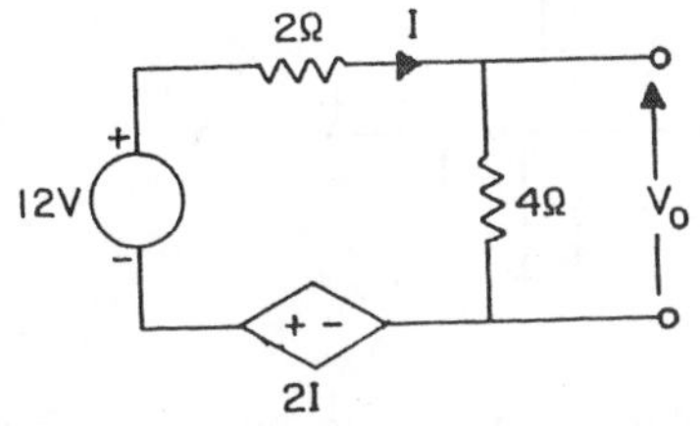

Solution: Using KVL $12-2I-4I+2I = 0$. Hence $I = 3A$ and therefore $V_o = 12V$.

43. Determine the current I in the circuit below.

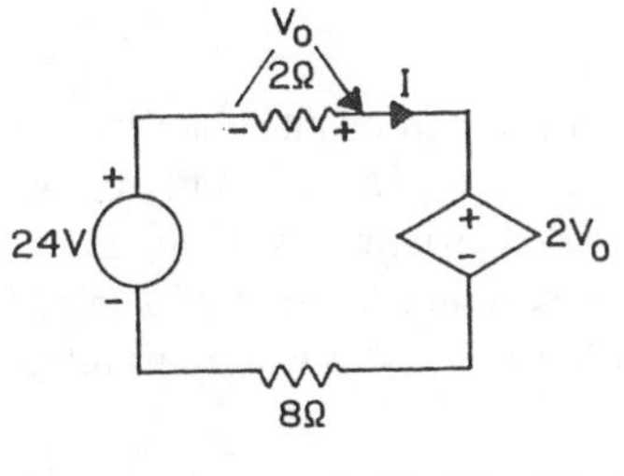

Solution: Using KVL $24-2I-2V_o-8I = 0$ and $V_o = -2I$. Therefore $I = 4A$ and $V_o = -8V$.

44. Determine I_o in the following circuit.

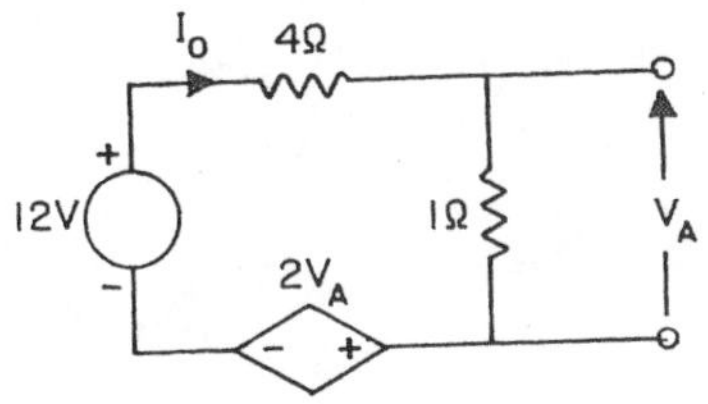

Solution: Using KVL $12-4I_o-I_o-2V_A = 0$ and $V_A = 1I_o$. Therefore $I_o = \frac{12}{7}$ A.

45. Find the current I_o in the circuit below.

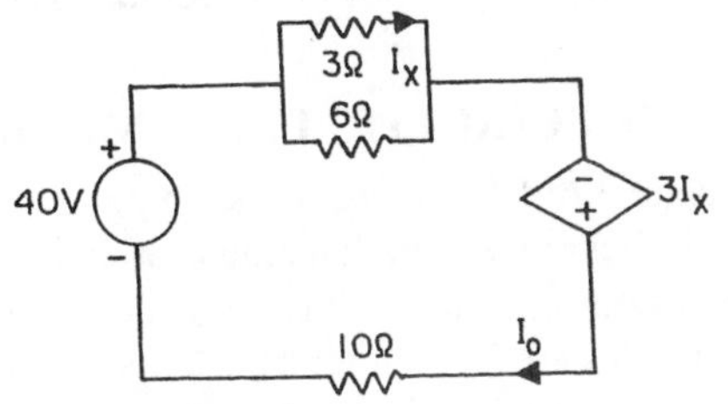

Solution: Using KVL $40-(3||6)I_o+3I_x-10I_o = 0$ and $I_x = \frac{(I_o)(6)}{3+6}$. Solving these equations yields $I_o = 4A$.

46. Determine the value of I_o in the circuit shown below. (a) With the op-amp present
(b) Without the op-amp

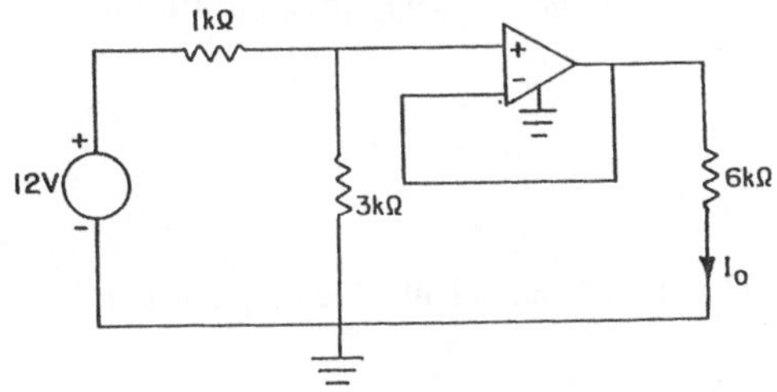

Solution: (a) The op-amp input is $V_{in} = \frac{(12)(3K)}{1K+3K} = 9V$ therefore $I_o = \frac{3}{2}$ ma. (b) Without the op-amp the 3KΩ and 6KΩ resistors are in parallel and $I_o = \frac{12}{1K+3K||6K}\left(\frac{3K}{6K}\right) = \frac{4}{3}$ ma.

47. Find I_o in the following network.

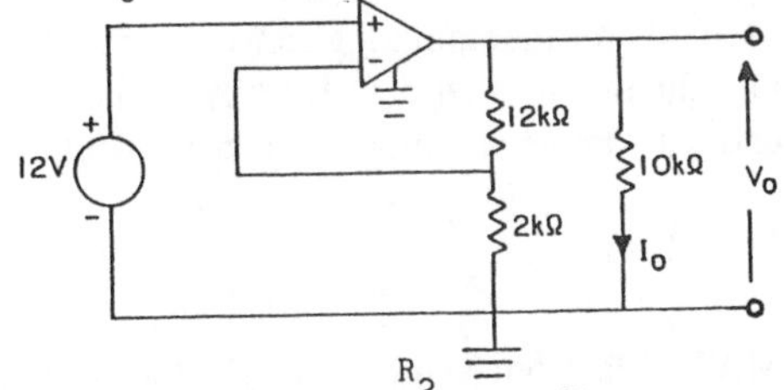

Solution: $V_o = (1+\frac{R_2}{R_1})V_1 = 84V$. Hence $I_o = \frac{84}{10K} = 8.4$ ma.

Chapter 3 NODAL & LOOP ANALYSIS TECHNIQUES

SOLVING CIRCUITS USING MATRICES

In chapter 3 of BECA, you are learning loop and mesh analysis techniques. Both methods produce a set of n simultaneous equations with n variables. In loop analysis, the variables are loop currents, and node voltages are the variables in nodal analysis. As with any set of simultaneous equations, they can be solved using matrices. Consider the circuit in Figure 3.1 where loop current have been defined. The resulting loop equations are

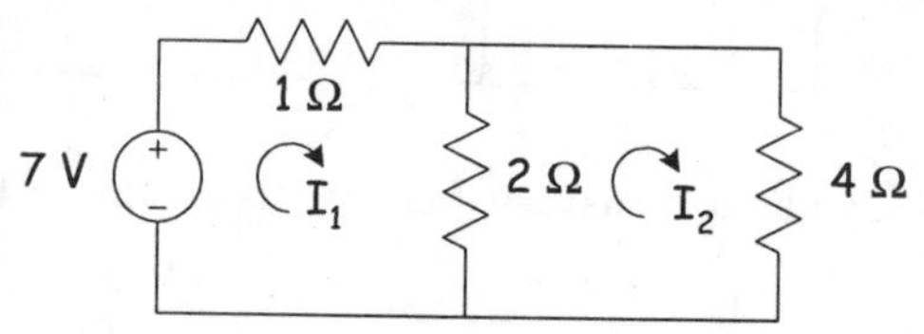

Figure 3.1 A simple dc circuit.

$$7 = 3I_1 - 2I_2$$
$$0 = -2I_1 + 6I_2$$

which can be written in matrix form as

$$\begin{bmatrix} 3 & -2 \\ -2 & 6 \end{bmatrix}\begin{bmatrix} I_1 \\ I_2 \end{bmatrix} = \begin{bmatrix} 7 \\ 0 \end{bmatrix} \qquad (3.1)$$

Using Cramers's Rule, the loop currents are found to be

$$I_1 = \frac{\begin{vmatrix} 7 & -2 \\ 0 & 6 \end{vmatrix}}{\begin{vmatrix} 3 & -2 \\ -2 & 6 \end{vmatrix}} = \frac{42-0}{18-4} = 3 \text{ A} \quad \text{and} \quad I_2 = \frac{\begin{vmatrix} 3 & 7 \\ -2 & 0 \end{vmatrix}}{\begin{vmatrix} 3 & -2 \\ -2 & 6 \end{vmatrix}} = \frac{0+14}{18-4} = 1 \text{ A} \qquad (3.2)$$

Methods for manipulating matrices, including Cramer's Rule, will not be discussed here but can be found in any introduction to matrices. Luckily, matrix operations are easily converted to computer code. In fact, most high-end calculators can perform matrix calculations.

MATLAB

MATLAB, from The MathWorks Inc., is a very powerful and popular piece of technical software. Originally a matrix manipulator (MATLAB stands for MATrix LABoratory) it has matured significantly to include numeric computation, advanced graphics and visualization, and its own high-level programming language. The details of MATLAB programming will not be treated here. We will instead list several examples of using MATLAB to solve the matrices that result from loop and mesh analyses. Each of the MATLAB examples in this study guide are on the CD-ROM.

MATLAB SIMULATIONS

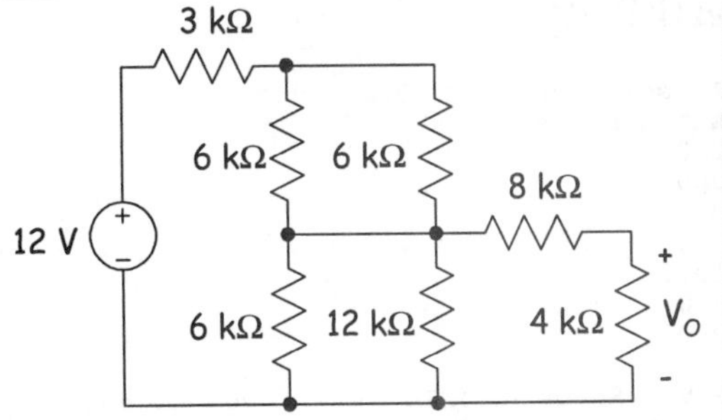

Figure 3.2. Circuit for first MATLAB simulation.

SIMULATION 1:

```
% MATLAB solution to Figure 3.2
% resistors and voltage sources
R1 = 3000;
R2 = 6000;
R3 = 6000;
R4 = 6000;
R5 = 12000;
R6 = 8000;
R7 = 4000;
Vs = 12.0;

% Solve V = RI where V is a vector of the voltage sources in each loop, R is a 4x4 matrix of
% resistances derived from applying KVL around each loop and I is a vector of loop currents.

R(1,1) = R1 + R2 + R4;
R(1,2) = -R2;
R(1,3) = -R4;
R(1,4) = 0;
R(2,1) = -R2;
R(2,2) = R2 + R3;
R(2,3) = 0;
R(2,4) = 0;
R(3,1) = -R4;
R(3,2) = 0;
R(3,3) = R4 + R5;
R(3,4) = -R5;
R(4,1) = 0;
R(4,2) = 0;
R(4,3) = -R5;
R(4,4) = R5 + R6 + R7;

% Voltage sources in each loop
V = [Vs 0 0 0]';

% Find the solution to V = RI, where I is the unknown, using MATLAB's matrix solution
function.

I = R\V;

% apply equations to find node voltages Vn
Vn(1) = I(4)*R7;
Vn(2) = Vn(1) + I(4)*R6;
Vn(3) = Vn(2) + I(2)*R3;

% print the output in a pretty form

fprintf ('The voltage labeled V0 = %g volts\n\n', Vn(1));

for j=1:4
        fprintf ('Loop Current #%d = %g milliamps\n',j,1000*I(j));
end;

fprintf ('\n');
for j=1:3
        fprintf ('Node Voltage #%d = %g Volts\n',j,Vn(j));
end;
```

The MATLAB solution is,

```
The voltage labeled V0 = 1.33333 volts

Loop Current #1 = 1.33333 milliamps
Loop Current #2 = 0.666667 milliamps
Loop Current #3 = 0.666667 milliamps
Loop Current #4 = 0.333333 milliamps

Node Voltage #1 = 1.33333 Volts
Node Voltage #2 = 4 Volts
Node Voltage #3 = 8 Volts
```

SIMULATION 2:

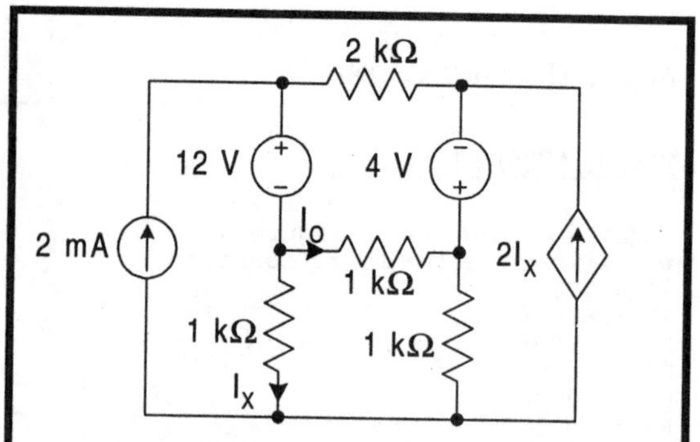

Figure 3.3. Circuit for second MATLAB simulation.

```
% MATLAB solution to Figure 3.3
% resistors and sources
R1 = 2000;
R2 = 1000;
R3 = 1000;
R4 = 1000;
Vs1 = 12.0;
Vs2 = 4.0;
Is1 = 0.002;

% Solve V = RI using KVL.  Only need to solve
% two equations for the two unknowns, I2 and I3.

R(1,1) = R1 + R2;
R(1,2) = -R2;
R(2,1) = -R2;
R(2,2) = R2 + R3 - R4;

% the voltage sources in each loop, with appropriate polarity
V = [(Vs1+Vs2) (R3*Is1 - 2*R4*Is1)]';

% Find the solution to the system V = RI where I is the unknown
% Use MATLAB's matrix solution function

I = R\V;

% Find all the loop currents
I_Loop(1) = Is1;
I_Loop(2) = I(1);
I_Loop(3) = I(2);
I_Loop(4) = -2*(I_Loop(1)-I_Loop(3));

% apply equations to find node voltages Vn
Vn(2) = (I_Loop(1) - I_Loop(3)) * R3;
Vn(1) = Vn(2) + Vs1;
Vn(4) = (I_Loop(3) - I_Loop(4))*R4;
Vn(3) = Vn(4) - Vs2;

% print the output in a pretty form
fprintf ('The branch current labeled Ix = %g milliamps\n',1000*(I_Loop(1) - I_Loop(3)));
fprintf ('The branch current labeled IO = %g milliamps\n\n',1000*(I_Loop(3) - I_Loop(2)));

for j=1:4
        fprintf ('Loop Current #%d = %g milliamps\n',j,1000*I_Loop(j));
end;

fprintf ('\n');
for j=1:4
        fprintf ('Node Voltage #%d = %g Volts\n',j,Vn(j));
end;
```

The MATLAB solution is,

```
The branch current labeled Ix = -3 milliamps
The branch current labeled IO = -2 milliamps

Loop Current #1 = 2 milliamps
Loop Current #2 = 7 milliamps
Loop Current #3 = 5 milliamps
Loop Current #4 = 6 milliamps

Node Voltage #1 = 9 Volts
Node Voltage #2 = -3 Volts
Node Voltage #3 = -5 Volts
Node Voltage #4 = -1 Volts
```

SIMULATION 3:

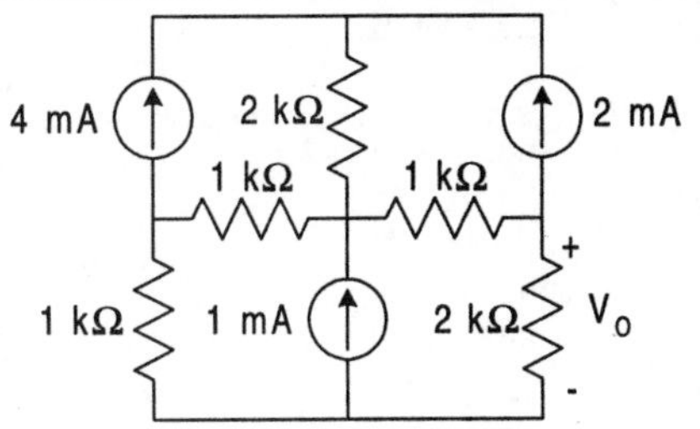

Figure 3.4. Circuit for third MATLAB simulation.

```
% MATLAB solution to Figure 3.4
% resistors and current sources
R1 = 2000;
R2 = 1000;
R3 = 1000;
R4 = 1000;
R5 = 2000;
Is1 = 0.004;
Is2 = 0.002;
Is3 = 0.001;

% Solve V = I*R where V is a vector of the node voltages,
% I is a vector of the branch currents and R is a matrix
% which describes the circuit topology.  Since we are
% using nodal analysis, the system is really V * 1/R = I

Rinv(1,1) = -1/R1;
Rinv(1,2) = 0;
Rinv(1,3) = 1/R1;
Rinv(1,4) = 0;
Rinv(2,1) = 0;
Rinv(2,2) = -1/R2 - 1/R4;
Rinv(2,3) = 1/R2;
Rinv(2,4) = 0;
Rinv(3,1) = 1/R1;
Rinv(3,2) = 1/R2;
Rinv(3,3) = -(1/R1 + 1/R2 + 1/R3);
Rinv(3,4) = 1/R3;
Rinv(4,1) = 0;
Rinv(4,2) = 0;
Rinv(4,3) = 1/R3;
Rinv(4,4) = -1/R3 - 1/R5;
Rinv(5,1) = 0;
Rinv(5,2) = 1/R4;
Rinv(5,3) = 0;
Rinv(5,4) = 1/R5;

% the source currents entering each node
IS = [-(Is1 + Is2) Is1 (-Is3) Is2 Is3]';

% Find the solution to the system with a matrix multiplication

V = Rinv \ IS;

% Compute the branch currents I once the node voltages are known
I(1) = (V(1) - V(3))/R1;
I(2) = (V(2) - V(3))/R2;
I(3) = (V(3) - V(4))/R3;
I(4) = V(2) / R4;
I(5) = V(4) / R5;

%print out the results in a pretty form
fprintf ('The voltage labeled Vo = %g volts\n\n',V(4));
        for j=1:5
                fprintf ('Branch Current #%d = %g milliamps\n',j,1000*I(j));
        end;
fprintf ('\n');
        for j=1:4
                fprintf ('Node Voltage #%d = %g Volts\n',j,V(j));
        end;
```

The MATLAB solution is,

```
The voltage labeled Vo = 1.6 volts

Branch Current #1 = 6 milliamps
Branch Current #2 = -4.2 milliamps
Branch Current #3 = 2.8 milliamps
Branch Current #4 = 0.2 milliamps
Branch Current #5 = 0.8 milliamps
Node Voltage #1 = 16.4 Volts
Node Voltage #2 = 0.2 Volts
Node Voltage #3 = 4.4 Volts
Node Voltage #4 = 1.6 Volts
```

SIMULATION 4:

```
% MATLAB solution to Figure 3.5
% resistors and sources
R1 = 2000;
R2 = 4000;
R3 = 4000;
R4 = 12000;
R5 = 2000;
VS1 = 6.0;
IS1 = 0.003;
IS2 = 0.001;

% Set up to solve the system V = RI.

R(1,1) = -R4;
R(1,2) = R3 + R4 + R5;
R(2,1) = R1 + R2;
R(2,2) = R3 + R5;
V(1) = VS1 + R3*(IS1 + IS2);
V(2) = V(1) + R1*IS1;

V = V';

% Find the solution to the system V = RI where I is the unknown
% Use MATLAB's matrix solution function

I = R\V;

I_Loop(2) = I(1);      % from the solved system of equations
I_Loop(4) = I(2);

I_Loop(1) = IS1;       % by observation
I_Loop(3) = I_Loop(4) - IS2;

% apply equations to find node voltages Vn
Vn(1) = VS1;
Vn(2) = Vn(1) + (I_Loop(1) - I_Loop(3)) * R3;
Vn(3) = Vn(2) + (I_Loop(1) - I_Loop(2)) * R1;
Vn(4) = I_Loop(4) * R5;

% print the output in a pretty form
fprintf ('The voltage labeled V0 = %g volts\n\n',Vn(3)-Vn(4));

for j=1:4
       fprintf ('Loop Current #%d = %g milliamps\n',j,1000*I_Loop(j));
end;

fprintf ('\n');
for j=1:4
       fprintf ('Node Voltage #%d = %g Volts\n',j,Vn(j));
end;
```

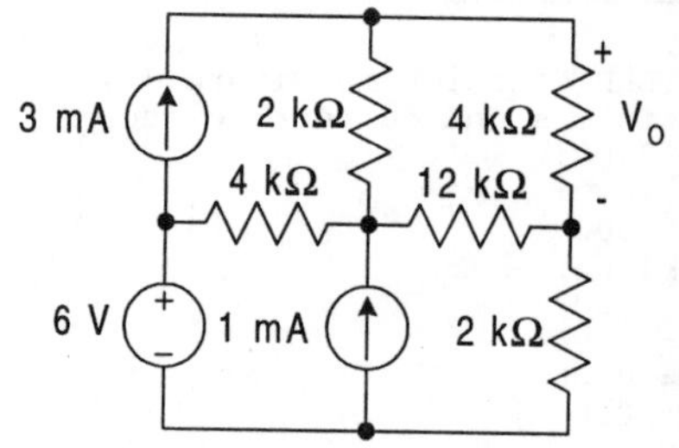

Figure 3.5. Circuit for fourth MATLAB simulation.

The MATLAB solution is,

```
The voltage labeled V0 = 5.89744 volts

Loop Current #1 = 3 milliamps
Loop Current #2 = 1.47436 milliamps
Loop Current #3 = 2.19231 milliamps
Loop Current #4 = 3.19231 milliamps

Node Voltage #1 = 6 Volts
Node Voltage #2 = 9.23077 Volts
Node Voltage #3 = 12.2821 Volts
Node Voltage #4 = 6.38462 Volts
```

PROBLEM SOLVING EXAMPLES

1. Find the current I_o in the following circuit using nodal equations.

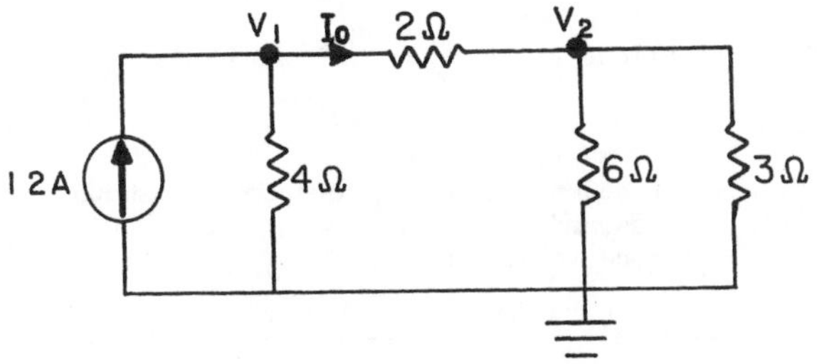

Solution: The nodal equations are

$$V_1(\tfrac{3}{4})-V_2(\tfrac{1}{2}) = 12$$

$$-V_1(\tfrac{1}{2})+V_2(1) = 0$$

Solving these equations yields $V_1 = 24V$, $V_2 = 12V$ and therefore $I_o = \frac{24-12}{2} = 6A$.

2. Use current division to find I_o in the previous problem.

Solution: The 2Ω resistor in series with the parallel combination of the 6Ω and 3Ω resistors yields a 4Ω resistor, therefore using current division $I_o = \frac{12(4)}{4+4} = 6A$.

3. Find V_o in the network below using nodal equations.

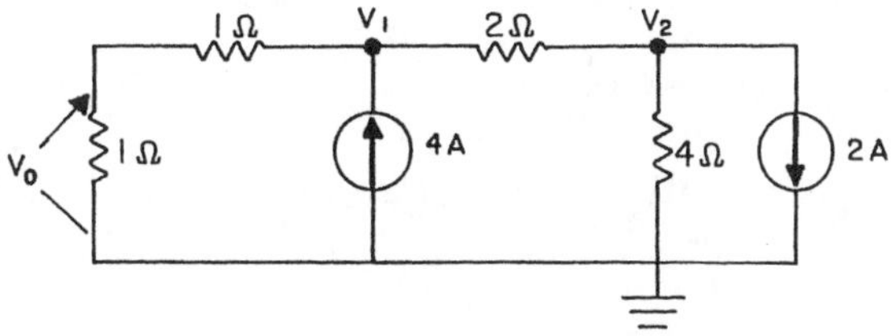

Solution: The nodal equations are

$$V_1(\tfrac{1}{2}+\tfrac{1}{2})-V_2(\tfrac{1}{2}) = 4$$

$$-V_1(\tfrac{1}{2})+V_2(\tfrac{1}{2}+\tfrac{1}{4}) = -2$$

Solving the equations for V_1 yields $V_1 = 4V$ and therefore using voltage division $V_o = V_1(\frac{1}{1+1}) = 2V$.

4. Use nodal equations to find I_o in the circuit shown below.

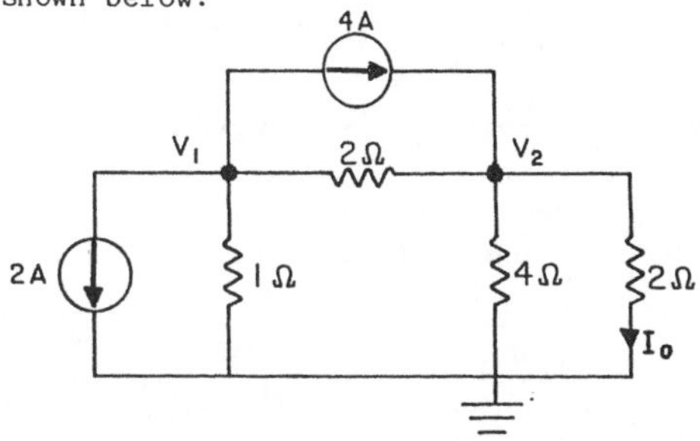

Solution: The nodal equations for the network are

$$V_1(\tfrac{1}{1}+\tfrac{1}{2})-V_2(\tfrac{1}{2}) = -2-4$$

$$-V_1(\tfrac{1}{2})+V_2(\tfrac{1}{2}+\tfrac{1}{4}+\tfrac{1}{2}) = 4$$

These equations yield $V_2 = \frac{24}{13}$ V and hence $I_o = \frac{12}{13}$ A.

5. Use node equations to find I_o in the following circuit.

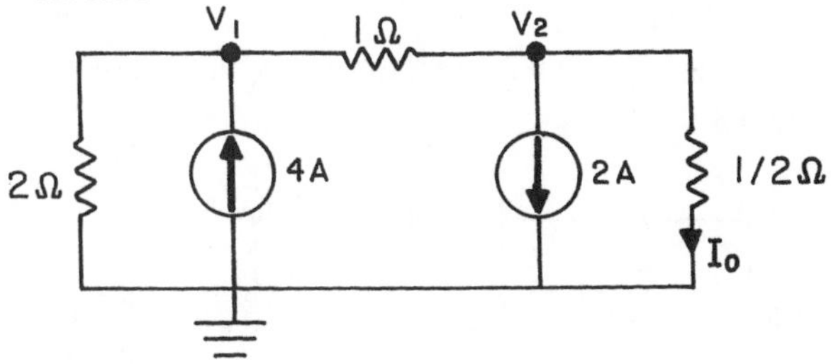

Solution: The equations which yield the node voltage are

$$V_1(\tfrac{1}{2}+\tfrac{1}{1})-V_2(\tfrac{1}{1}) = 4$$

$$-V_1(\tfrac{1}{1})+V_2(1+2) = -2$$

From these equations we obtain $V_2 = \frac{2}{7}$ V and hence $I_o = \frac{4}{7}$ A.

6. Find I_A in the network below using nodal analysis.

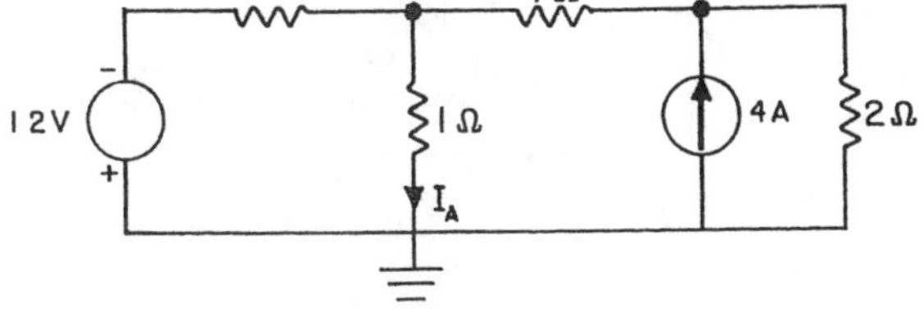

Solution: The nodal equations for the network are

$$\frac{V_A-(-12)}{2} + \frac{V_A}{1} + \frac{V_A-V_B}{1} = 0$$

$$\frac{V_B-V_A}{1} + \frac{V_B}{2} - 4 = 0$$

Solving these equations we obtain $V_A = -1.82V$ and hence $I_A = -1.82A$.

7. Use node equations to find I_o in the following network.

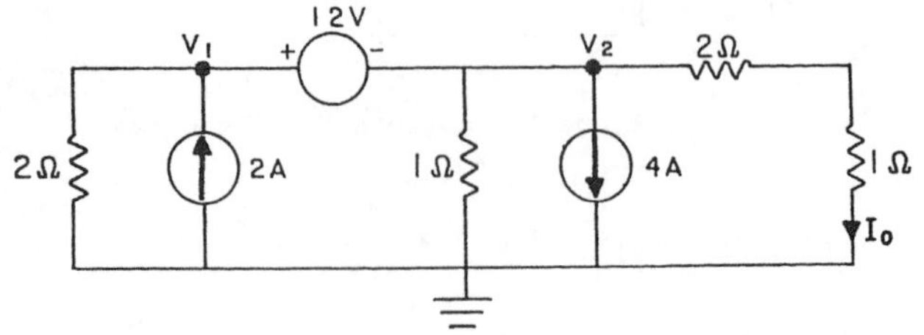

Solution: The constraint equation for the supernode is $V_1-V_2 = 12$ and KCL at the supernode yields $\frac{V_1}{2} -2+ \frac{V_2}{1} +4+ \frac{V_2}{3} = 0$. Solving these equations we obtain $V_2 = -\frac{48}{11}$ V and therefore $I_o = -\frac{16}{11}$ A.

8. Use nodal equations to find I_o in the circuit shown below.

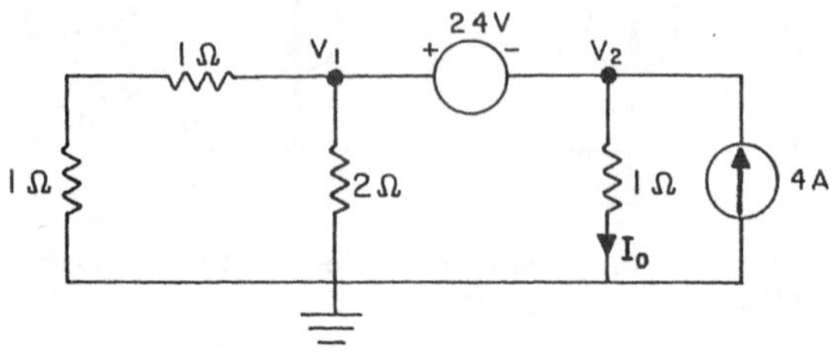

Solution: KCL for the supernode containing the voltage source is $\frac{V_2}{1} + \frac{V_2+24}{2} + \frac{V_2+24}{2} = 4$. Solving this equation we obtain $V_2 = -10V$ and therefore $I_o = -10A$.

9. Use nodal equations to find I_o in the following network.

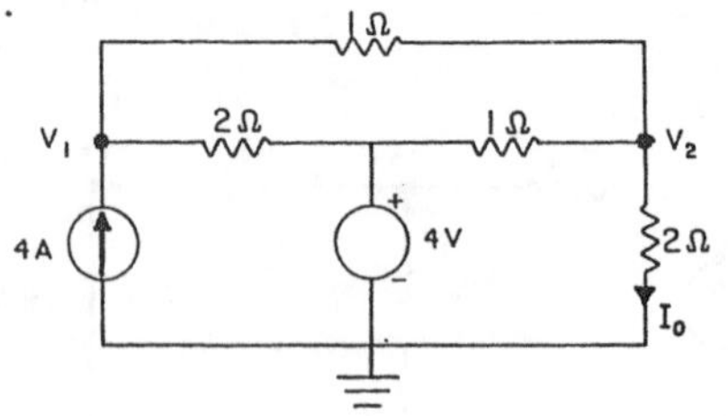

Solution: The nodal equations for the network are

$$\frac{V_1-4}{2} + \frac{V_1-V_2}{1} = 4$$

$$\frac{V_2-V_1}{1} + \frac{V_2-4}{1} + \frac{V_2}{2} = 0$$

These equations yield $V_2 = \frac{48}{11}$ V and hence $I_o = \frac{24}{11}$ A.

10. Write the node equations for the circuit shown below in standard form, which when solved, will yield all node voltages.

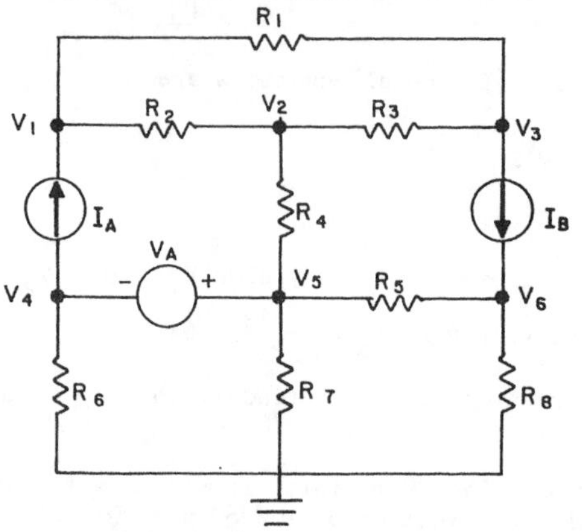

Solution: The equations for the network are

$$V_1(\frac{1}{R_1} + \frac{1}{R_2})-V_2(\frac{1}{R_2})-V_3(\frac{1}{R_1}) = I_A$$

$$-V_1(\frac{1}{R_2})+V_2(\frac{1}{R_2} + \frac{1}{R_3} + \frac{1}{R_4})-V_3(\frac{1}{R_3})-V_5(\frac{1}{R_4}) = 0$$

$$-V_1(\frac{1}{R_1})-V_2(\frac{1}{R_3})+V_3(\frac{1}{R_1} + \frac{1}{R_3}) = -I_B$$

$$-V_2(\frac{1}{R_4})+V_4(\frac{1}{R_6})+V_5(\frac{1}{R_4} + \frac{1}{R_5} + \frac{1}{R_7})-V_6(\frac{1}{R_5}) = -I_A$$

$$-V_4+V_5 = V_A$$

$$-V_5(\frac{1}{R_5})+V_6(\frac{1}{R_5} + \frac{1}{R_8}) = I_B$$

Note that the 4th and 5th equations are for the supernode containing V_A.

11. Find the current I_x in the network shown below using nodal analysis.

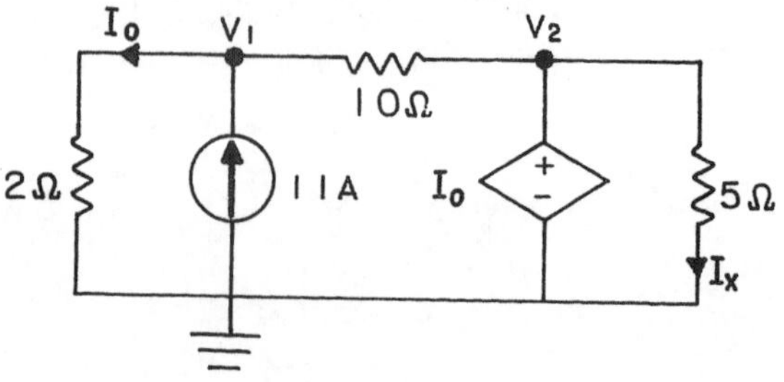

Solution: The equations for the network are

$$V_1\left(\frac{1}{2}+\frac{1}{10}\right)-V_2\left(\frac{1}{10}\right) = 11$$

$$V_2 = I_o = \left(\frac{V_1}{2}\right)$$

Solving these equations yields $V_1 = 20V$ and hence $V_2 = 10V$ and $I_x = 2A$.

12. Find the current I_o in the following network using nodal analysis.

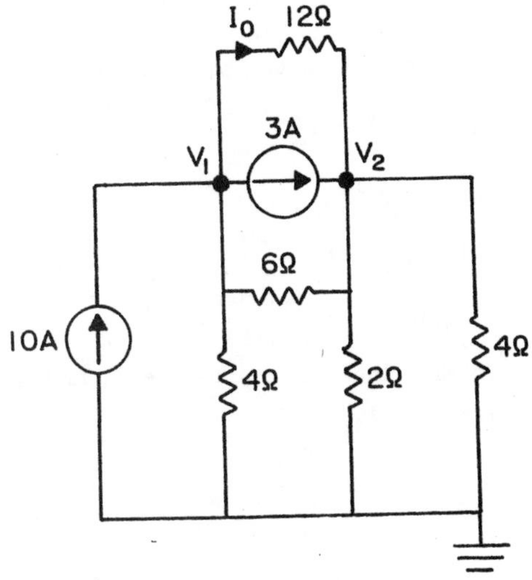

Solution: The nodal equations are

$$V_1\left(\frac{1}{12}+\frac{1}{6}+\frac{1}{4}\right)-V_2\left(\frac{1}{12}+\frac{1}{6}\right) = 10-3$$

$$-V_1\left(\frac{1}{12}+\frac{1}{6}\right)+V_2\left(\frac{1}{12}+\frac{1}{6}+\frac{1}{2}+\frac{1}{4}\right) = 3$$

Solving the equations yields $V_1 = 17.71V$, $V_2 = 7.43V$ and therefore $I_o = \frac{17.71-7.43}{12} = 0.857A$.

13. Write the nodal equations for the circuit shown below.

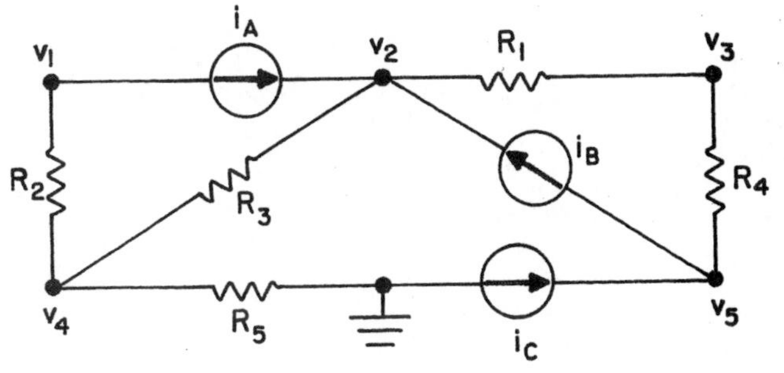

Solution: The nodal equations are

$$V_1\left(\frac{1}{R_2}\right)-V_4\left(\frac{1}{R_2}\right) = -i_A$$

$$V_2\left(\frac{1}{R_1}+\frac{1}{R_3}\right)-V_3\left(\frac{1}{R_1}\right)-V_4\left(\frac{1}{R_3}\right) = i_A+i_B$$

$$-V_2\left(\frac{1}{R_1}\right)+V_3\left(\frac{1}{R_1}+\frac{1}{R_4}\right)-V_5\left(\frac{1}{R_4}\right) = 0$$

$$-V_1\left(\frac{1}{R_2}\right)-V_2\left(\frac{1}{R_3}\right)+V_4\left(\frac{1}{R_2}+\frac{1}{R_3}+\frac{1}{R_5}\right) = 0$$

$$-V_3\left(\frac{1}{R_4}\right)+V_5\left(\frac{1}{R_4}\right) = i_C-i_B$$

14. Find the power absorbed by the 3Ω resistor in the network below using loop equations.

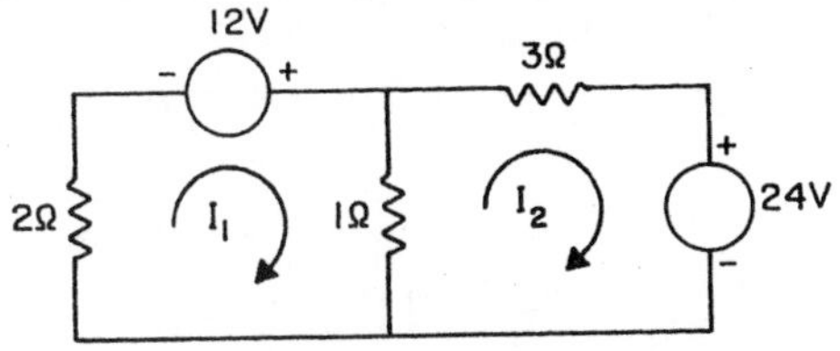

Solution: The loop equations are

$$3I_1-I_2 = 12$$

$$-I_1+4I_2 = -24$$

Solving these equations yields $I_2 = -\frac{60}{11}A$.

Therefore $P_{3\Omega} = \left(\frac{60}{11}\right)^2(3) = 89.25W$.

15. Write the loop equations for the network shown below.

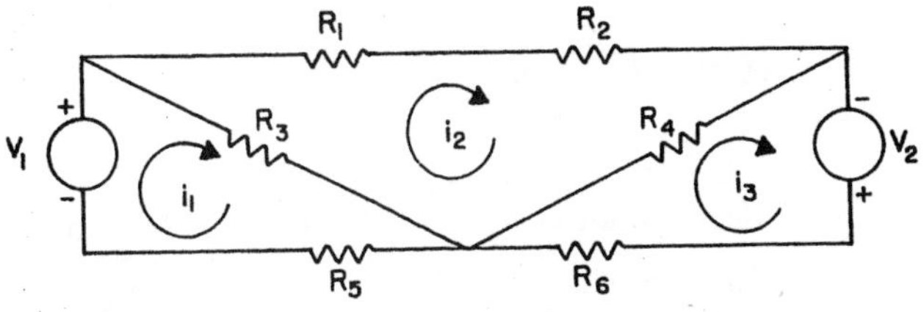

Solution: The loop equations for the network are

$$i_1(R_3+R_5)-i_2(R_3)-i_3(0) = V_1$$

$$-i_1(R_3)+i_2(R_1+R_2+R_3+R_4)-i_3(R_4) = 0$$

$$-i_1(0)-i_2(R_4)+i_3(R_4+R_6) = V_2$$

16. Write the mesh equations in standard form for the following network.

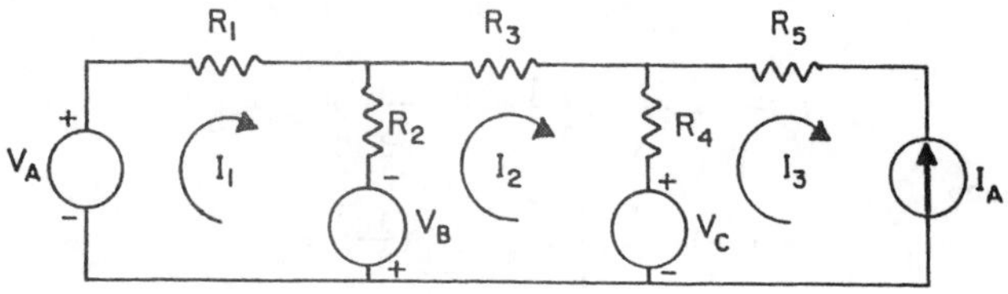

Solution: The mesh equations in standard form are

$I_1(R_1+R_2)-I_2(R_2)-I_3(0) = V_A+V_B$

$-I_1(R_2)+I_2(R_2+R_3+R_4)-I_3(R_4) = -V_B-V_C$

$I_3 = -I_A$

17. Use loop equations to determine V_o in the following circuit.

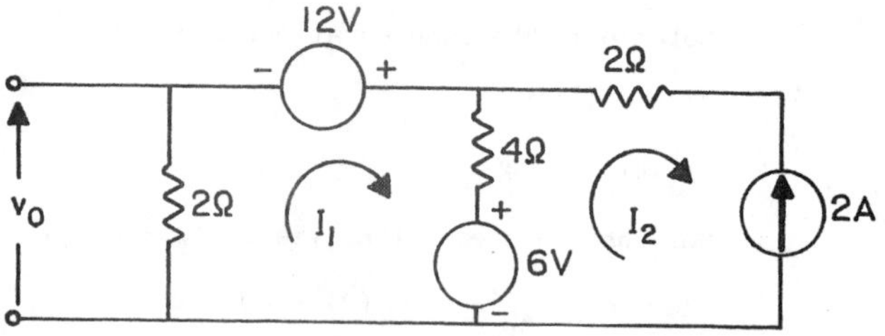

Solution: The loop equations for the network are

$12-6 = 6I_1-4I_2$

$I_2 = -2$

Solving these equations yields $I_1 = -\frac{1}{3}$ A and $V_o = -2I_1 = \frac{2}{3}$ V.

18. Determine the loop currents in the following network.

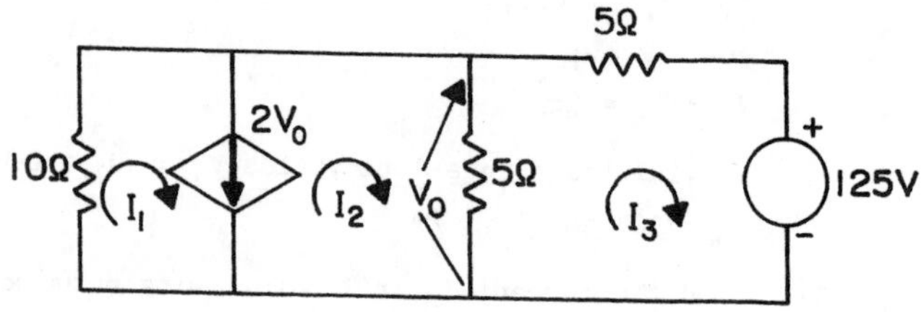

Solution: The loop equations are

$I_1-I_2 = 2V_o$

$V_o = 5(I_2-I_3)$

$10I_1+5(I_2-I_3) = 0$

$-5I_2+10I_3 = -125$

Solving these equations yields $I_2 = -21$A, $I_3 = -23$A and $I_1 = -1$A.

Chapter 4 ADDITIONAL ANALYSIS TECHNIQUES

SIMULATING DC CIRCUITS USING PSPICE

In the mid 1960's, Dr. Donald Pederson at the University of California at Berkeley began work on a program for simulating circuit performance which came to be known as SPICE (Simulation Program with Integrated Circuit Emphasis). At Pederson's insistence, the program has remained public domain software. However, several companies have written their own simulators, based on the SPICE engine, with advanced front-end interfaces and graphics. One such product is PSpice, formerly produced by MicroSim but recently acquired by OrCAD.

WHAT IS PSPICE

In this study guide, we will be discussing PSpice version 8.0 from MicroSim. OrCAD's Version 9.0 is scheduled for release in late 1998/early 1999 with significant changes.

Strictly speaking, PSpice is a circuit simulator only. However, most engineers use two other programs that make PSpice easier to use, *Schematics* and PROBE. *Schematics* is a schematic-entry program that allows you to draw the circuit you wish to simulate. PROBE, a graphing utility that interfaces directly to PSpice and *Schematics*, creates plots from the simulation results. Instructions on using all three of these programs in unison are given in the BECA text.

DISPLAY RESULTS ON THE SCHEMATIC

One of the advances in Version 8.0 is the ability to easily display dc node voltages and branch currents directly on the circuit diagrams you will draw in *Schematics*. To display voltages, open *Schematics* and go to the Analysis menu. Select Display Results on Schematics option and choose Enable Voltage Display as shown in Figure 4.1. To display currents, repeat the process and choose Enable Current Display.

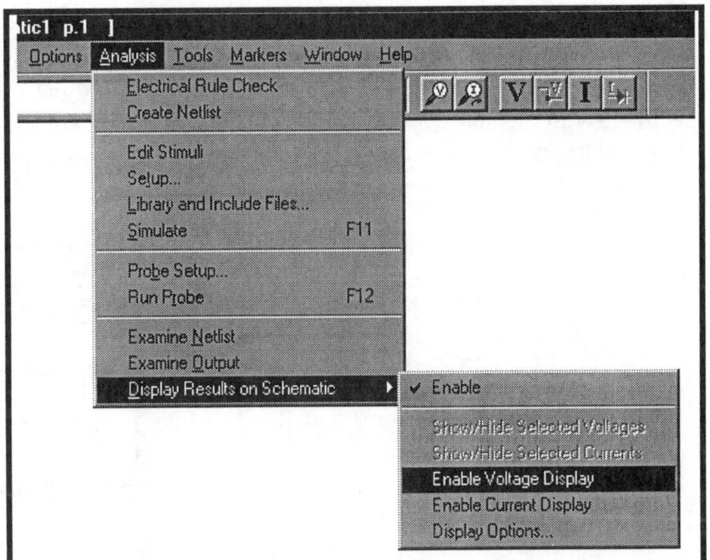

Figure 4.1. Using the menus to Enable Voltage Display.

Currents can also be displayed using *Schematics*' dc ammeter part called IPROBE. Simply insert the IPROBE part into any branch, just like a real ammeter, and the value will be displayed on the screen. Unfortunately, the IPROBE part in Version 8.0 has a bug - the displayed current has no sign! So, you know the current's magnitude but not its direction. We recommend you use the Display Results on Schematic feature instead.

DC VISUAL TUTORIAL

Throughout this guide we will refer to Pspice Visual Tutors. These are video files that document the step-by-step processes in circuit creation and simulation. All of the visual tutors are on the CD-ROM and, since they are executable files, you can run any of them at any time. The first tutor is called DC_TUTOR.EXE. Give it a try.

THE BECA LIBRARY

On the Study Guide CD-ROM you will also find a directory called BECA LIBRARIES which contains four PSpice library files. These libraries contain custom parts that have been created for your convenience and edification. Text files describing each part can be found in the BECA PART READMES directory on the CD-ROM. These text files can be opened using the PSpice Text Editor, Notepad or Wordpad.

Custom Parts

Shown in Figure 4.2 are two parts in the BECA library that may be of some use in dc simulations - the CPL part and the instantaneous wattmeter. The CPL part is a constant power load. As the name implies, it draws constant power regardless of the voltage across it. Details on using the CPL part is given in the files CPL.TXT in the BECA PART READMES directory.

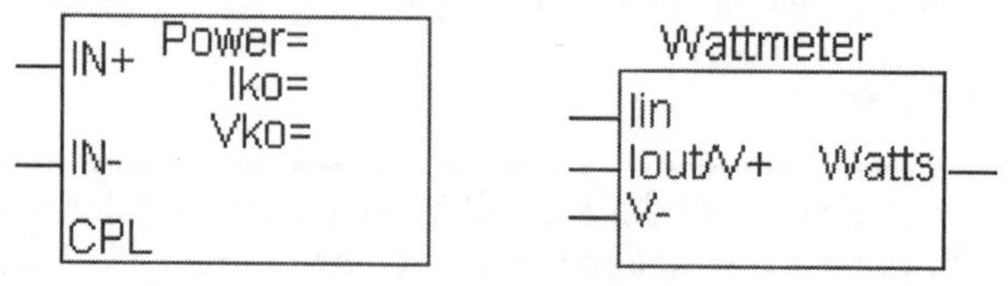

Figure 4.2. The CPL (Constant Power Load) and Wattmeter parts in the BECA library.

Power can be measured using the instantaneous wattmeter. It has three connections just like its real world counterpart. Look at the READMES file WATTMETER.TXT for instructions on using this part.

In the BECA library, you'll also find a quasi-ideal operational amplifier. Its *Schematics* symbol, equivalent circuit and Attributes box are shown in Figure 4.3. You can specify the op-amp's gain, input and output resistances and supply voltages. VCC is the positive supply and VEE is the negative supply, which have been set to $\pm$ 5 V in Figure 4.3c. As in a real op-amp, the output voltage in our quasi-ideal op-amp cannot exceed the supply voltages. Note that the output voltage will be referenced to the ground node (node #0) in the *Schematics* diagram.

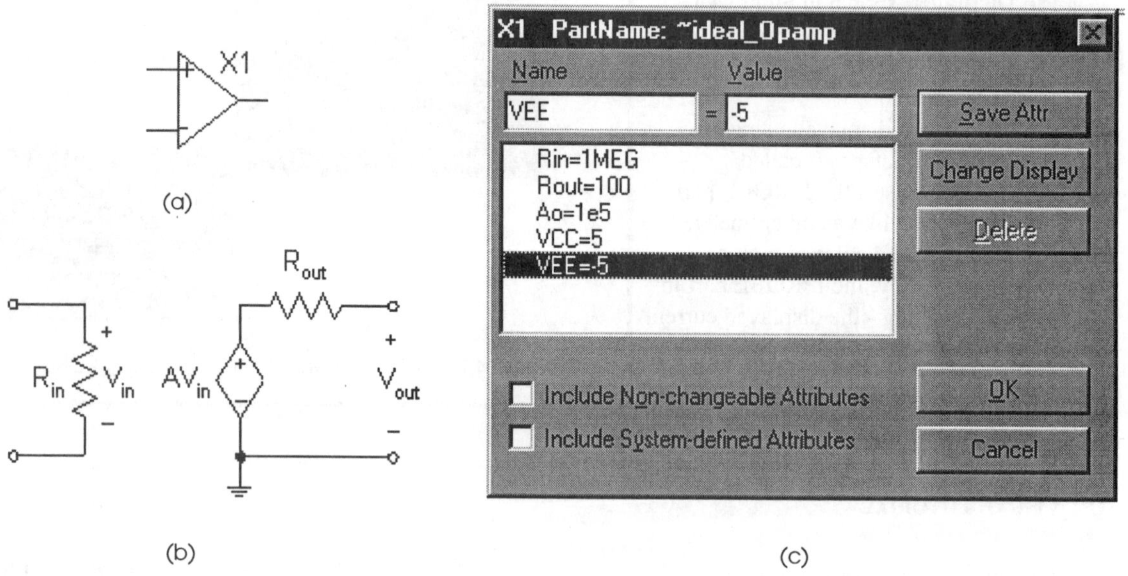

Figure 4.3. The quasi-ideal op-amp in the BECA library: (a) the Schematics symbol, (b) the equivalent circuit and (c) the Attributes box.

Getting the BECA Library

To get the BECA libraries into PSpice, copy all four library files from the CD-ROM to the MsimEv_8 subdirectory called lib.

Including the BECA Library

To use the BECA parts, *Schematics* must be permitted to access them. This process is called "including" the library. To include the BECA libraries:

1. Open *Schematics*, go to the Analysis menu and choose Library and Include Files. The dialog box in Figure 4.4a will appear.
2. Using the Browse button, select BECA.lib in the MsimEv_8/lib folder.
3. Click on the Add Library* button. Click OK. Now PSpice knows where the electrical models for the parts are located.
4. Where the are BECA part symbols? In the Options menu, select Editor Configuration. The window in Figure 4.4b will open.
5. Enter C:\MSimEV_8\lib in the Library Path field and select Library Settings to open the Library Settings window in Figure 4.4c.
6. Next, use Browse to select the file BECA.slb in the Msim_Ev.8/lib.
7. Back in the Library Settings, click OK.
8. Back in the Editor Configuration window, click on Add Library*, then OK.

You now have full access to the BECA parts.

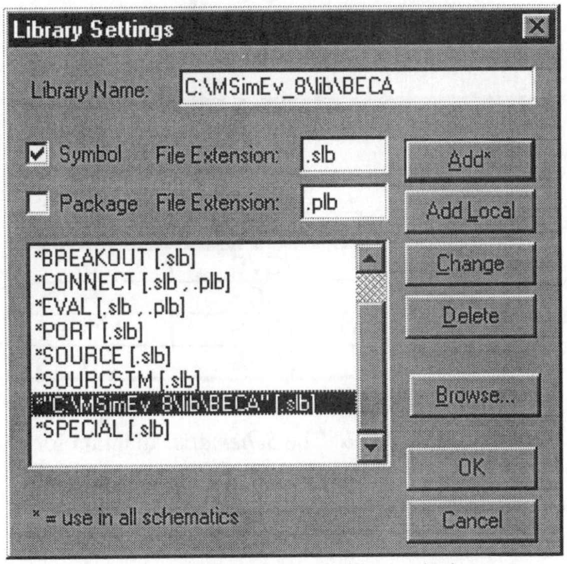

Figure 4.4c. The Library Settings window.

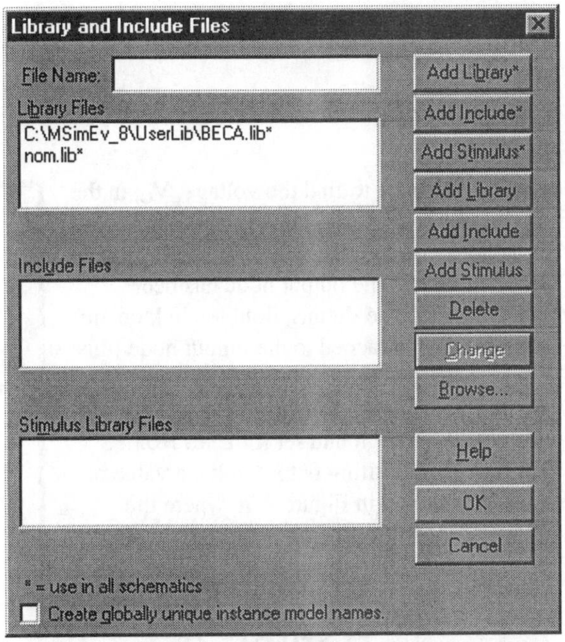

Figure 4.4a. Including the BECA parts library.

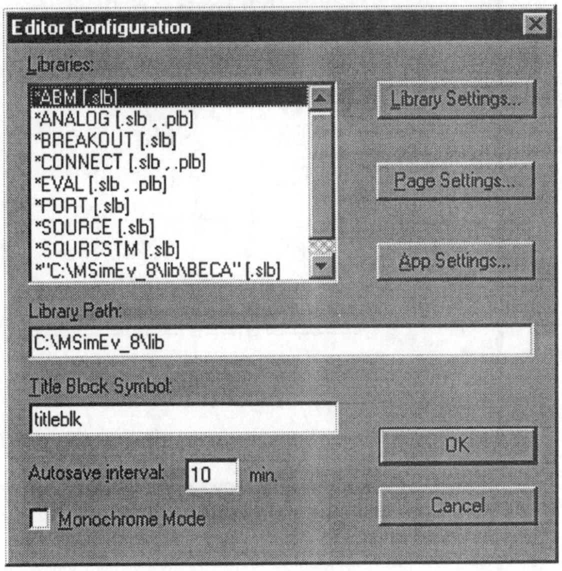

Figure 4.4b. The Editor Configuration window.

PSPICE SIMULATIONS

Simulation One

Let's use PSpice to find the voltage, V_O, in the circuit in Figure 4.6. The corresponding *Schematics* circuit diagram is shown in Figure 4.5a. Note that the output node has been labeled "Vo". To do this, double-click on any wire segment attached to the output node (this opens a simple dialog box) and type the desired node name. Perform the simulation with the SETUP menu set for Bias Point Detail. The resulting output voltage value of 1.33 V is shown in Figure 4.5b where the Display Results on Schematic option has been selected.

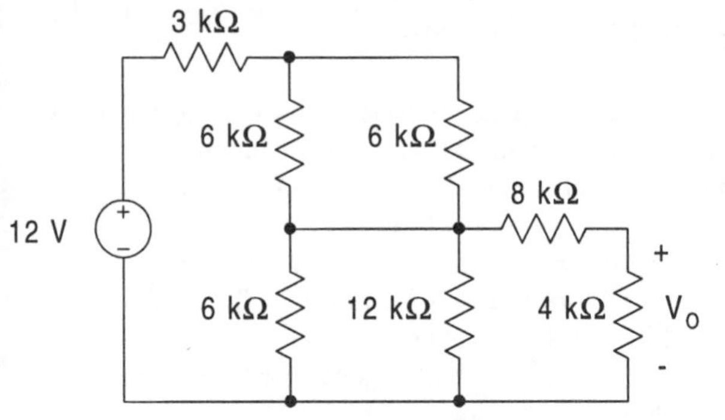

Figure 4.6. Circuit for first PSICE simulation.

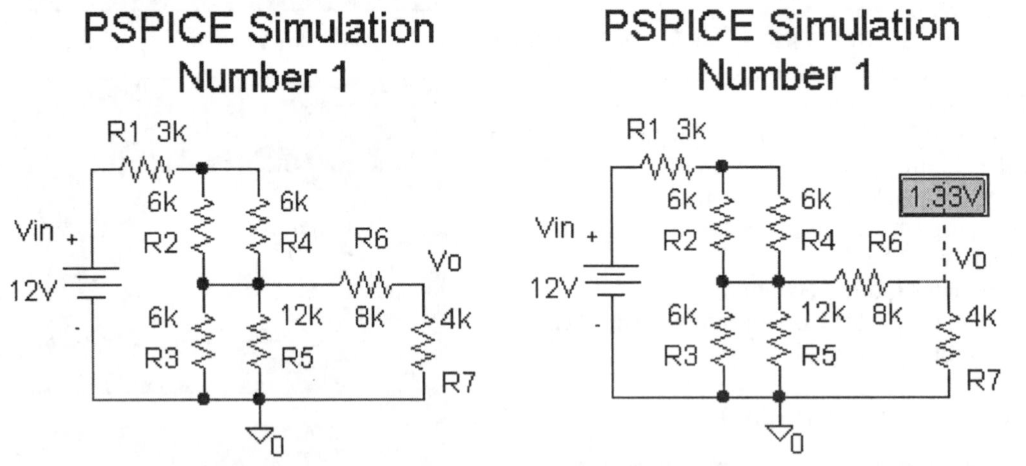

Figure 4.5. *Schematics* diagrams for the circuit in Figure 4.4, (a) pre-simulation and (b) post-simulation.

Simulation Two

Since PSpice contains parts for all four dependent source types; current-controlled current sources, voltage-controlled current sources, current-controlled voltages and voltage-controlled voltage sources, simulating dependent source circuits is quite easy. All dependent sources are in the ANALOG library. TABLE 4.1 lists the dependents sources and their PSpice parts names.

TABLE 4.1

PSpice DEPENDENT SOURCES

Source Type	Part Name
VCVS	E
CCCS	F
VCCS	G
CCVS	H

Figure 4.7a shows a dependent-source rich circuit we will simulated in PSPICE. The *Schematics* diagram for the circuit is shown in Figure 4.7b. To set the gain (multiplicative factor) of a dependent source, double-click on the source to open its Attributes Dialog box, type in the gain and select OK. Simulation results (V_{OUT} = - 24 V) are shown in Figure 4.7b where the Display Results on Schematic option has been selected.

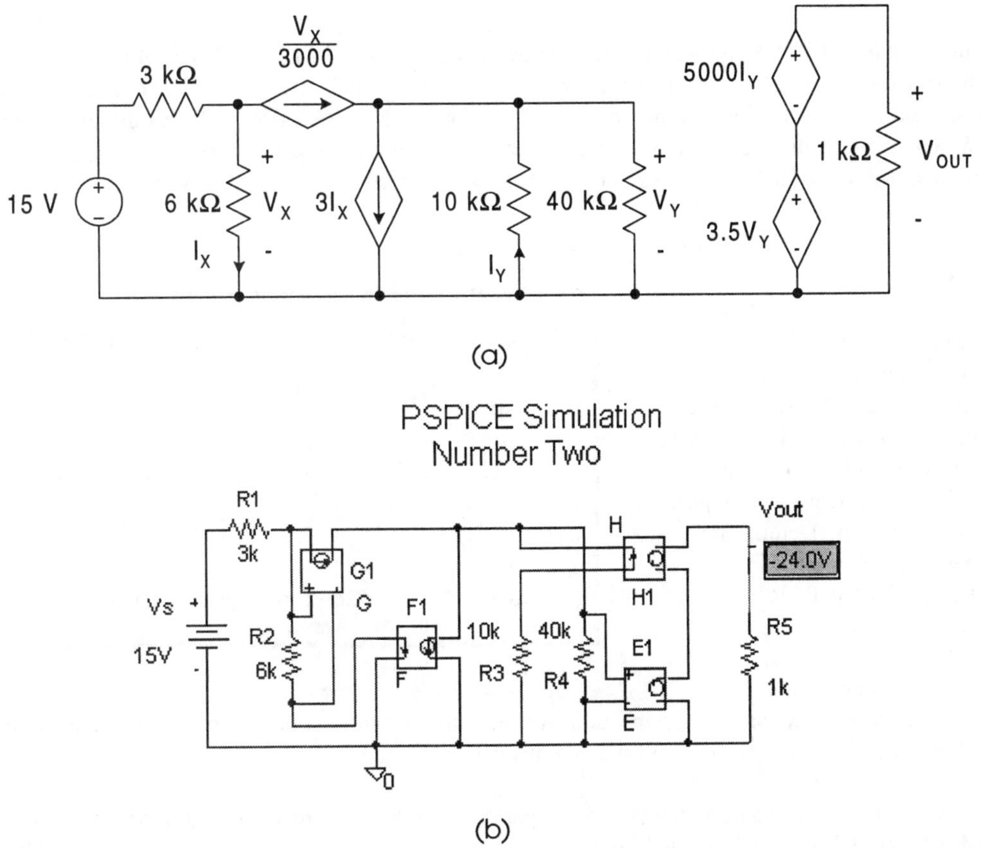

Figure 4.7. Schematics diagrams for the circuit in Figure 4.6, (a) pre-simulation and (b) post-simulation.

Simulation 3 - Using the Quasi-ideal Op-amp

Let's use the quasi-ideal op-amp in the BECA library to simulate a circuit that finds the average value of two voltages. The circuit in Figure 4.8 will perform this task. The output of the first op-amp is

$$V_{O1} = -\frac{R_3}{R_1}V_1 - \frac{R_3}{R_2}V_2$$

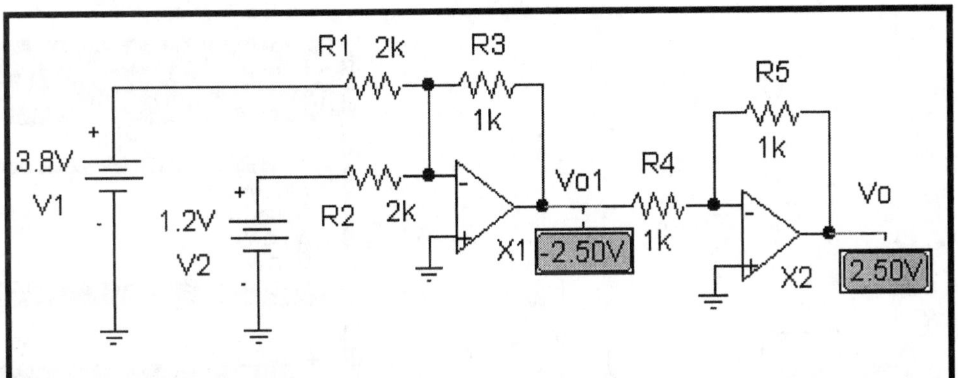

Figure 4.8. A averaging circuit simulating using the quasi-ideal op-amp in the BECA library.

For obvious reasons, the first op-amp subcircuit is called a summer. If we make $R_1 = R_2 = 2\,R_3$, then we have the negative of the average of V_1 and V_2 - namely -2.5 V. The second op-amp subcircuit has a gain of -1 which yields a positive average at V_O.

THE BECA LIBRARY

On the Study Guide CD-ROM you will also find a directory called BECA LIBRARIES which contains four PSpice library files. These libraries contain custom parts that have been created for your convenience and edification. Text files describing each part can be found in the BECA PART READMES directory on the CD-ROM. These text files can be opened using the PSpice Text Editor, Notepad or Wordpad. To use the custom parts, you must install the BECA libraries into *Schematics*. Instructions are listed on page 29.

Custom Parts

Shown in Figure 4.2 are two parts in the BECA library that may be of some use in dc simulations - the CPL part and the instantaneous wattmeter. The CPL part is a constant power load. As the name implies, it draws constant power regardless of the voltage across it. Details on using the CPL part is given in the files CPL.TXT in the BECA PART READMES directory.

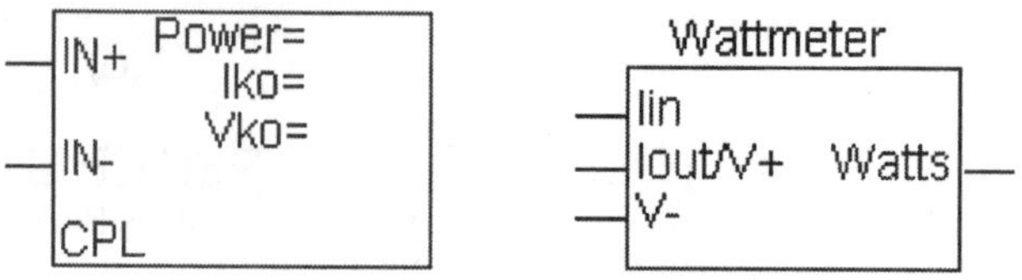

Figure 4.2. The CPL (Constant Power Load) and Wattmeter parts in the BECA library.

Power can be measured using the instantaneous wattmeter. It has three connections just like its real world counterpart. Look at the READMES file WATTMETER.TXT for instructions on using this part.

In the BECA library, you'll also find a quasi-ideal operational amplifier. Its *Schematics* symbol, equivalent circuit and Attributes box are shown in Figure 4.3. You can specify the op-amp's gain, input and output resistances and supply voltages. VCC is the positive supply and VEE is the negative supply, which have been set to ± 5 V in Figure 4.3c. As in a real op-amp, the output voltage in our quasi-ideal op-amp cannot exceed the supply voltages. Note that the output voltage will be referenced to the ground node (node #0) in the *Schematics* diagram.

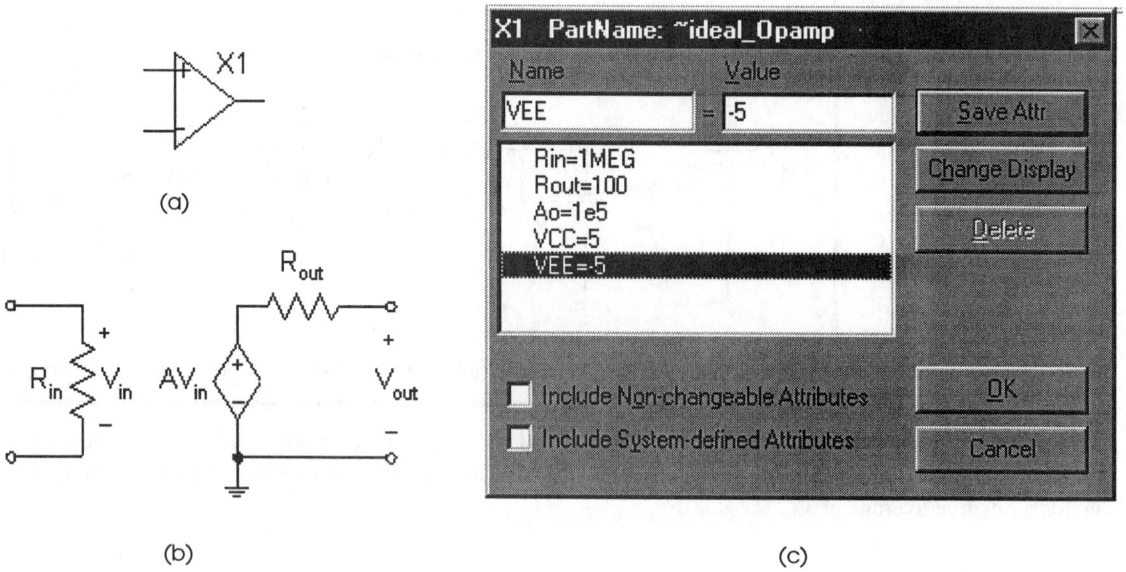

Figure 4.3. The quasi-ideal op-amp in the BECA library: (a) the Schematics symbol, (b) the equivalent circuit and (c) the Attributes box.

Getting the BECA Library

To get the BECA libraries into PSpice, copy all four library files from the CD-ROM to the MsimEv_8 subdirectory called UserLib.

Including the BECA Library

To use the BECA parts, *Schematics* must be permitted to access them. This process is called "including" the library. To include the BECA libraries:

1. Open *Schematics*, go to the Analysis menu and choose Library and Include Files. The dialog box in Figure 4.4a will appear.
2. Using the Browse button, select BECA.lib in the MsimEv_8/UseLlib folder. Click OPEN.
3. Back in the Library and Include Files window, click on the Add Library* button. Click OK. Now PSpice knows where the electrical models for the parts are located.
4. Where are the BECA part symbols? In the Options menu, select Editor Configuration. The window in Figure 4.4b will open.
5. Select Library Settings to open the Library Settings window in Figure 4.4c.
6. Next, use Browse to select the file BECA.slb in the Msim_Ev.8/UserLib folder. Click open.
7. Back in the Library Settings, click on Add*, then click OK.
8. Back in the Editor Configuration window, click OK.

You now have full access to the BECA parts.

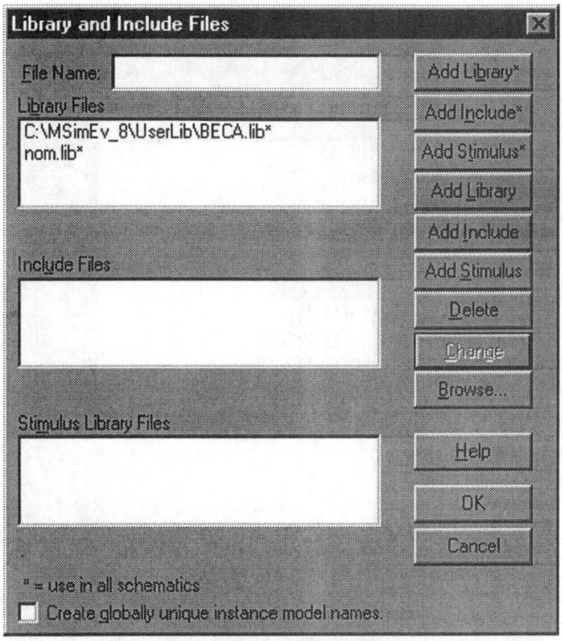

Figure 4.4a. Including the BECA parts library.

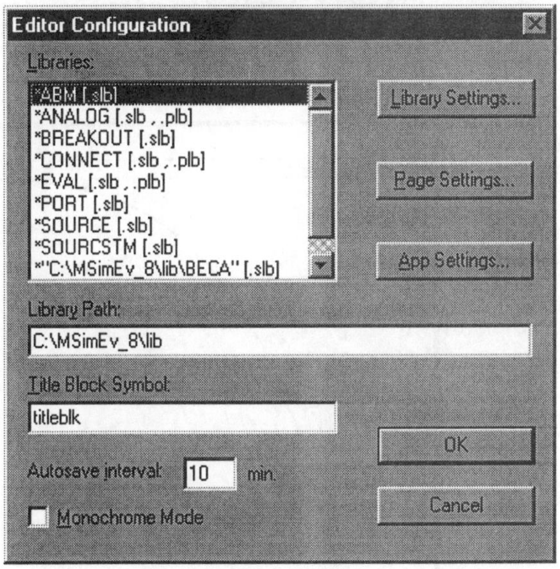

Figure 4.4b. The Editor Configuration window.

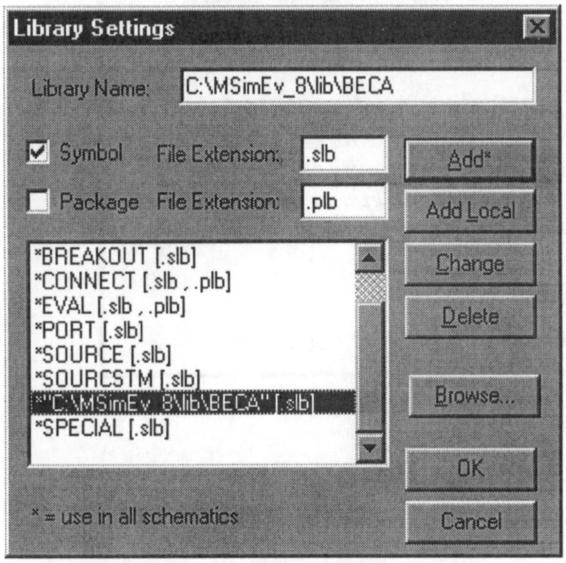

Figure 4.4c. The Library Settings window.

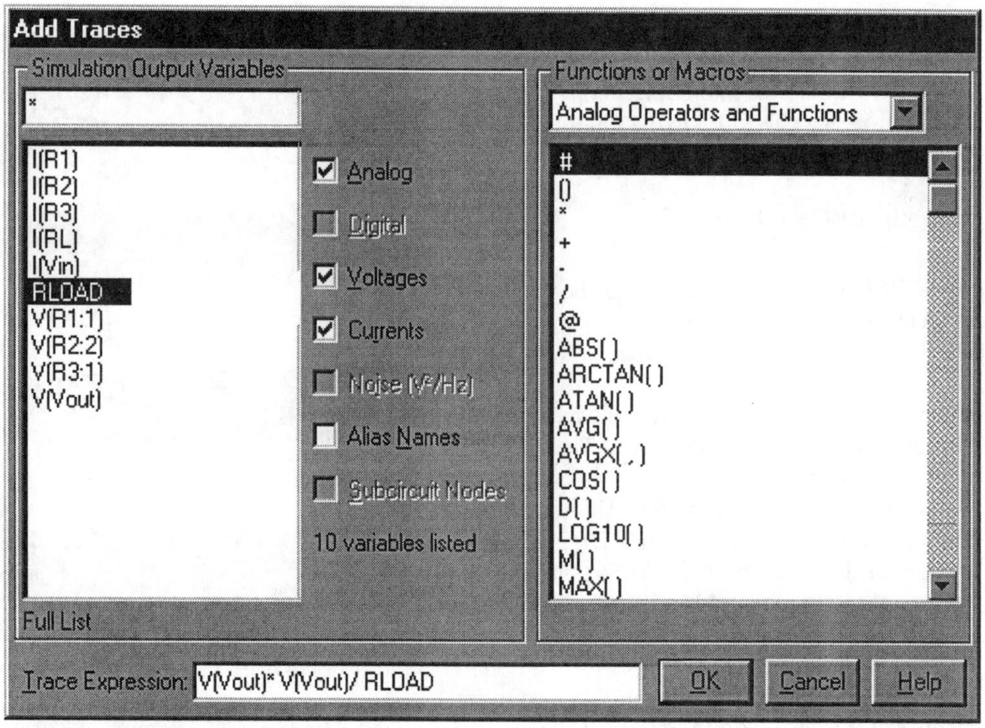

Figure 4.14. The Add Traces dialog box.

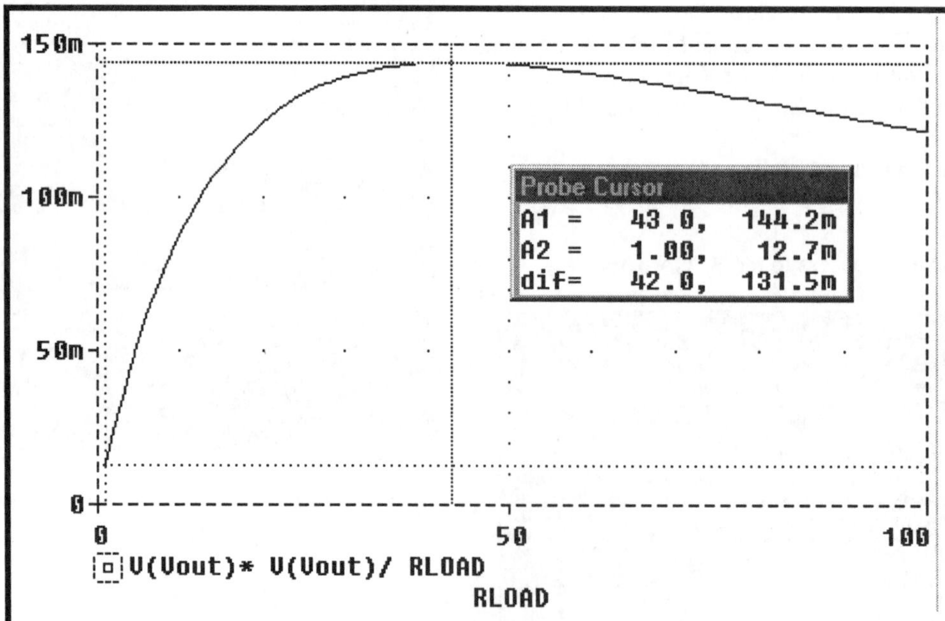

Figure 4.15. PROBE plot showing maximum power transfer at about 43 Ω.

Documentation and the README Part

Good engineers always document their work with commentary and informative graphics. *Schematics* provides three methods of documentation. First, terse messages can be added by selecting either **Text** or **Text Box** from the **Draw** menu. Second, graphics can be added by selecting **Insert Picture** from the **Draw** menu. And third, extensive commentary can be appended using the **README** part from the **SPECIAL** library. Simply add the part to the schematic and double click on it. When its attribute box opens, type in the name of the text file that contains your comments (*filename*.txt). If the file does not exist, you will be prompted to create it. Your text file can be written in any text editor. Most people use Notepad, Wordpad or the PSpice Text Editor.

Upon finishing your commentary, save it and close the text editor. Now, if at a later date you return to that particular schematic, the README part and the text file reference are included. To open the file, just double-click on the file's name. Note that the *Schematics* file (.sch extension) and the text file must be in the same directories they were in when you specified the README part file name reference. This is easily done if you keep all file relating to a schematic in the same directory.

SIMULATING CIRCUITS USING ELECTRONICS WORKBENCH

What is Electronics Workbench (EWB)

Electronics Workbench, from Interactive Image Technologies, is another SPICE-based simulator with a schematic entry user interface and plotting utilities. The most unique and popular feature of EWB is its virtual instruments which include a voltmeter, ammeter, oscilloscope as well as other instruments too advanced to mention at this point. While EWB is not quite as prevalent as PSpice, being quite a bit less expensive has made it a very popular entry-level simulator for individuals. In this guide, we will not discuss how to use EWB. However, we will include circuit simulation examples for those of you who do.

EWB Simulation One

Let's simulate the circuit in Figure 4.16 to find the voltage, V_{OUT}. The EWB schematic is shown in Figure 4.17a. The voltage display part is the above mentioned voltmeter. Figure 4.17b shows the simulation results, namely, $V_{OUT} = 3.556$ V.

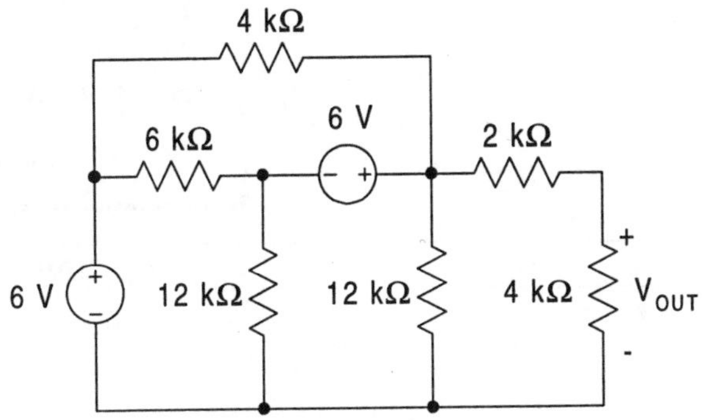

Figure 4.16. Circuit diagram for the first EWB simulation.

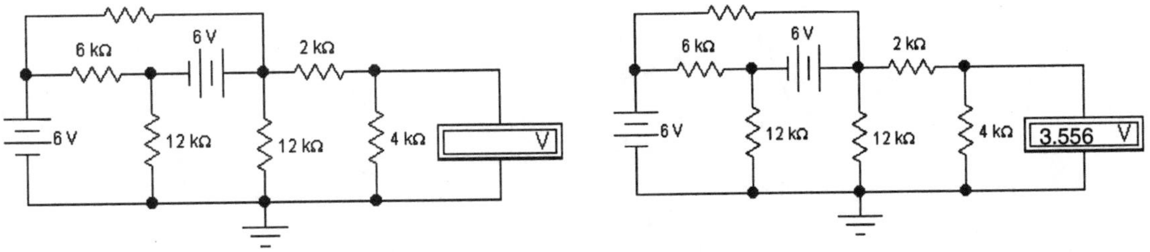

(a) (b)

Figure 4.17. EWB circuit diagrams for the circuit in Fig. 4.16, (a) before and (b) after simulation.

EWB Simulation Two

For our second EWB example, we'll simulate the dependent source circuit in Figure 4.18 to find the current, I_O. The EWB schematic is shown in Figure 4.19a where we have used the virtual multimeter in EWB to meter I_O. Figure 4.19b shows the "faceplate" of the meter set to measure dc amps. The simulation results are $I_O = -631.6\ \mu A$.

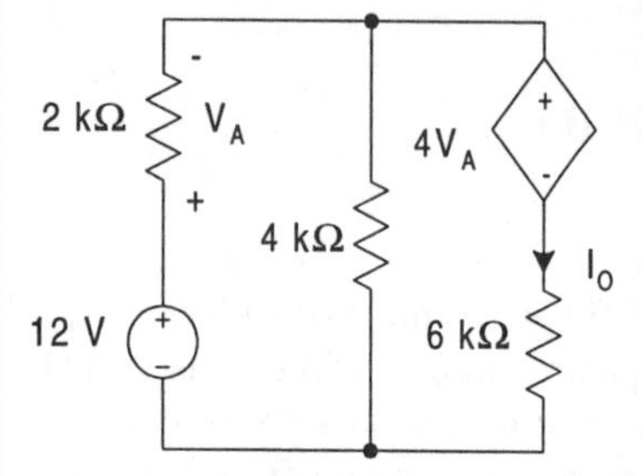

Figure 4.18. Circuit diagram for the second EWB simulation.

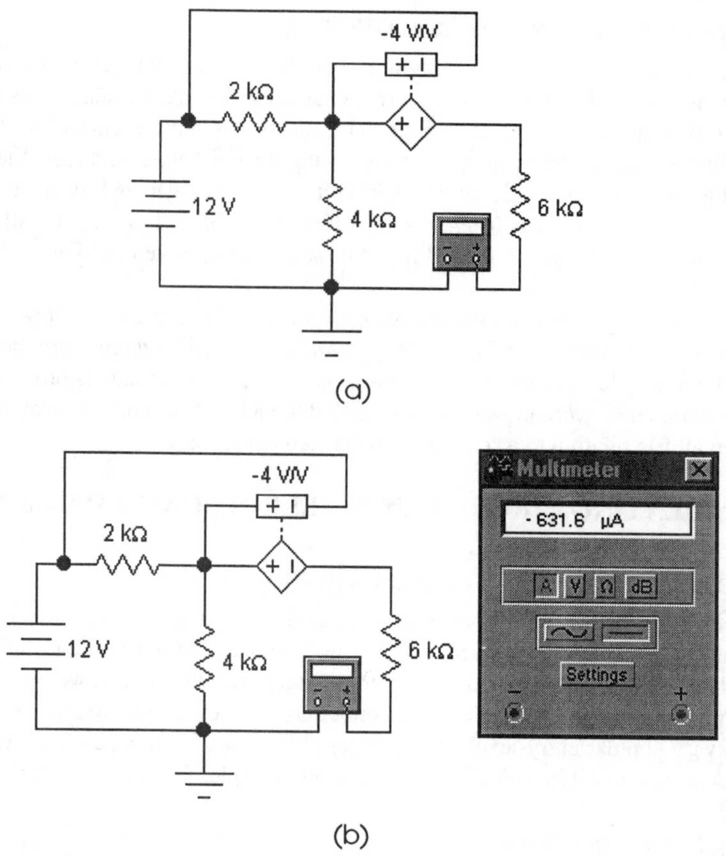

Figure 4.19. EWB schematics for the circuit in Figure 4.17, (a) before and (b) after simulation. The virtual multimeter in EWB is used to meter Io.

NOTES

PROBLEM SOLVING EXAMPLES

1. Assume I_o = 1A and use linearity to find the actual value of I_o in the following network.

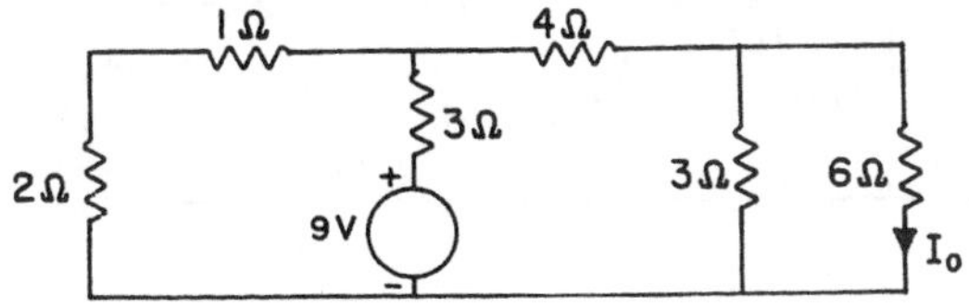

Solution: If we label the network as follows,

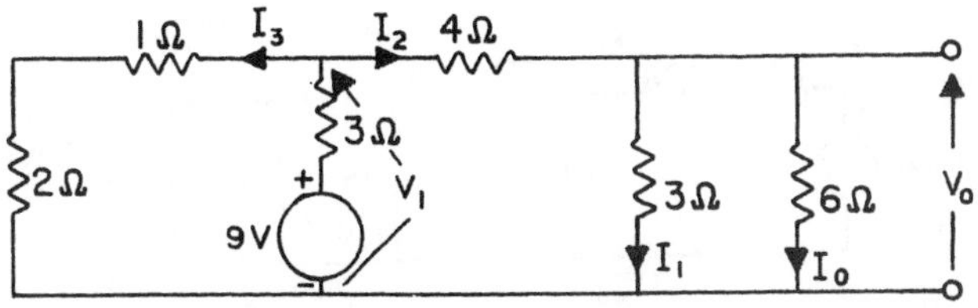

Then $V_o = (1)(6) = 6V$ and $I_1 = \frac{6}{3} = 2A$. $I_2 = I_1+I_o = 3A$ and hence $V_1 = 4I_2+V_o = 18V$. $I_3 = \frac{V_1}{2+1} = 6A$ and therefore the source voltage is $V_S = 3(I_2+I_3)+V_1 = 45V$. Using linearity $\frac{45}{1} = \frac{9}{I_o}$.

Therefore $I_o = \frac{1}{5}$ A.

2. Find I_o in the following network using superposition.

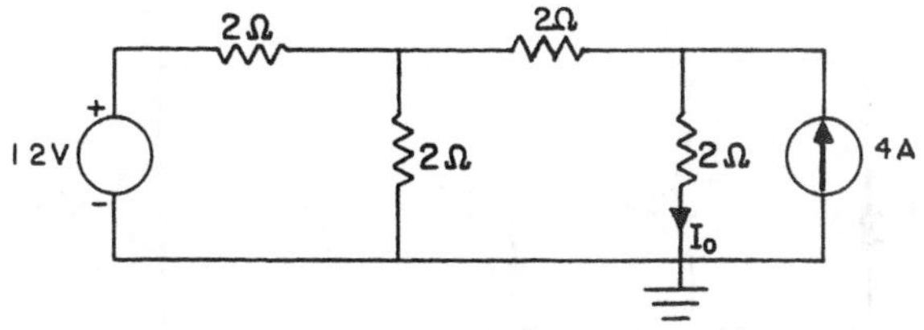

Solution: Consider the following networks.

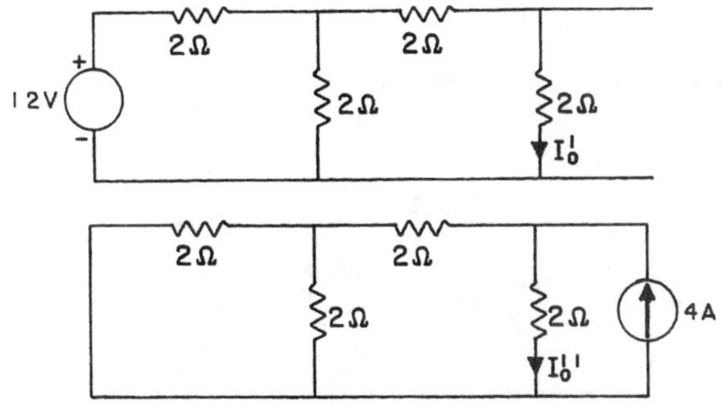

$$I_o' = \frac{12}{2+\frac{(2)(4)}{2+4}}\left(\frac{2}{6}\right) = \frac{12}{10} \text{ A}$$

$$I_o'' = \frac{(4)(2+1)}{2+2+1} = \frac{12}{5} \text{ A}$$

Therefore $I_o = I_o'+I_o'' = \frac{18}{5}$ A.

3. Use superposition to determine I_o in the circuit below.

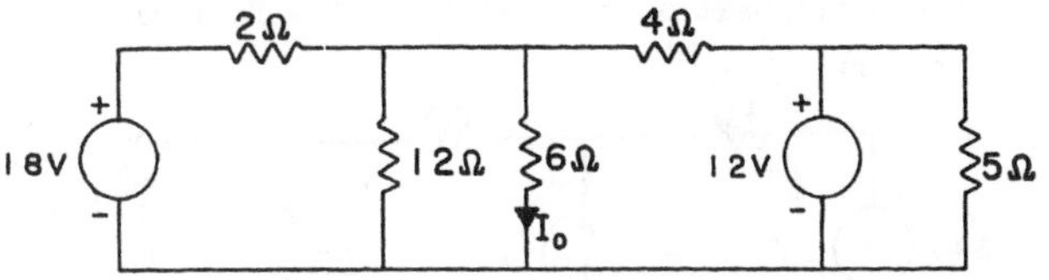

Solution: Consider the following networks.

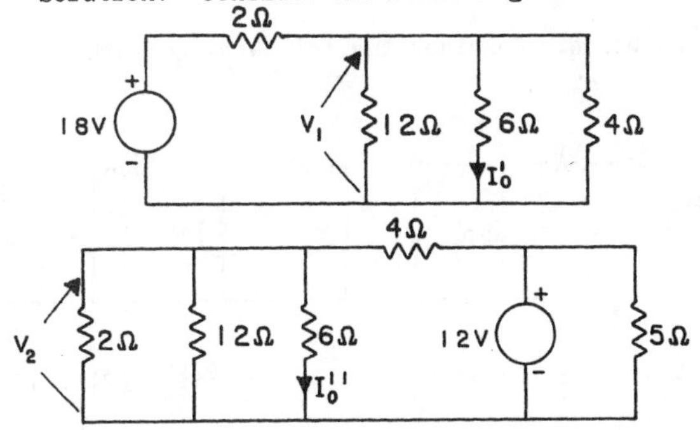

$$V_1 = \frac{18}{2+2}(2) = 9V \qquad V_2 = \frac{12}{4+\frac{8}{6}}\left(\frac{8}{6}\right) = 3V$$

$$I_o' = \frac{9}{6} \text{ A} \qquad I_o'' = \frac{1}{2} \text{ A}$$

$$I_o = I_o'+I_o'' = 2A$$

4. Determine the current I_x in the circuit shown below, using superposition.

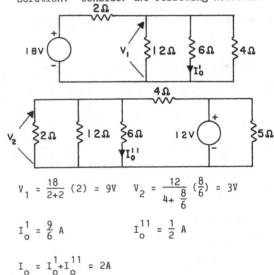

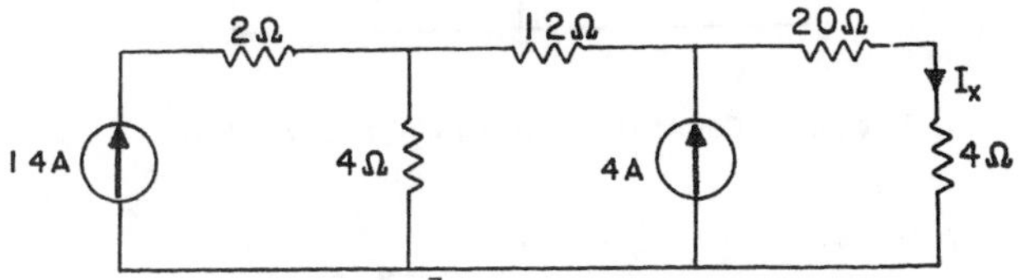

Solution: Consider the two networks below.

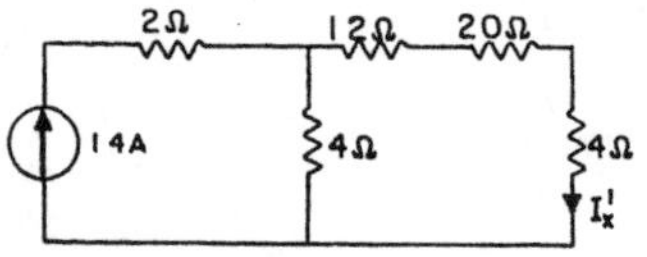

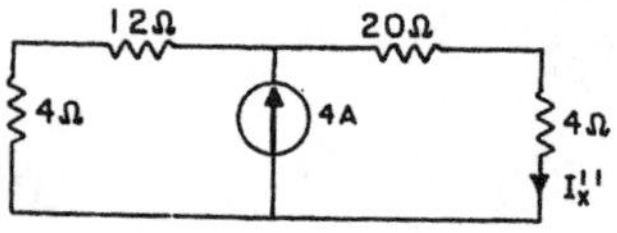

$I_x' = \frac{(14)(4)}{4+12+20+4} = 1.4A$ $\quad I_o'' = \frac{(4)(16)}{16+24} = 1.6A$

$I_o = I_o' + I_o'' = 3A.$

5. Use superposition to find V_o in the following network.

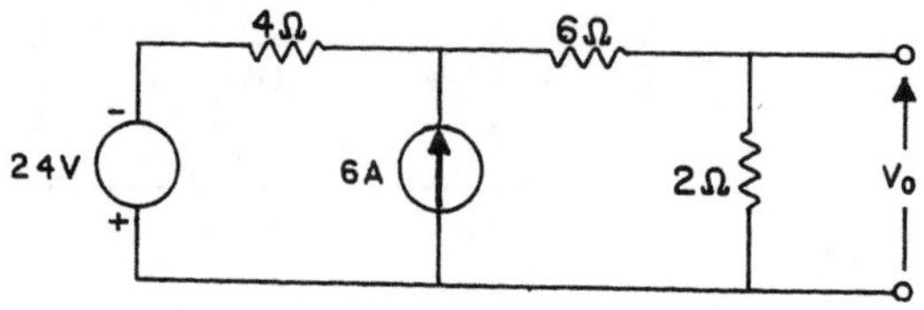

Solution: Consider the following networks.

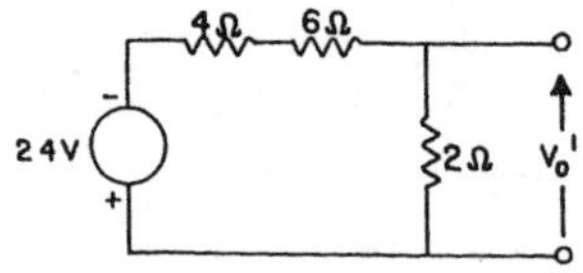

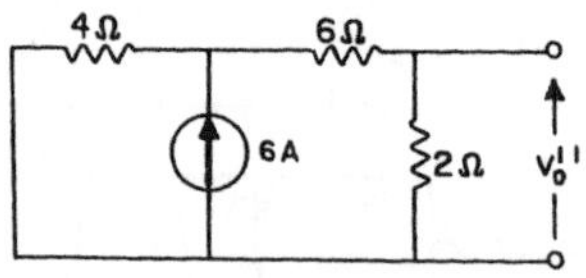

$V_o' = -\frac{24}{12}(2) = -4V$ $\quad V_o'' = \frac{(6)(4)}{12}(2) = 4V$

$V_o = V_o' + V_o'' = 0V$

6. Use superposition to find V_o in the following network.

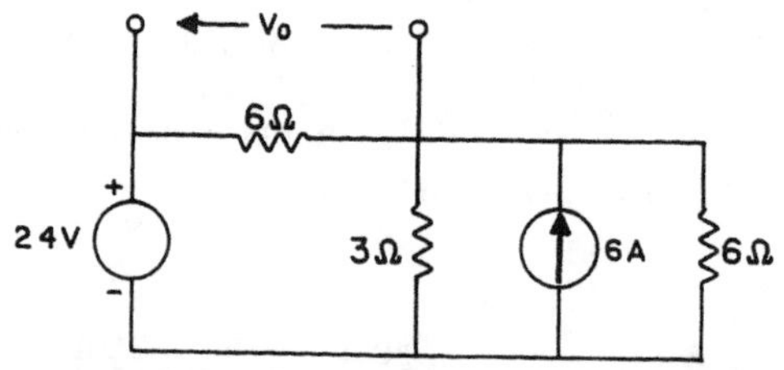

Solution: Consider the two networks below.

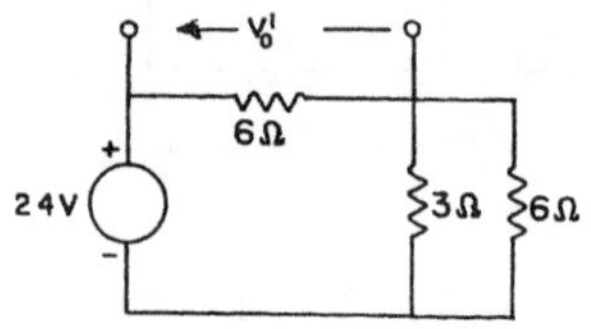

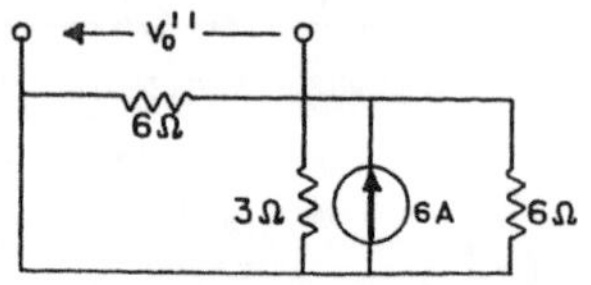

$V_o' = \frac{24}{6+2}(6) = 18V$ $\quad V_o'' = -\frac{(6)(2)}{2+6}(6) = -9V$

$V_o = V_o' + V_o'' = 9V$

7. Use source transformation to find I_o in the following circuit.

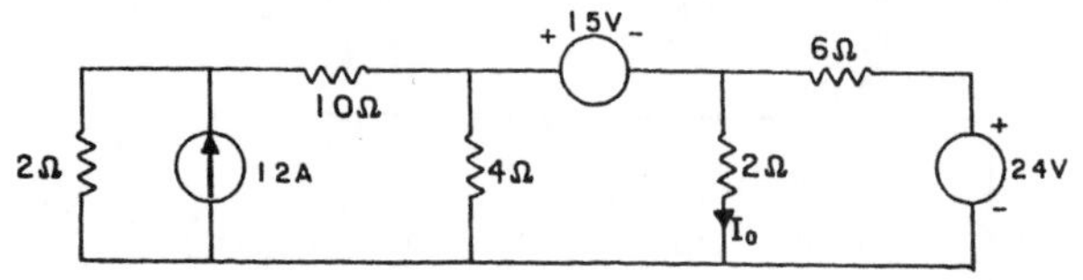

Solution: The following sequence of figures illustrates the transformation.

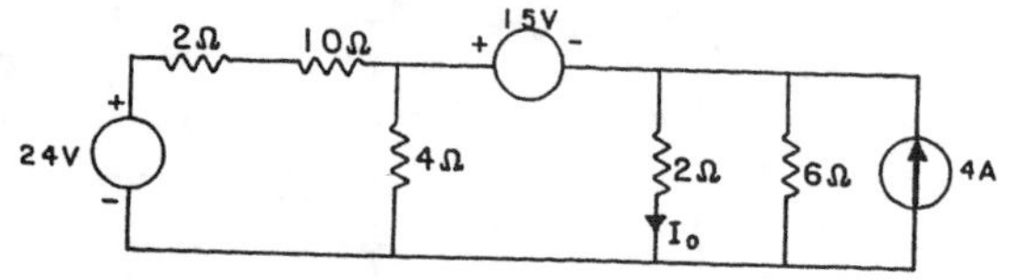

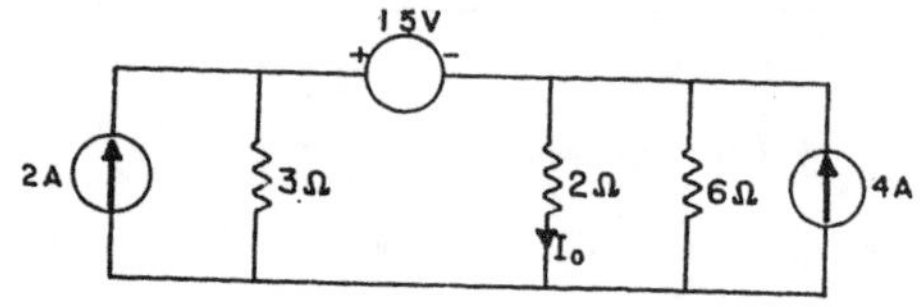

Converting the 2A source and 3Ω resistor into a 6V source in series with a 3Ω resistor; adding the 15V source and converting back to a current source in parallel with a resistor yields

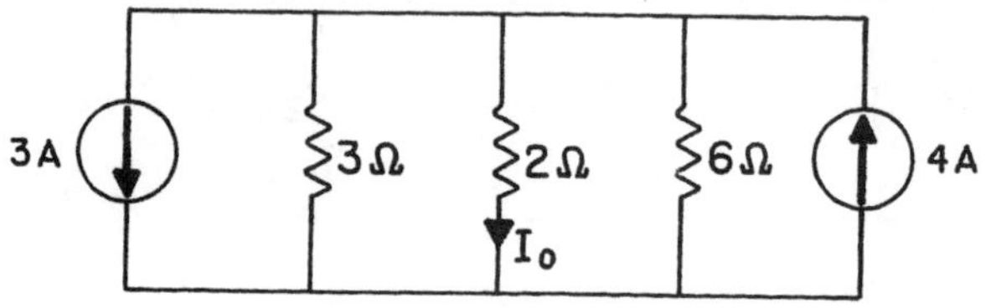

Hence $I_o = \frac{1}{2}$ A

8. Use source transformation to find I_o in the network below.

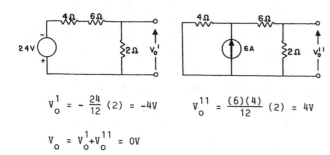

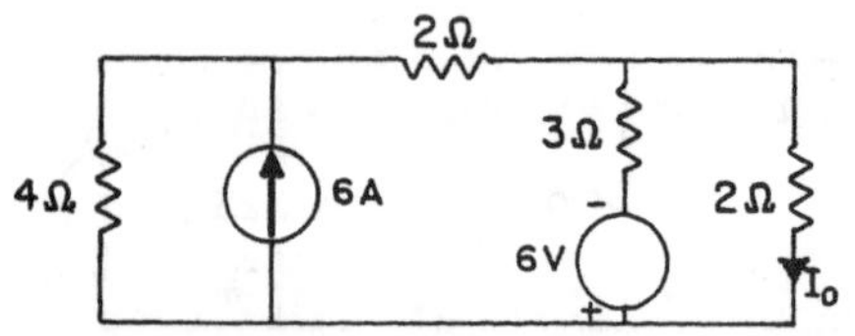

Solution: The following figures illustrate the transformation.

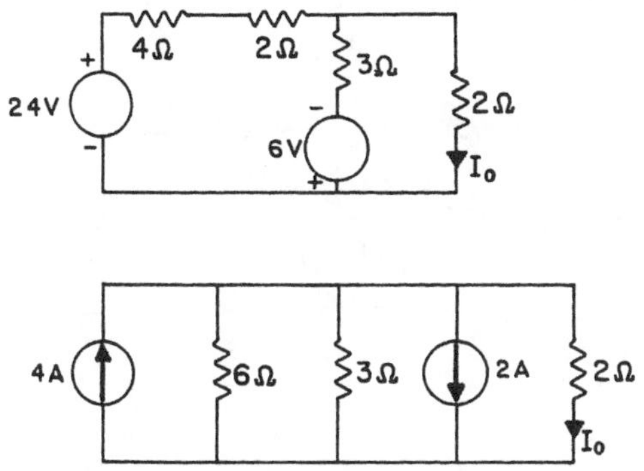

Hence $I_o = \frac{2(2)}{2+2} = 1A$.

9. Use source transformation to find I_o in the following network.

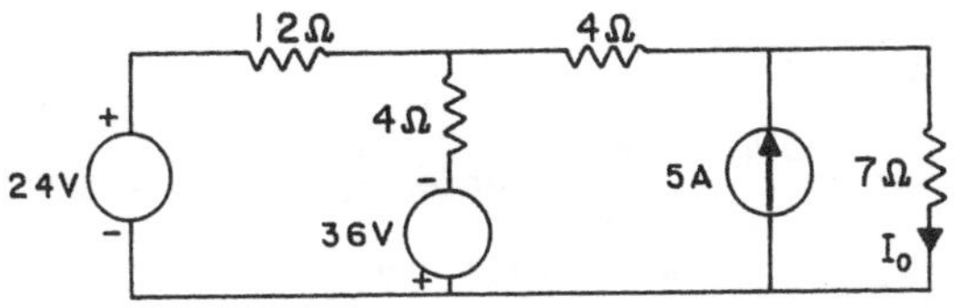

Solution: The following figures illustrate the transformation.

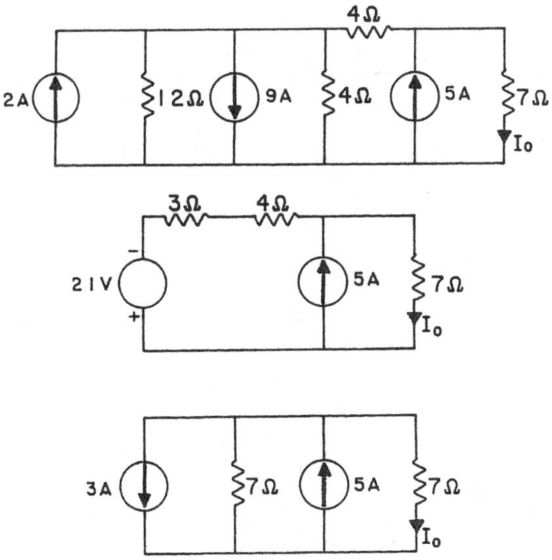

Hence $I_o = \frac{(2)(7)}{7+7} = 1A$.

10. Determine I_o in the network below, using source transformation.

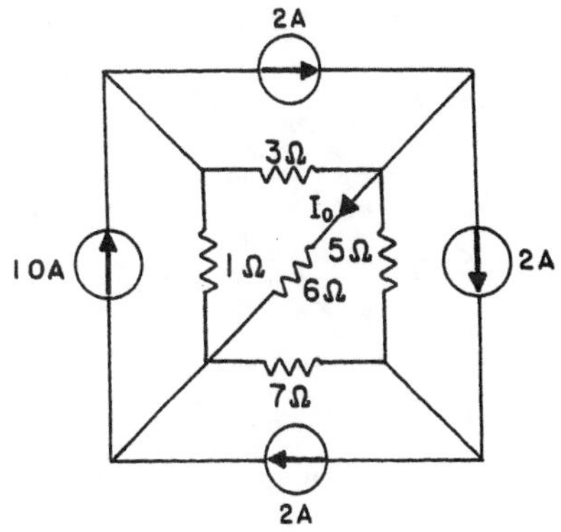

Solution: The sequence of figures illustrates the transformation.

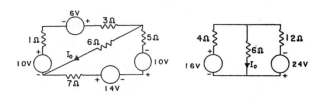

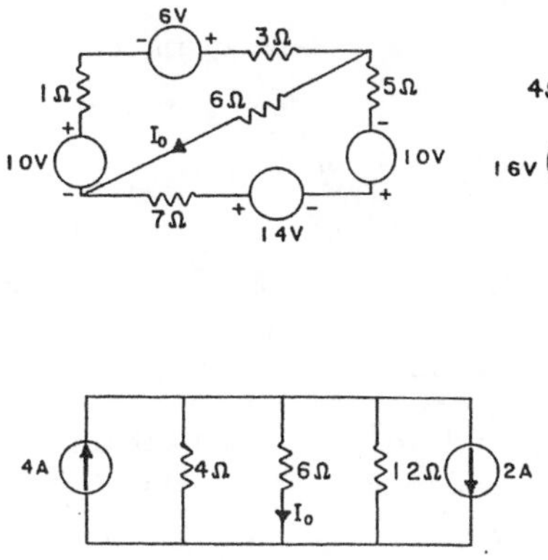

$I_o = \frac{(2)(3)}{6+3} = \frac{2}{3}$ A

11. Use Thevenin's theorem to find V_o in the following circuit.

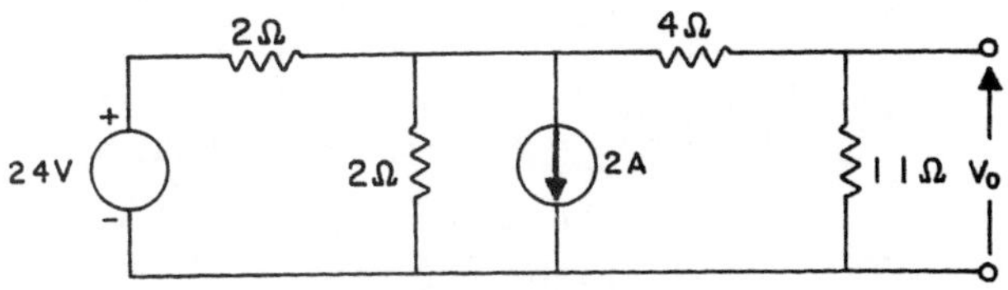

Solution: The open-circuit voltage is found from

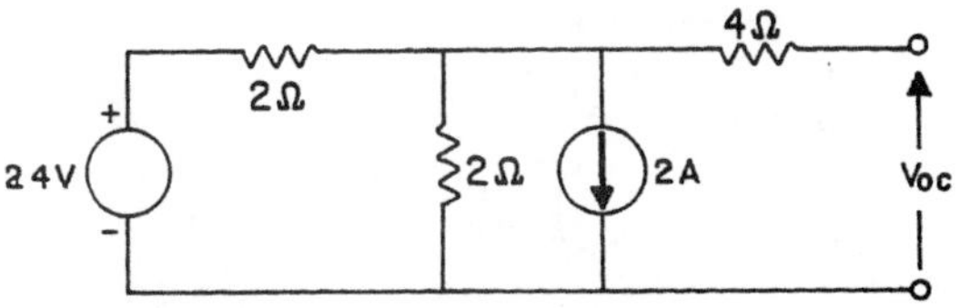

The node equation for V_{oc} is

$$\frac{V_{oc}-24}{2} + \frac{V_{oc}}{2} + 2 = 0$$

Therefore $V_{oc} = 10V$. R_{TH} is found from the network

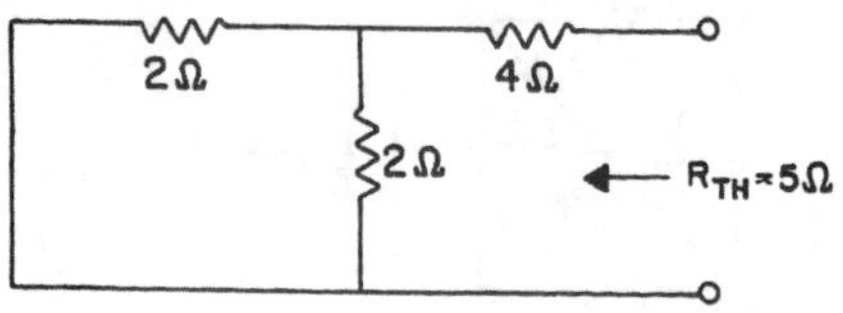

Hence

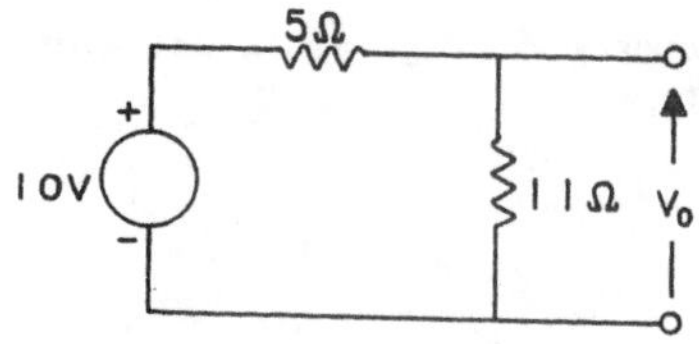

$$V_o = \frac{(10)(11)}{16} = 6.88V$$

12. Use Thevenin's theorem to find I_o in the following circuit.

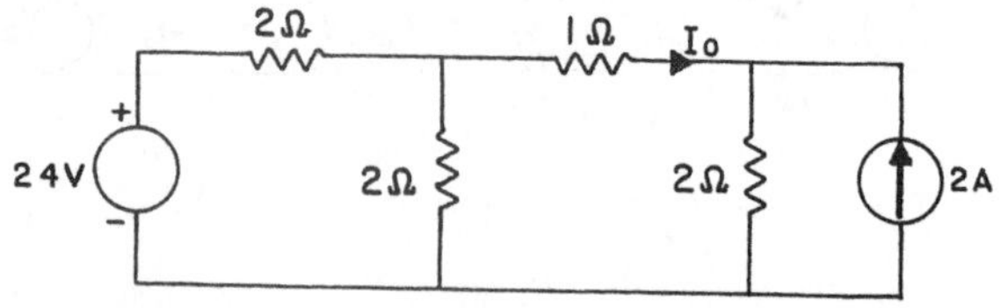

Solution: V_{oc} is found from the following network.

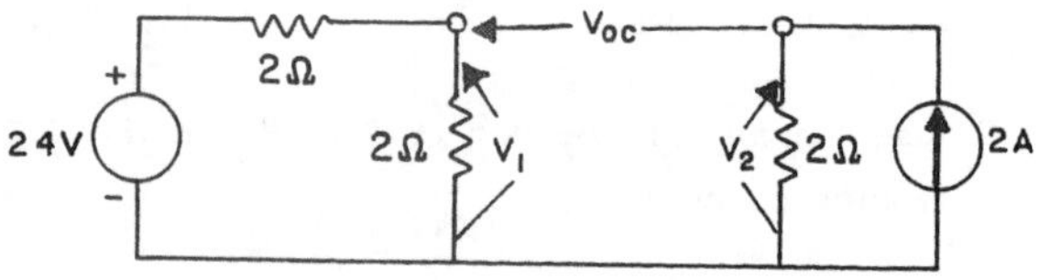

$V_1 = 12V$ and $V_2 = 4V$, hence $V_{oc} = V_1 - V_2 = 8V$.

R_{TH} is obtained from the network

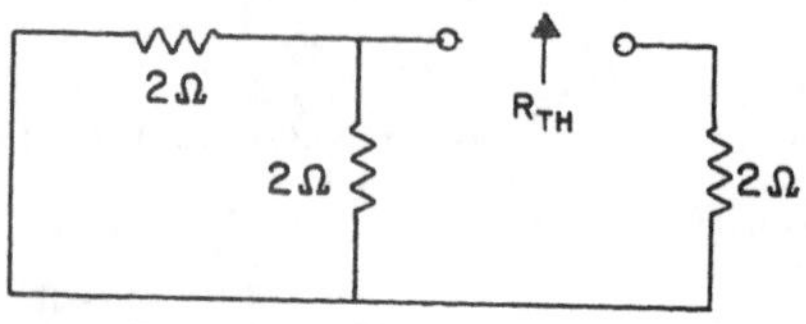

$R_{TH} = 3\Omega$ and hence

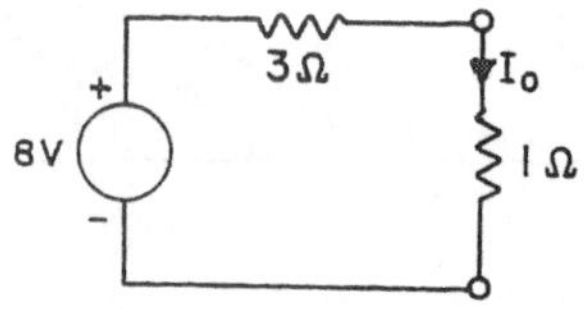

$$I_o = \frac{8}{4} = 2A$$

13. Use Thevenin's theorem to find V_o in the circuit below.

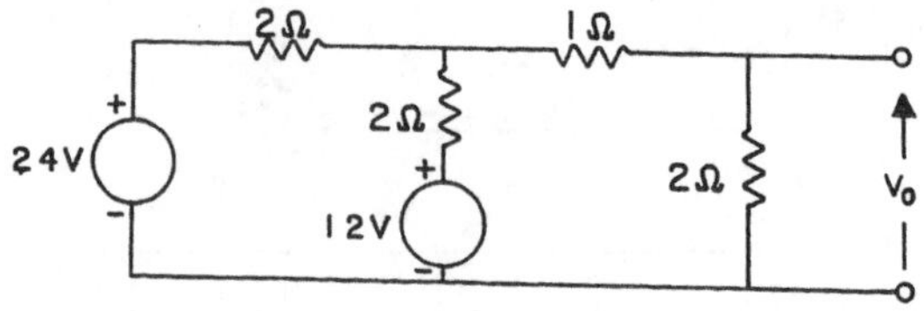

Solution: V_{oc} is derived from the following network.

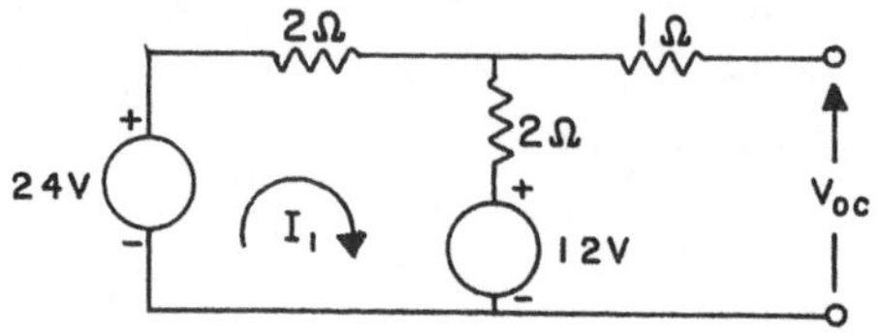

$I = \frac{24-12}{4} = 3A$ and hence $V_{oc} = (2)(3)+12 = 18V$.

R_{TH} is obtained from the circuit

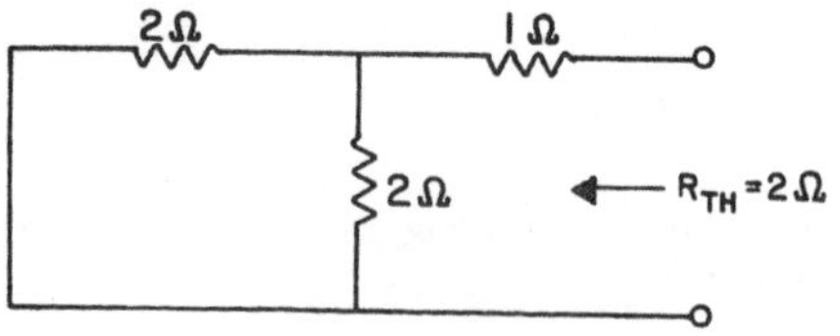

Hence

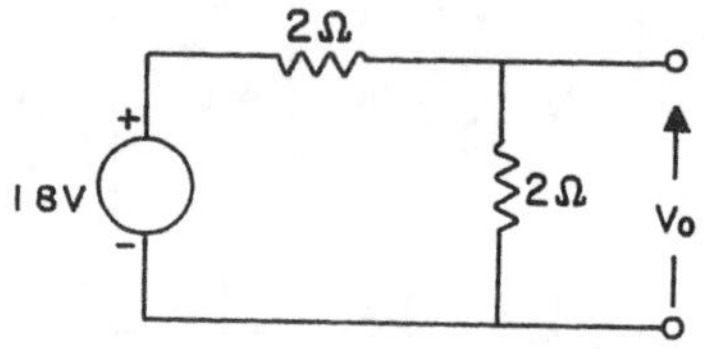

$V_o = 9V$

14. Use Norton's theorem to find V_o in the network below.

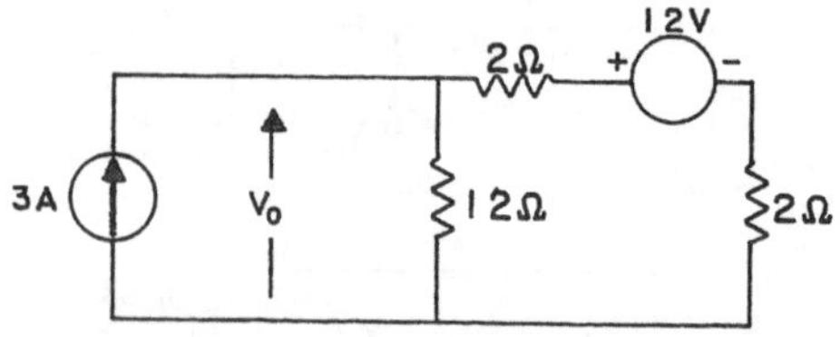

Solution: I_{sc} can be obtained from the following network.

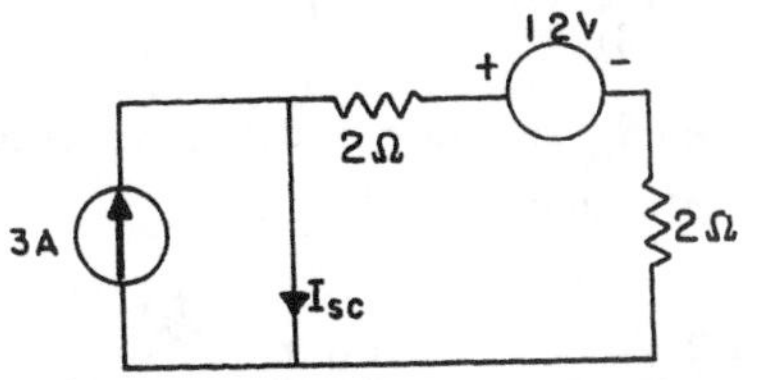

$I_{sc} = 3 + \frac{12}{4} = 6A$. Then R_{TH} is

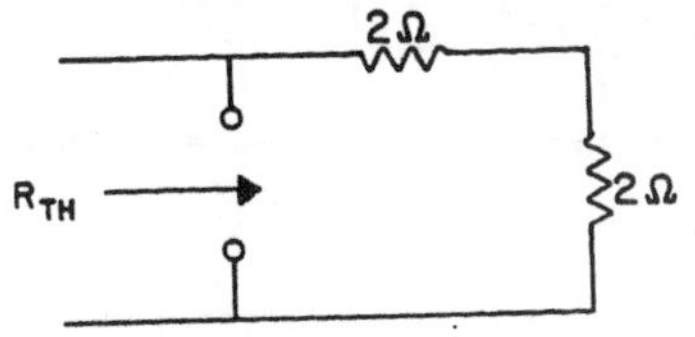

$R_{TH} = 4\Omega$. Hence

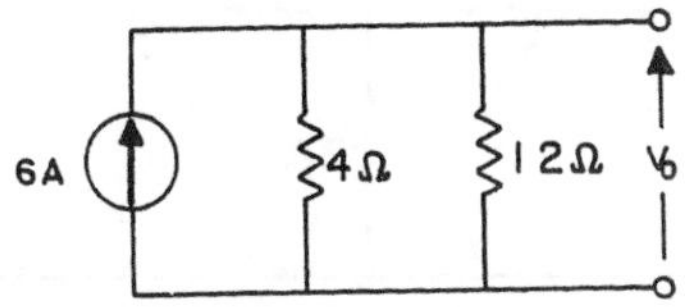

$V_o = 6(3) = 18V$

15. Form a Norton equivalent at A-B and use it to find I_o in the following circuit.

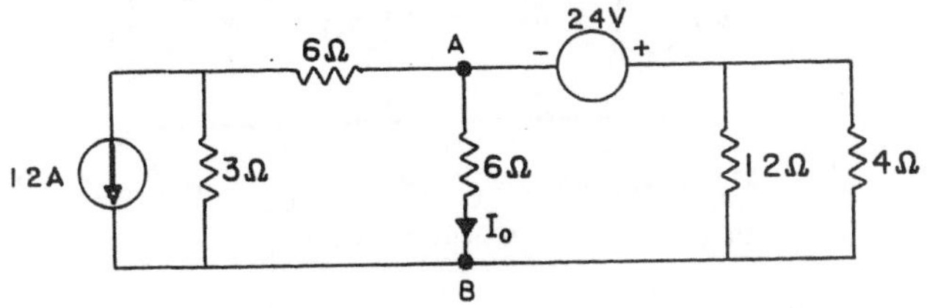

Solution: I_{sc} is obtained from the circuit

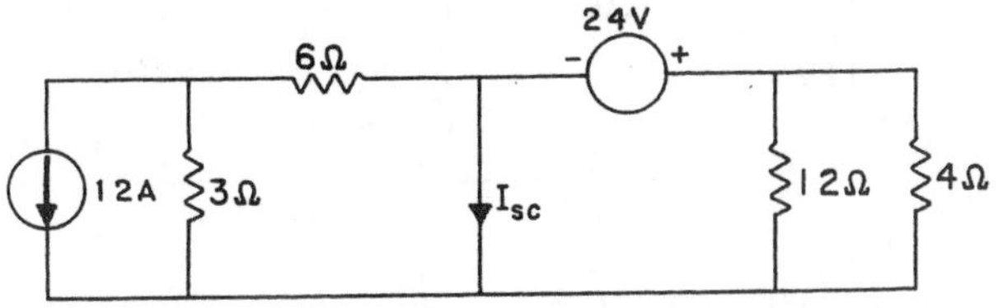

$I_{sc} = -\frac{12(3)}{6+3} - \frac{24}{3} = -12A$. R_{TH} is derived from the network

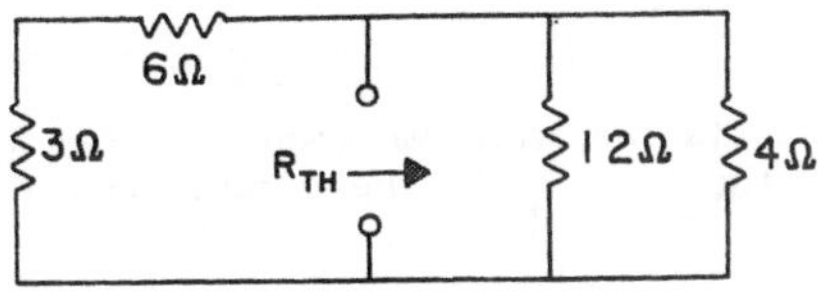

$R_{TH} = \frac{(3)(9)}{3+9} = \frac{27}{12}\ \Omega$, hence

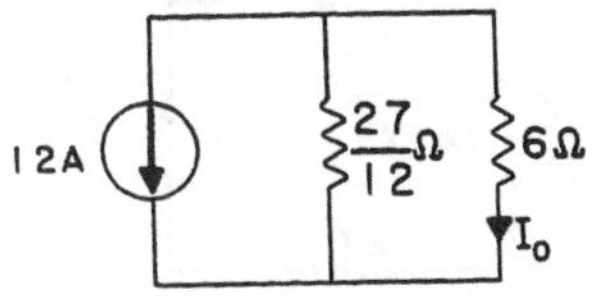

$$I_o = \frac{(-12)(\frac{27}{12})}{\frac{27}{12}+6} = -3.27A$$

16. Use Thevenin's theorem to find V_o in the following network.

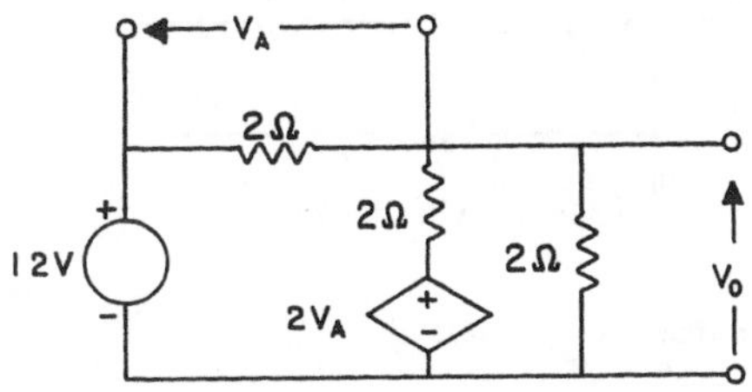

Solution: V_{oc} is found as follows.

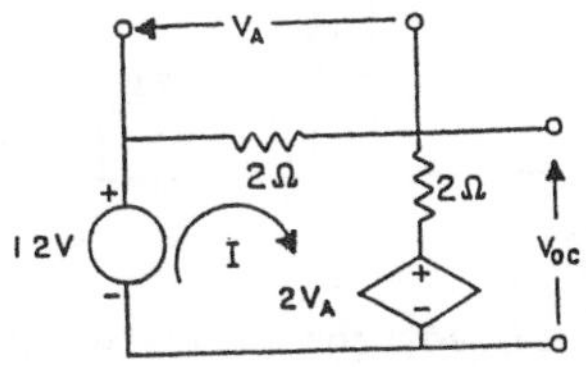

$I = \frac{12-2V_A}{4}$ and $V_A = 2I$.

Hence $I = \frac{3}{2}$ A and

$V_{oc} = 12-2(\frac{3}{2}) = 9V$.

I_{sc} is derived from the network

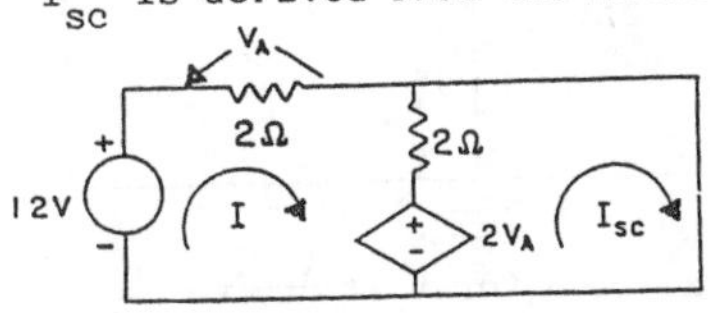

$4I-2I_{sc} = 12-2V_A$

$V_A = 2I$

$I_{sc} = 18A$ and hence $R_{TH} = \frac{V_{oc}}{I_{sc}} = \frac{1}{2}\ \Omega$.

Hence

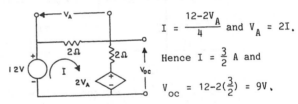

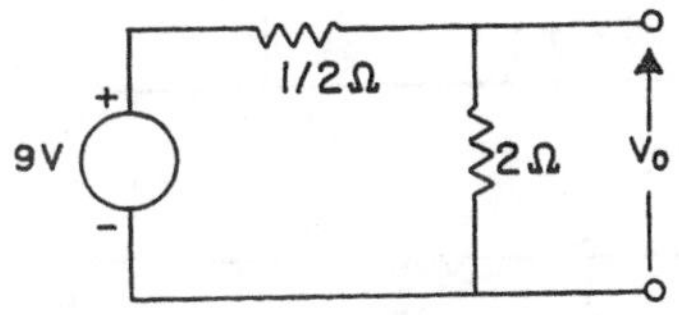

$V_o = \frac{36}{5}$ V

17. Use Thevenin's theorem to find the power in the 4Ω resistor in the circuit below.

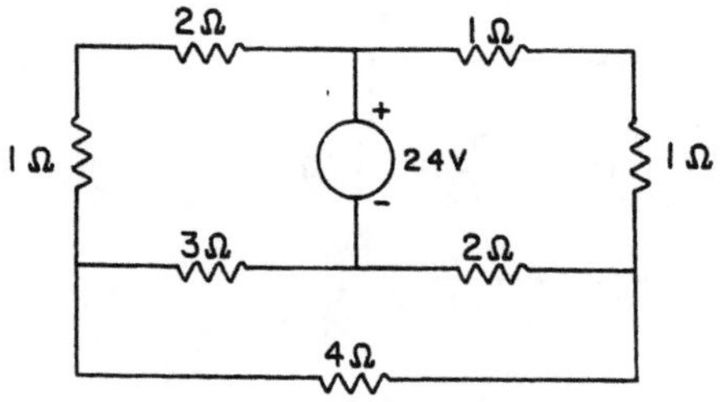

Solution: V_{oc} is obtained from the network

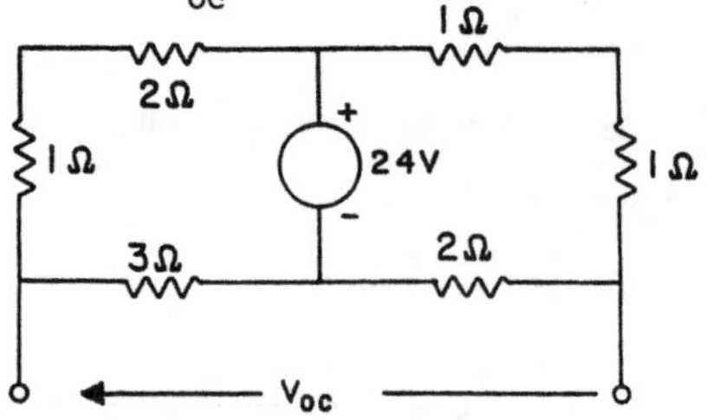

$$V_{oc} = \frac{24}{6}(3) - \frac{24}{4}(2) = 0V$$

Therefore $P_{4\Omega} = 0W$.

18. Use Norton's theorem to find V_o in the network below.

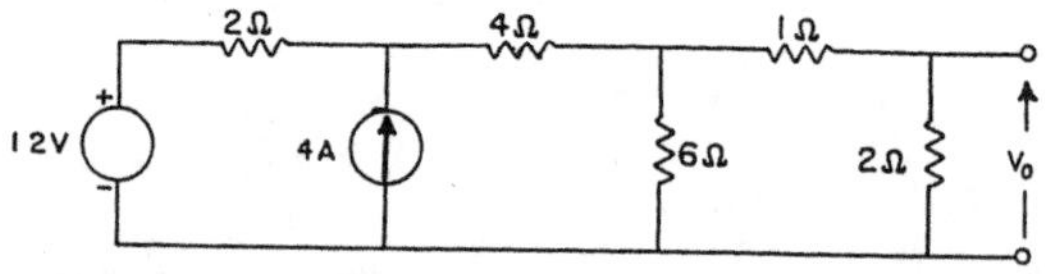

Solution: I_{sc} is obtained from the network

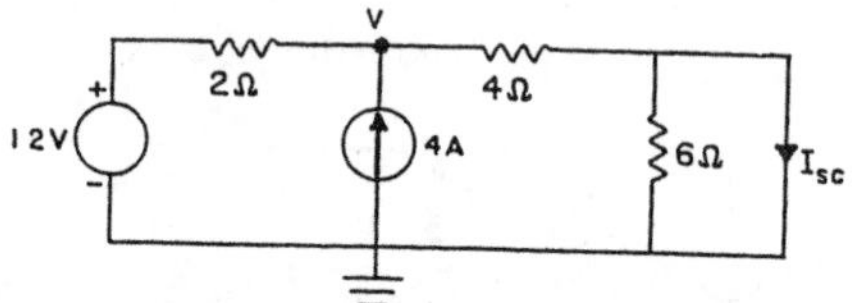

The node equation for V is $\frac{V-12}{2} + \frac{V}{4} = 4$ and

hence $V = \frac{40}{3}$ V and $I_{sc} = \frac{\frac{40}{3}}{4} = \frac{10}{3}$ A. R_{TH} is

obtained from the circuit

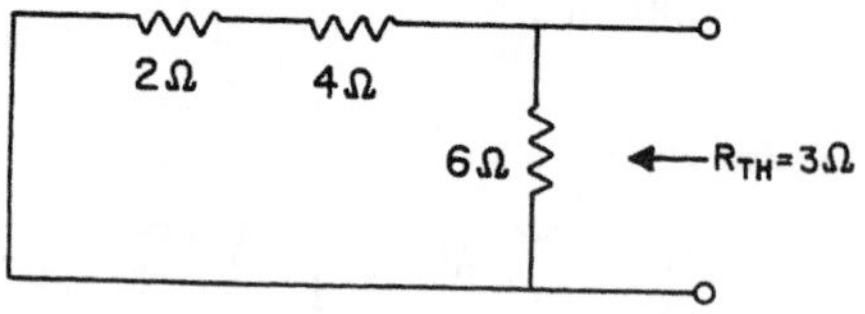

Therefore

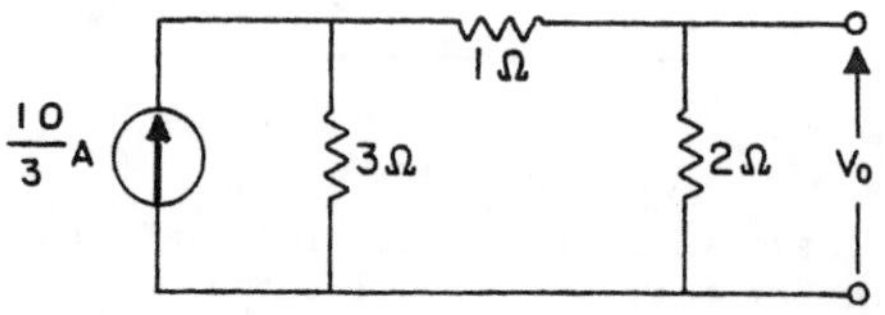

$$V_o = \left(\frac{10}{3}\right)\left(\frac{3}{3+3}\right)(2) = \frac{10}{3} \text{ V}$$

19. Determine V_o in the circuit shown below, using Thevenin's theorem.

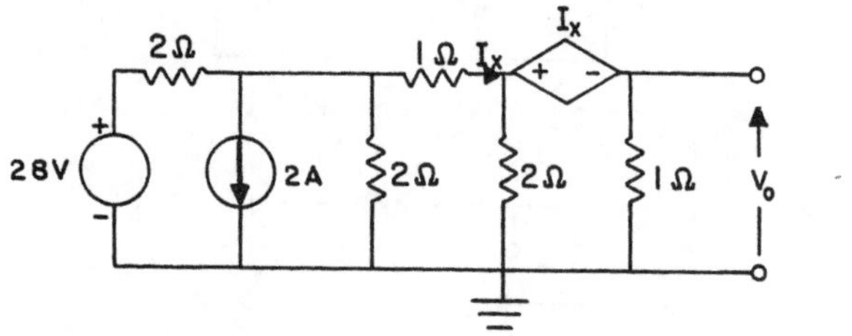

Solution: Using source transformation the network can be reduced to

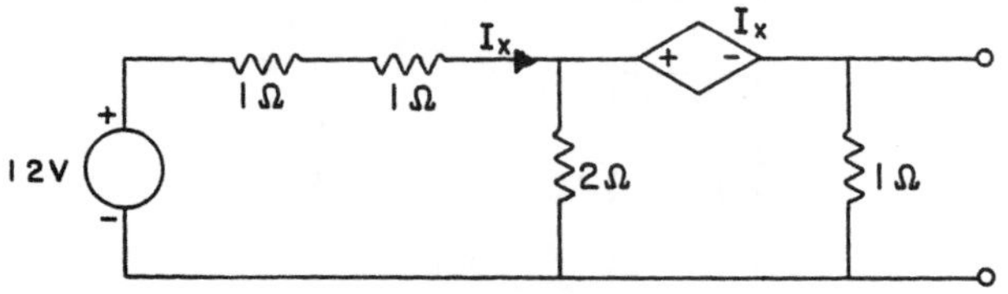

V_{oc} can be obtained from the network

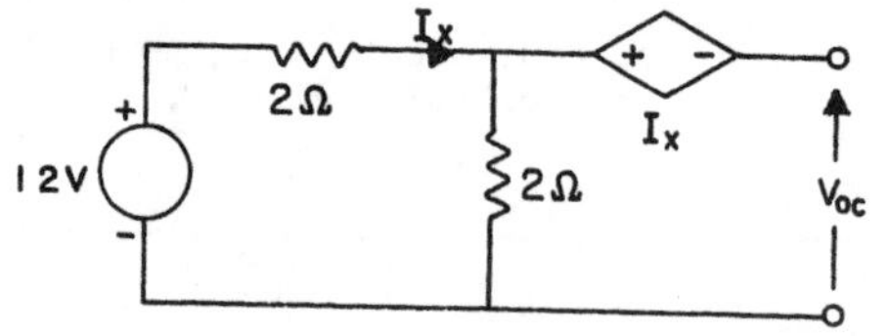

$4I_x = 12$, hence $I_x = 3$ and $V_{oc} = 6-3 = 3V$. R_{TH} is derived from the network

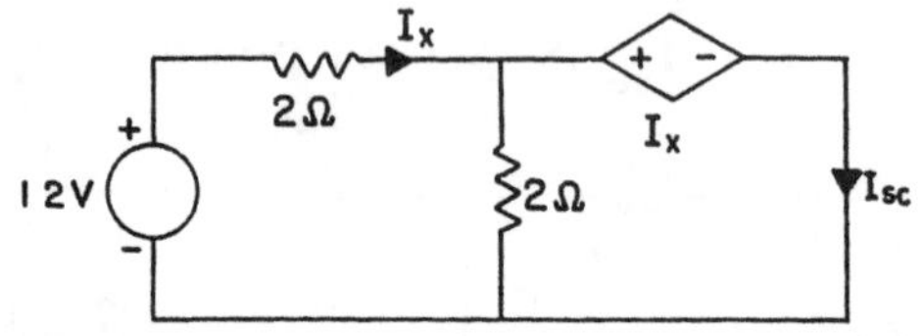

$12-2I_x = 2I_x$ hence $I_x = 3A$ and $I_{sc} = 2A$.

Therefore $R_{TH} = \frac{V_{oc}}{I_{sc}} = \frac{3}{2}\ \Omega$. Hence

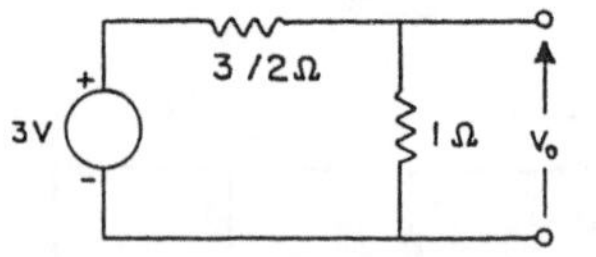

$$V_o = \frac{6}{5} \text{ V}$$

20. What is the maximum power that can be delivered to the load R_L in the circuit shown below.

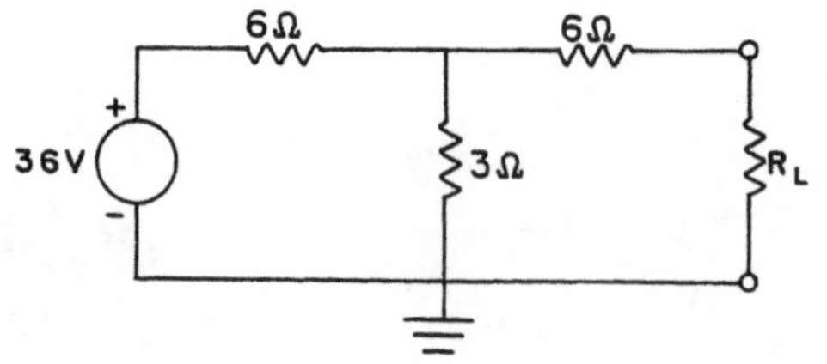

Chapter 5 CAPACITANCE & INDUCTANCE

CAPACITORS

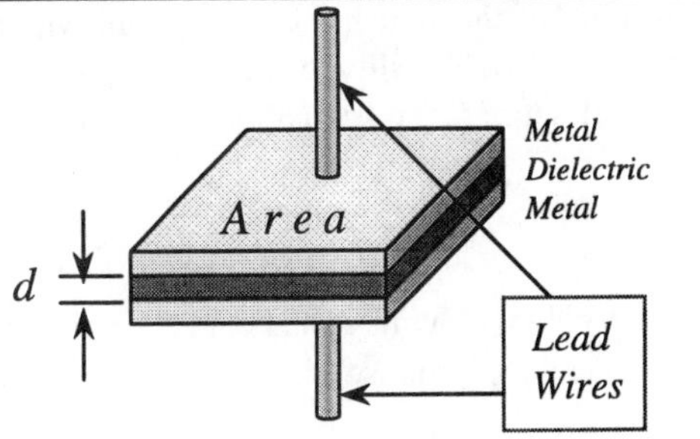

Figure 5.1. An inside view of a prototypical capacitor.

As mentioned in the BECA text and depicted Figure 5.1, capacitors consist of two sheets of metal, called plates, that are separated by a thin sheet of dielectric material. When a voltage is applied between the plates, an electric field must exist inside the dielectric. Since the dielectric is an insulator, no current flows through it and energy is stored in the electric field. Let's compare this situation to that of a resistor. When a voltage is applied to a resistor, a electric field exists in the resistor. Also, in agreement with Ohm's Law, current flows through the resistor. We know the power consumed by the resistor is $P = IV$ and energy usage is the integral of the power. So, in the resistor, an electric field exists, same as in the capacitor, but since current flows, energy is lost as heat rather than stored. The *I-V* relationship for the capacitor is

$$i(t) = C\frac{dv(t)}{dt} \tag{5.1}$$

where *i(t)* and *v(t)* obey the passive sign convention and *C* is the capacitance in Farads (F). Since *i(t)* is the derivative of charge with respect to time, *C* can be expressed as,

$$C = \frac{q(t)}{v(t)} \tag{5.2}$$

where *q(t)* and *v(t)* are the charge and voltage on the capacitor respectively. Also, it is shown in the BECA text that the energy stored in the capacitor is

$$w(t) = \frac{C}{2}v^2(t) \tag{5.3}$$

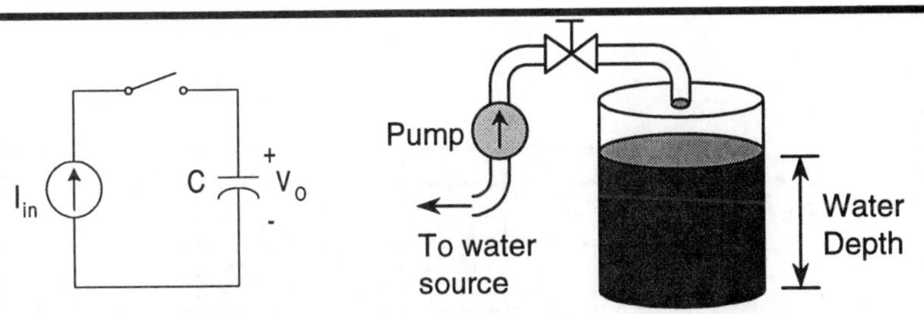

Figure 5.2. An diagrammatic analogy between energy storage in a capacitor versus a vat of water.

A simple analogy can be made between electrical energy storage in a capacitor and energy storage in a hydraulic system as seen in Figure 5.2. The switch is analogous to the valve where an open switch (no current) corresponds to a closed valve (no water flow). The charge on the capacitor is analogous to the gallons of water in the vat and the height of the water in the vat as similar to the voltage across the capacitor. The pump speed sets the water flow rate just as the current source sets the charge flow rate. When the valve(switch) is opened(closed) more water(current) flows to the vat(capacitor). This increases the water height(voltage) value. For a given water(current) flow rate, the rate of change in water height(voltage) value depends on the area of the vat(capacitor value). So, the capacitance is analogous to the vat's area! Just as a thin vat fills quickly and a wide vat fills slowly, a small capacitor's voltage changes quickly while a larger capacitor's voltage changes slower.

A capacitor's value is related to the it's dimensions and materials as defined in Figure 5.1 by the equation,

$$C = \frac{\varepsilon}{d} A \tag{5.4}$$

where A is the area and ε is the permitivity of the dielectric material. Permitivity, a material constant, quantifies the dielectric's ability to store energy in an electric field. (The permitivity of dielectrics used in capacitor manufacture range from about 0.1 pF/cm to 1000 pF/cm). This can be seen by combining (5.3) and (5.4) as

$$w(t) = \frac{A\varepsilon}{2d} v^2(t) \tag{5.5}$$

So, for a given voltage and capacitor dimensions, a larger permitivity means more stored energy. Combining (5.2) and (5.4) and solving for ε yields

$$\varepsilon = \frac{d}{A} \frac{q(t)}{v(t)} \tag{5.6}$$

We see that for a given voltage, more permitivity means more stored charge and thus more stored energy. Since physical circuits operate at specified voltages (i.e. automobile circuits use the 12-V battery) it makes sense to think of higher permitivity yielding more stored charge and more energy per applied volt. From (5.4), capacitors employing dielectrics with large ε values will have large capacitances. For this reason, capacitor manufacturer's are always searching for materials with higher permitivities.

SPECIFICATIONS

At this point in your studies, there are three capacitor specifications you should know about; capacitance value, working voltage and tolerance.

Capacitance Value

Standard capacitor values range from a few pF (10^{-12} F) to about 50 mF. Capacitor larger than 1 F are available but will not be discussed here. Table 5.1 lists the standard capacitor values. Manufacturers always list capacitance values as pF or μF. For example, a 1 nF capacitor will be referred to as a 1000 pF capacitor.

Table 5.1

Standard Capacitor Values

pF	pF	pF	pF	μF	μF	μF	μF	μF	μF	μF
1	10	100	1000	0.010	0.10	1.0	10	100	1000	10000
	12	120	1200	0.012	0.12	1.2	12	120	1200	12000
1.5	15	150	1500	0.015	0.15	1.5	15	150	1500	15000
	18	180	1800	0.018	0.18	1.8	18	180	1800	18000
2	20	200	2000	0.020	0.20	2.0	20	200	2000	20000
	22	220	2200	0.022	0.20	2.2	22	220	2200	22000
	27	270	2700	0.027	0.27	2.7	27	270	2700	27000
3	33	330	3300	0.033	0.33	3.3	33	330	3300	33000
4	39	390	3900	0.039	0.39	3.9	39	390	3900	39000
5	47	470	4700	0.047	0.47	4.7	47	470	4700	47000
6	51	510	5100	0.051	0.51	5.1	51	510	5100	51000
7	56	560	5600	0.056	0.56	5.6	56	560	5600	56000
8	68	680	6800	0.068	0.68	6.8	68	680	6800	68000
9	82	820	8200	0.082	0.82	8.2	82	820	8200	82000

Larger values up to several dozen Farads are available.

Working Voltage

From Figure 5.1, it should be apparent that if the voltage across the dielectric is allowed to increase, eventually the dielectric will break down, with thinner dielectrics failing at lower voltages. Since it is critical to keep the applied voltage below the breakdown point, manufacturers specify the working voltage rating of the capacitor. Simply put, the dc voltage across a capacitor should not exceed the voltage rating! Working voltage also goes under the alias, DC voltage rating and just plain voltage rating. In general, for capacitors of the same physical size and manufacturing process, as the working voltage increases, the capacitance decreases. This is logical since increasing the voltage rating requires a thicker dielectric which, from (5.4), results in a smaller capacitance. Standard working voltage ratings range from 6.3 V to 500 V. Both smaller and larger rating are, of course, available.

Tolerance

An adjunct to the capacitance value is the tolerance, usually listed as a percentage of the nominal value. Standard tolerance values are ±5%, ±10% and ±20%. Occasionally, tolerances for single-digit pF capacitors are listed in pF. For example, 5 pF ± 0.25 pF.

CONSTRUCTION TECHNIQUES

While there are many materials and construction processes for capacitor manufacture, we will focus on the most common: rolled film (paper and plastic) multi-layer (ceramic and mica) and aluminum electrolytic capacitors.

Rolled Construction Capacitors

The capacitor in Figure 5.1 is ill-suited for packaging, especially when C is large. An alternate scheme, called rolled construction, takes narrow strips of dielectric and metal and rolls them up like a jelly roll. This is sketched in Figure 5.3 for a single dielectric layer although many dielectric - metal laminations can be used. To further decrease the size of the capacitor, the metal can be deposited in super-thin layers on either side of the dielectric. This process is growing in popularity. After rolling the strips, lead wires and an external case are added to complete the process.

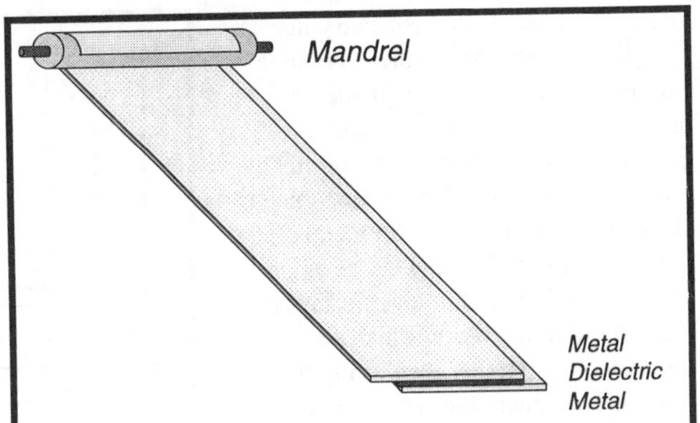

Figure 5.3. Sketch of a rolled construction capacitor layup.

Paper

One of the original dielectrics used in commercial rolled capacitors was paper soaked in oil. While paper has been replaced by plastic dielectrics in the most applications, paper caps are still used in high voltage/high power systems.

Plastic

Plastic dielectric capacitors are the most popular capacitors on the market today. They are available in a wide range of capacitance values and working voltages, and they're inexpensive. A partial list of the various plastics used as dielectrics is given in Table 5.2. Specifications for air and paper are included for completeness.

Table 5.2

Partial Listing of Paper and Plastic Dielectrics

Material	Relative Permativity*	Other Uses
Air	1.0	
Kraft paper**	4.4	
Polyester	2.4	wiring insulation
Polyethylene (PET)	3.1	wiring insulation
Polypropolene (PP)	2.2	wiring insulation
Polytetrafluoroethylene (PTFE)	2.1	wiring insulation

Relative Permitivity is the ratio of ε for material to that of air.

Ceramic Capacitors

A ceramic capacitor looks very much like the prototypical capacitor in Figure 5.1. The dielectric is made of ceramic (aluminum oxide) and the metal layers are deposited. Ceramic caps are available in the three configurations shown in Figure. The lead wires on the button capacitor in a make it is quite versatile. It can be used on printed circuit boards or in a hobbyist's hand-soldered project. The single and multi-layer ceramic caps (sometimes called chip capacitors) are packaged for surface mount PC boards. Having no lead wires make these configurations very small (less than 10 mm on a side).

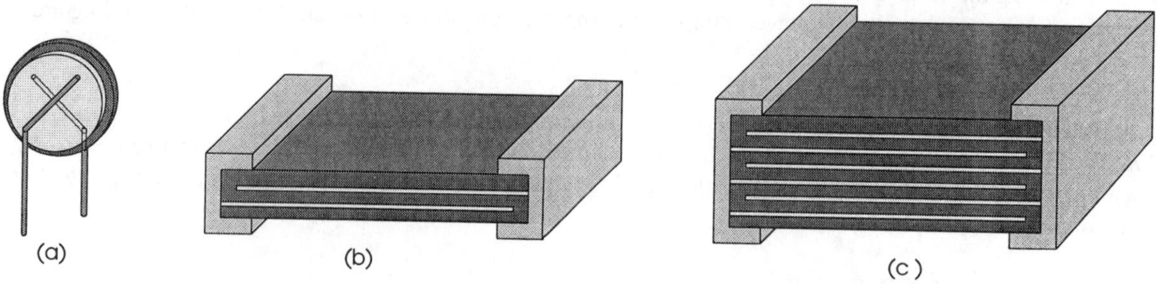

Figure 5.4. Ceramic capacitor configurations. (a) button,, (b) chip and(c) multi-level ceramic capacitor.

Aluminum Electrolytic Capacitors

Of the three construction techniques discussed here, electrolytic capacitors have the highest capacitance per cubic centimeter. This is accomplished by drastically reducing the dielectric thickness. Let's begin by explaining the materials inside the electrolytic which is sketched in Figure 5.6. The metal plates are made of aluminum. One plate is oxidized to form a thin layer (a few 100 nm) of aluminum oxide which serves as the dielectric. Rather than trying to intimately connect the second plate to the

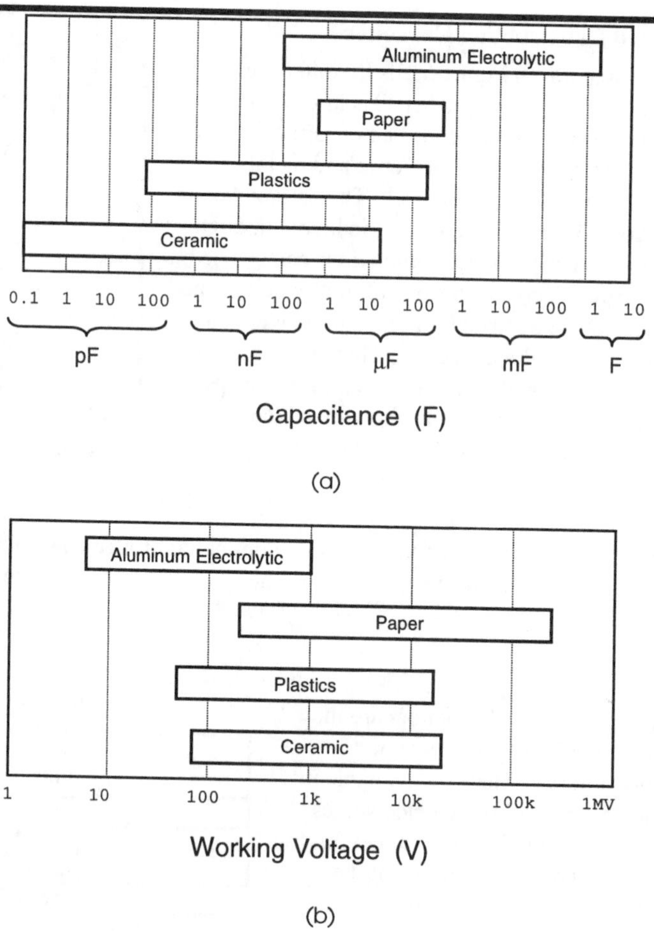

Figure 5.5. Capacitance and working voltage ranges for several types of capacitors.

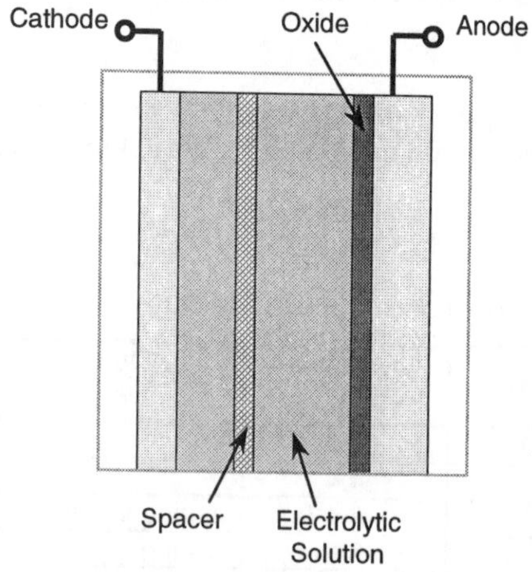

Figure 5.6. An internal sketch of an aluminum electrolytic capacitor.

oxide, a electrolyte solution (usually glycol-based) is added that serves as an intermediate connection. Finally, a porous spacer is added between the plates to keep them apart. To package the capacitor, the plates and spacer are rolled and placed in a cylindrical case.

An unfortunate feature of the electrolytic is that it is sensitive to the polarity of the applied voltage. When the oxidized electrode is positive, the capacitor works great. However, the oxide actually conducts when the voltage polarity reverses! Since the oxidized electrode must be kept positive, it is called the anode and the other plate is the cathode. A mark on the capacitor's case, a positive on the anode or a negative sign on the cathode, identifies the terminals.

Now that we've covered the different materials used in capacitor manufacture, we can compare their performance. Capacitor value and working voltage ranges for a variety of dielectrics is given in Figure 5.5.

MARKINGS

Manufacturers mark specification values on a capacitor in one of three ways: plain English, a color code or a numeric code. When the capacitor case is large, capacitance, tolerance and working voltage are just printed on the side. Units are in pF or μF. However, some companies use "mfd" or "MFD" when they mean micro-Farads. Some capacitor's use the resistor color code shown in Table 1of Chapter 2, where the capacitance is always in pF. Color codes do not include working voltage. Physically small capacitors use a numeric code to specify value and tolerance. Capacitance value is listed in a three digit code where the first two digits are magnitude and the third digit is a multiplier. Be forewarned that the same code is used for both pF and μF! So, use common sense in deciding whether a particular code should be interpreted as pF or μF. In general, disk capacitors smaller than a pencil's diameter are coded in pF and larger packages are coded in μF. The letters M ($\pm$20%), K ($\pm$10%) and J ($\pm$5%) specify the tolerance. As an example, consider a very small capacitor that is coded, 473K. The capacitance is 47×10^3 pF, equals 47,000 pF or 47 nF and the tolerance is $\pm$ 10%.

INDUCTORS

As mentioned in the BECA text, whenever current flows in a wire, a magnetic field is produced. When the wire is arranged in a coil, as shown in Figure 5.7, the magnetic field is focused inside the coil. Thus, energy can be stored in the magnetic field of a coil of wire carrying current. Inductors are little more than coils of wire designed for this purpose.

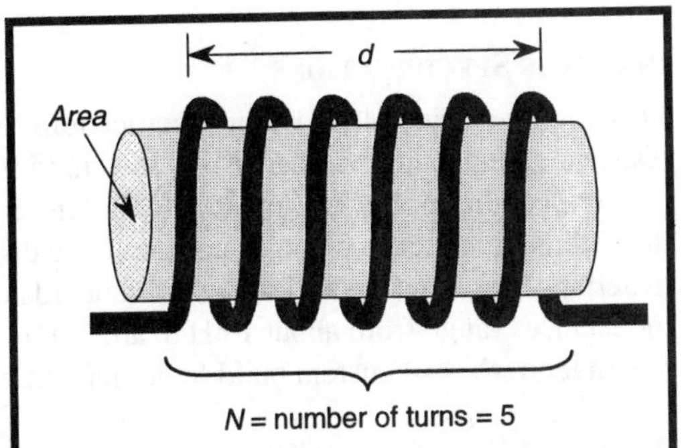

Figure 5.7. A conceptual view of the inductor.

The *I-V* relationship for the inductor is

$$v(t) = L\frac{dv(t)}{dt} \qquad (5.7)$$

where L is the inductance in Henries (H). The energy stored in the inductor's magnetic field is

$$w(t) = \frac{L}{2}i^2(t) \qquad (5.8)$$

Inductance is related to the dimensions of the inductor, defined in Figure 5.7, by the equation,

$$L = \frac{AN^2}{d}\mu \qquad (5.9)$$

where N is the number of wraps, or turns, in the coils and μ is the permeability of the material inside the coil. Permeability is a material constant that quantifies how much magnetic field is created per amp of current in the coil. To see how μ affects the inductor performance, combine (5.7) and (5.8) to yield

$$w(t) = \left[\frac{AN^2}{2d} i^2(t)\right] \mu \tag{5.10}$$

Equation (5.10) indicates that increasing μ will, for a given current, increase the stored energy. For this reason, inductor manufacturers put high permeability materials, called magnetic cores, inside inductor coils.

As seen in Figure 5.8,.an analogy exists between energy storage in the magnetic field of an inductor and energy storage in a flywheel. The energy stored in a spinning flywheel depend on its the mass, *m*, and angular speed, v, (radians/sec.) and is given by

$$w(t) = Iv^2(t) \tag{5.11}$$

where I, the moment of inertia, is directly related to mass. Comparing (5.8) and (5.11), it is clear that inductance is analogous to mass. Just as a heavier wheel stores more energy than a lighter wheel, a larger inductance will store more energy in its magnetic field.

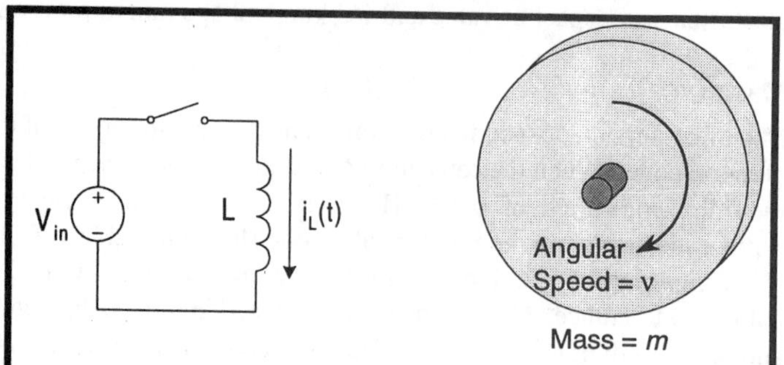

Figure 5.8. A pictorial energy storage analogy between the inductor and a flywheel.

INDUCTOR SPECIFICATIONS

There are two principle inductor specifications, inductance and resistance. Inductance value will, as seen in (5.9), depends on the permeability of the core material. Therefore, we will save the details of inductance ranges until we have discussed core materials. For now, we will state that standard commercial inductances range from about 1 nH to around 0.1 H. Of course, manufacturer's will custom build larger inductors - for a price.

If we recall from Chapter 2 that wire-wound resistors are just coils of wire, it is logical that inductors will have resistance. The major difference between wire-wound resistors and inductors is the wire material. High resistance materials like Nichrome™ are used in resistors while low resistance copper is used in inductors. Inductor resistance, measured at dc, is nothing more than the wiring resistance. Given that copper wire is used, the resistance depends only on the length and diameter of the wire. Typical resistance values are 10 to 500 mΩ. Table 5.3 lists the American Wire Gauge (AWG) standard wire diameters and the resulting resistance per foot of copper wire.

Table 5.3

Resistance per foot of solid copper wire.

AWG No.	Diameter (in.)	mΩ/ft
12	0.0808	1.59
14	0.0641	2.54
16	0.0508	4.06
18	0.0400	6.50
20	0.0320	10.4
22	0.0253	16.5
24	0.0201	26.2
26	0.0159	41.6
28	0.0126	66.2
30	0.0100	105
32	0.0080	167
34	0.0063	267
36	0.0049	428
38	0.0039	684
40	0.0031	1094

CORES

To increase inductance while maintaining a small component size, manufacturer's usually put high permeability materials (called magnetic cores) in their inductors. The most popular cores are iron alloys, powdered iron, molybdenum permalloy powder (MPP) and ferrites. The permeability and useful frequency ranges for each are listed in Table 5.4. Due to its bulk and weight, iron is used mostly in the power industry (60 Hz operation). Powered iron is exactly that, iron particles that have been compressed under high pressure. Molybdenum permalloy, an alloy of molybdenum, nickel and steel, is also formed by pressure. Ferrite cores are made of ceramics that contain particles of nickel, zinc, manganese and iron oxide that have been sintered at 2000C. Due to this manufacturing process, ferrite cores can be made in just about any shape imaginable.

Table 5.4

Permeability and Useful Frequency Ranges for Several Core Materials

Material	Relative Permeability*	Operating Frequency
Air	1	dc - GHz
Iron Alloys	250 - 2000	dc - 20 kHz
Iron Powder	5 - 80	2 kHz - 100 MHz
MPP	14 - 550	10 kHz - 1 MHz
MnZn Ferrite	750 - 15,000	10 kHz - 2 MHz
NiZn Ferrite	10 - 1500	200 kHz - 100 MHz

* Rel. permeability is normalized μ for air, $4\pi \times 10^{-7}$ H/m

There is a feature of these core materials that is not evident from the above discussion - core saturation. From the BECA text, we know that current and magnetic field are linearly related. This relationship holds for air core inductors. However, for magnetic materials, there is a maximum magnetic field strength that cannot be exceeded regardless of the current. This is shown in Figure 5.9. If the current increases beyond the core saturation limit, the additional energy is dissipated rather than stored in the magnetic field. Since it is important to know the maximum current an inductor can carry before saturation occur, manufacturers include a maximum current rating specification on magnetic core inductors. Maximum currents range from a few milliamps to tens of amps.

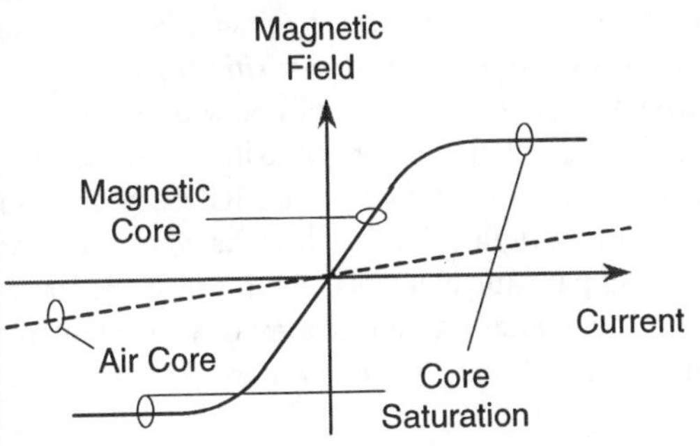

Figure 5.9. Plot of magnetic field vs. current showing core saturation.

MARKINGS

Inductor specifications for value and tolerance are marked on inductors in one of two ways; in plain English or in a color code. Maximum current must be gotten from the manufacturer's datasheets.

The color code, based on the resistor color code listed in Table 1 in Chapter 1, specifies value in μH and tolerance as a percentage. Since physically small inductors can be indistinguishable from resistors a double-wide silver band is added as shown in Figure . For example, a color code of YELLOW-VIOLET-RED-SILVER identifies a 47×10^2 = 4.7 nF + 10% capacitor. To identify inductances less than 10 μH, a gold band denotes a decimal point. To illustrate, a color code of YELLOW-GOLD-VIOLET-GOLD denotes a 4.7 $\pm$ 5% inductor.

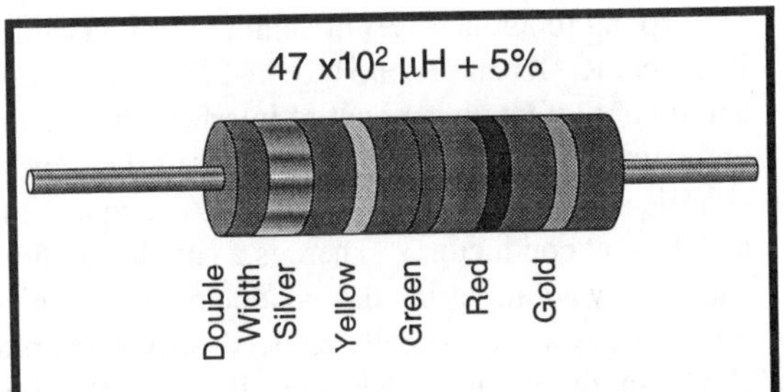

Figure 5.10. An example of the inductor color code.

CAPACITANCE AND INDUCTANCE AS MODELING ELEMENTS

An important application of resistance, capacitance and inductance is modeling things that are not necessarily resistors, capacitors or inductors. For, example, transistor performance is often modeled using equivalent R, L and C elements. Another example is modeling interference between two wires or two circuits. Interference occurs when an energy field generated at one point in a circuit affects voltages and/or currents in another circuit. The two circuits don't even have to be wired together! When the coupling is due to an electric field, we say the coupling is capacitive. When a magnetic field is the culprit, the interference is called inductive.

A Capacitive Coupling Example (Crosstalk)

Consider the four-conductor ribbon cable in Figure 5.11a that runs from a PC to some peripheral equipment. As mentioned in Table 5.2, many materials used as dielectrics in capacitors are also used as wiring insulation. So, we should not be surprised to find that capacitances exist between the cables. Here the wiring functions as the capacitor plates and the insulation as the dielectric. Since this capacitance is unintentional and unwanted, it is called *parasitic capacitance.* Obviously, a longer cable, or one with thinner insulation, will have more parasitic capacitance. An equivalent circuit diagram that MODELS this situation is shown in Figure 5.11b. Now the wires are connected through parasitic elements and it is possible for a signal on one conductor to induce a replica signal on another line. We call this condition *crosstalk.*

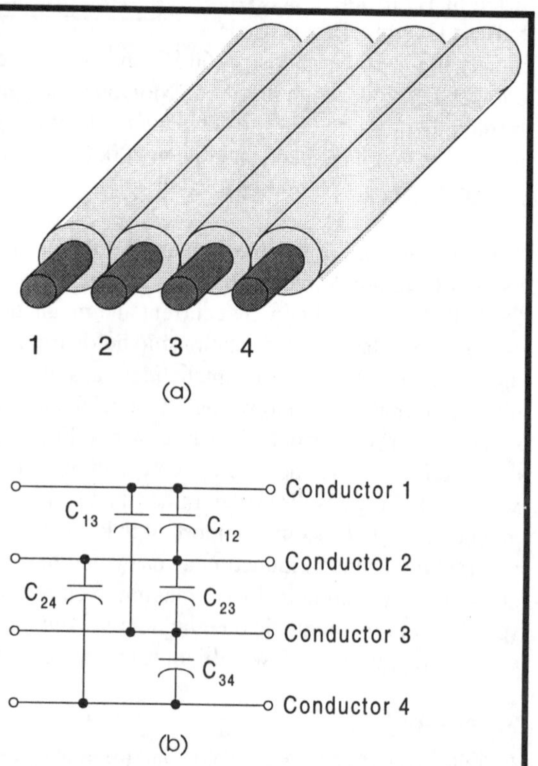

Figure 5.11. A four conductor cable with parasitic capacitances.

Let's investigate the amount of crosstalk that might occur by simulating the circuit in Figure 5.12 with PSpice. In particular, we want to see if the 5-V pulses on conductor 1 will affect conductors 2 and 3. The parasitic capacitance values of 10 fF and 5 fF correspond to about 3 feet of standard ribbon cable. Resistors R_1, R_2 and R_3 are required for PSpice to work and do not affect the amount of interference. The unfortunate results are shown in Figure 5.13. There is significant crosstalk, particularly, 49% at conductor 2 and 35% at conductor 3. There is a simple solution to this problem that is commonly employed in ribbon cables. By adding grounded conductors between conductors 1, 2 and 3 as shown in Figure 5.14a, the interference will drastically reduced. Given the number of parasitics in Figure 5.14b, we have chosen not to show you the PSpice Schematics diagram. But, the simulation results are in Figure 5.15. Crosstalk is < 16% at conductor 2, reduction by a factor of 3, and < 3 % at conductor 3, a factor of 12 reduction! Significant improvements indeed.

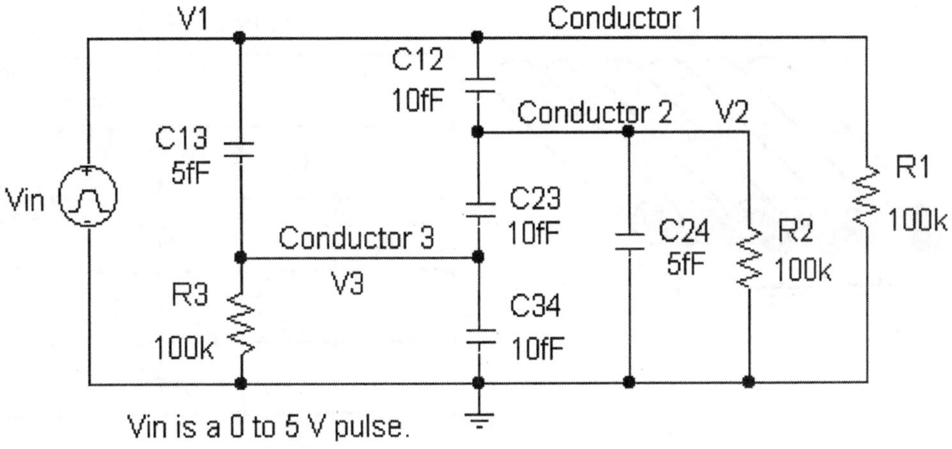

Figure 5.12. PSpice circuit for simulating crosstalk in 4-conductor ribbon cable.

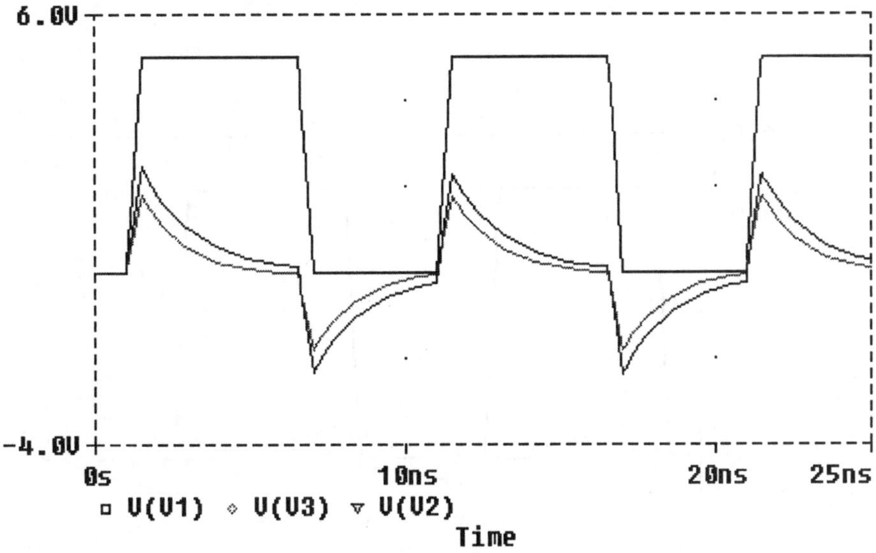

Figure 5.13. Crosstalk simulation results. Crosstalk is 49% at conductor 2 and 35% at conductor 3 - unacceptable.

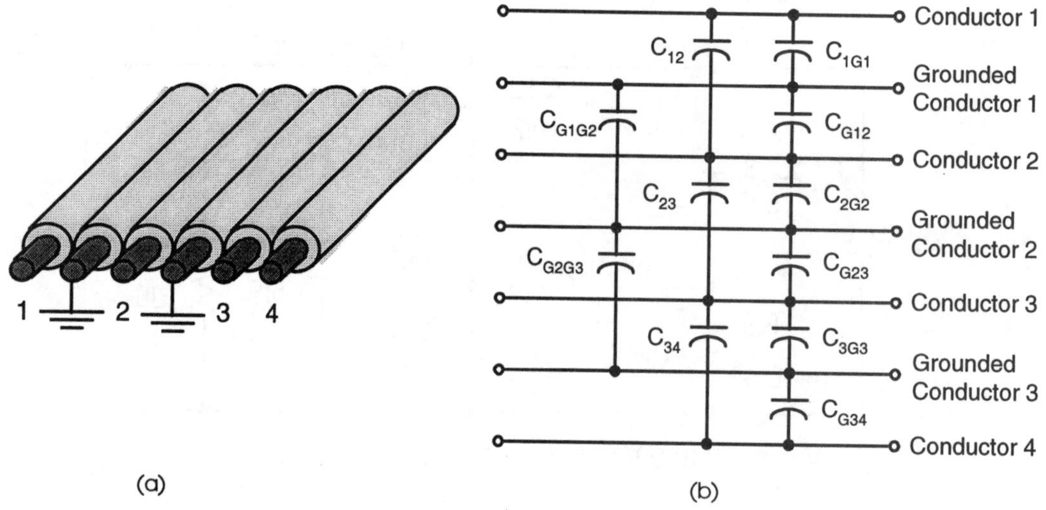

Figure 5.14. Solving the crosstalk problem by (a) adding grounded conductors between the signal-carrying conductors. The equivalent parasitic circuit is shown in (b).

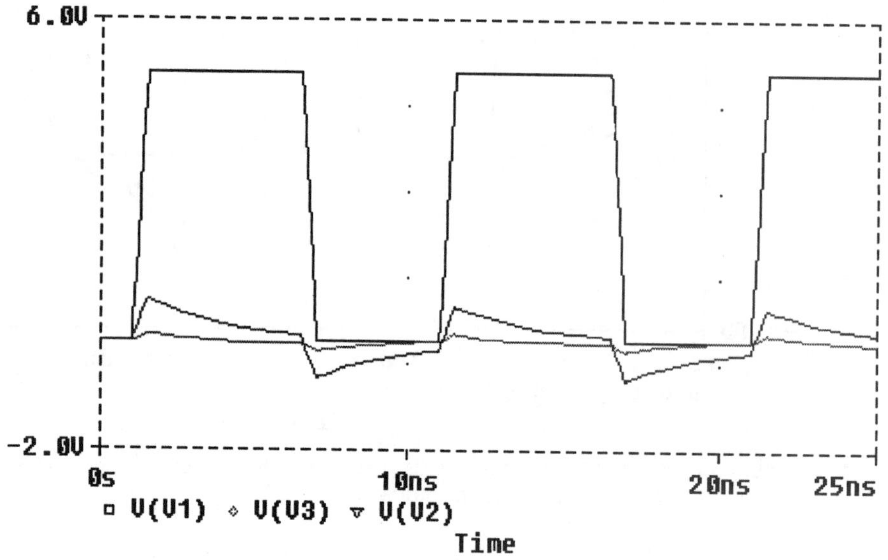

Figure 5.15. Simulation results of the crosstalk solution. Crosstalk is reduced to 16% at conductor 2 and 3% at conductor 3.

PROBLEM SOLVING EXAMPLES

1. What is the charge stored on a 20-μF capacitor that is charged to 12 V?

   Solution: $Q = CV = 240\mu C$

2. Find the voltage across and the charge stored by each capacitor, as well as the total energy stored if the following network is charged to 6V.

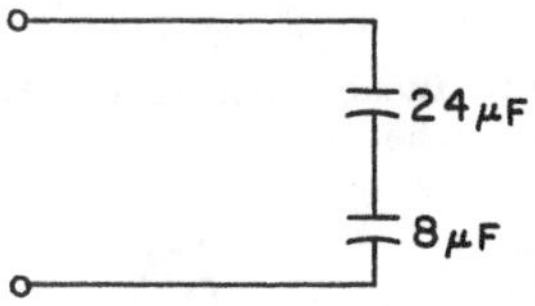

Solution: $V_{24} = (6)(\frac{8}{24+8}) = \frac{3}{2}$ V

$V_8 = (6)(\frac{24}{24+8}) = \frac{9}{2}$ V

$Q = CV = 36\mu C$

and

$W_{24} = \frac{1}{2} CV^2 = 27\mu J$

$W_8 = \frac{1}{2} CV^2 = 81\mu J$

3. Find the total capacitance of the network below.

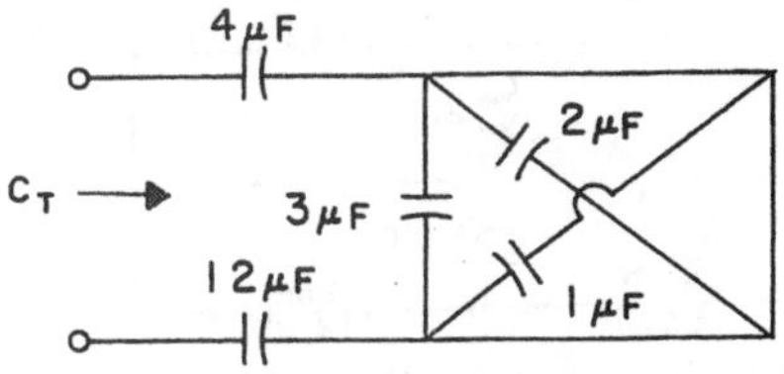

Solution: Because of the short circuit around the outer loop the network reduces to

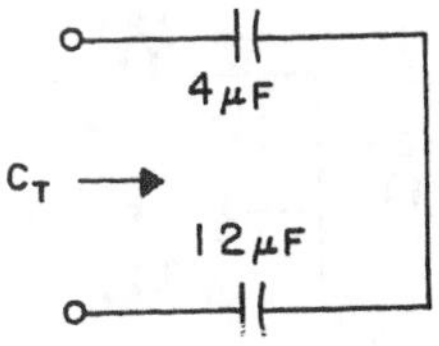

and $C_T = 3\mu F$

4. Find the total capacitance of the following network.

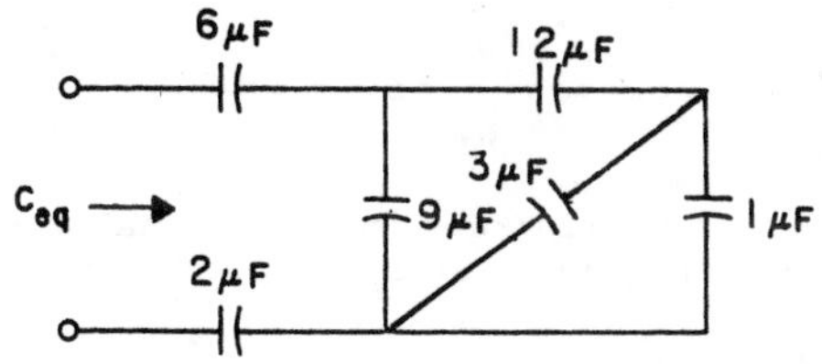

Solution: The network is reduced as follows.

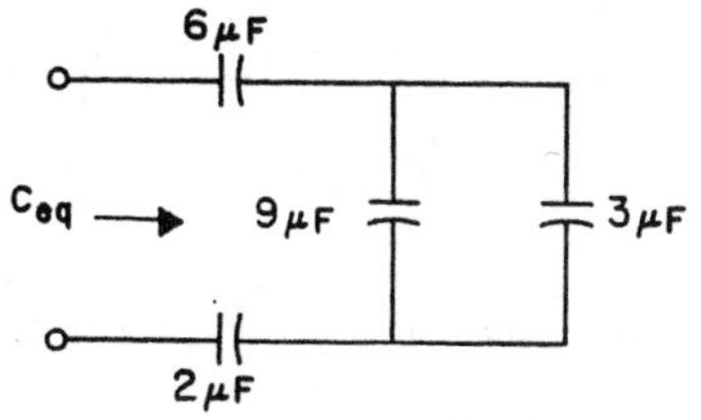

Hence $C_{eq} = \dfrac{\left[\dfrac{(12\mu F)(2\mu F)}{12\mu F+2\mu F}\right] 6\mu F}{\dfrac{(12\mu F)(2\mu F)}{12\mu F+2\mu F} + 6\mu F} = \frac{4}{3}\ \mu F.$

5. Determine the equivalent capacitance of the network below.

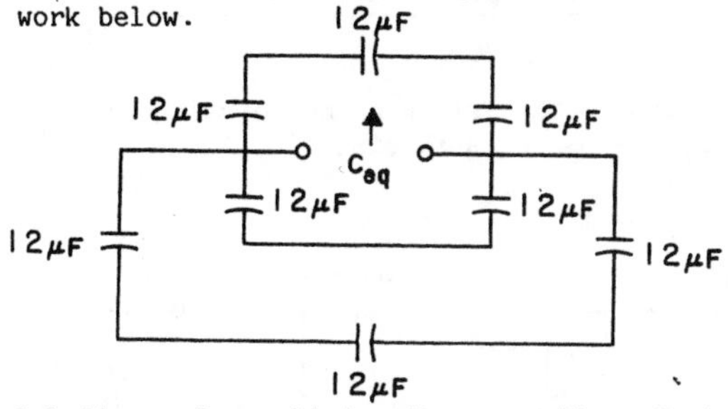

Solution: By combining the capacitors that are connected in series the network reduces to

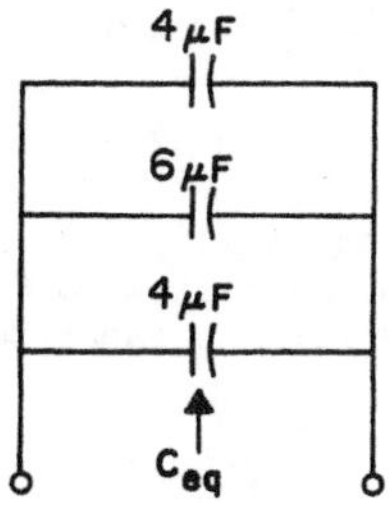

Hence $C_{eq} = 14\mu F$

6. Find the total capacitance of the network below.

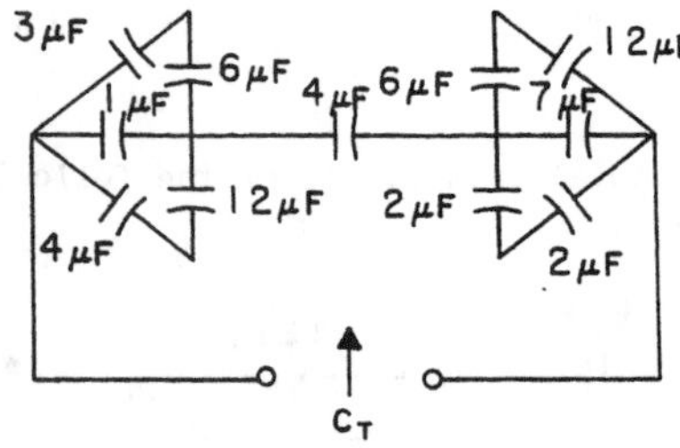

Solution: Combining the capacitors that are connected in series yields

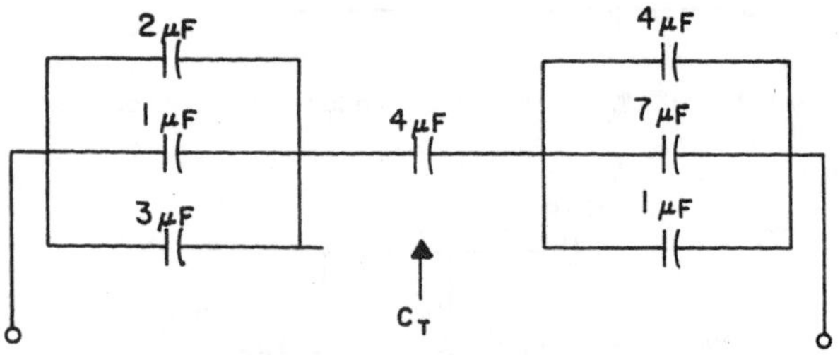

Now adding the parallel capacitors and combining the three remaining capacitors yields $C_T = 2\mu F$.

7. Find the total capacitance of the following network.

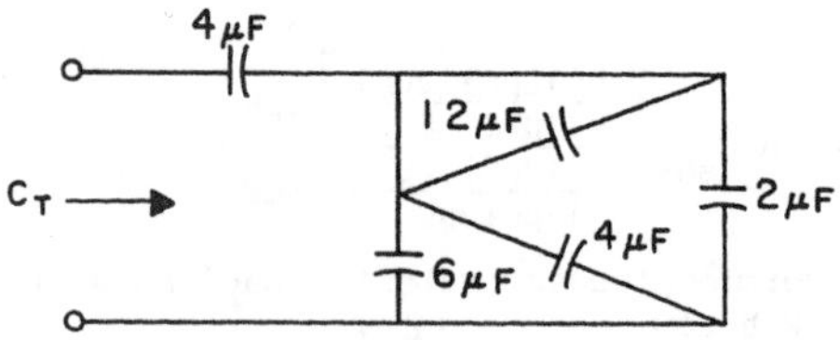

Solution: The following network sequence illustrates the analysis.

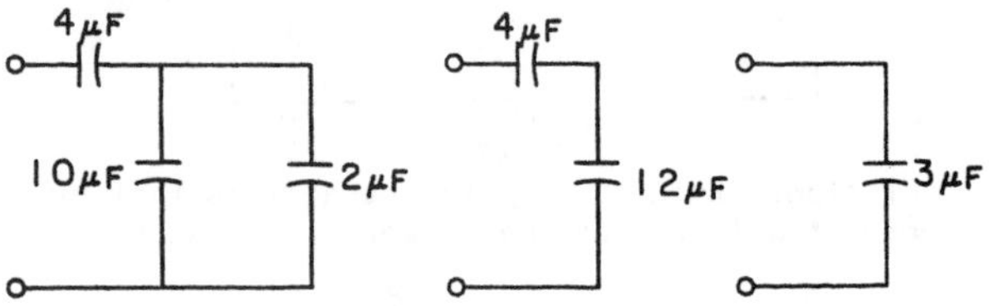

8. The voltage v(t) across a 10μF capacitor is shown in the following Figure. Determine the waveform for the capacitor current.

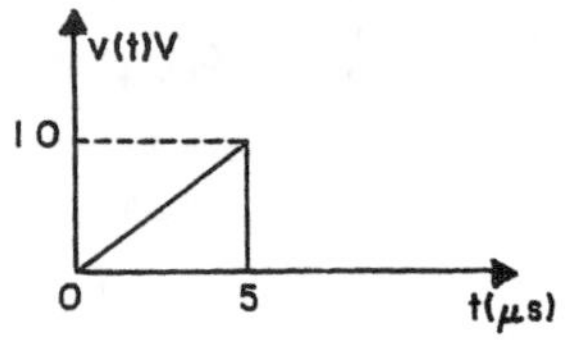

Solution: $v(t) = \frac{10}{5\times10^{-6}} t \quad 0 \leq t \leq 5\mu sec$

$= 0 \quad t > 5\mu sec$

$$i(t) = C\frac{dv(t)}{dt} = 10\times10^{-6}\left(\frac{10}{5\times10^{-6}}\right) = 20A$$

$0 \leq t \leq 5\mu sec$

Therefore

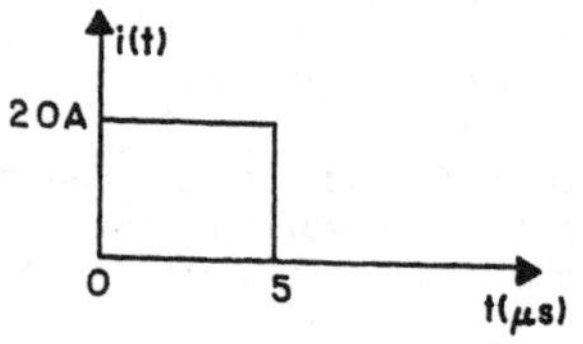

9. Determine the inductance at the terminals A-B in the network below.

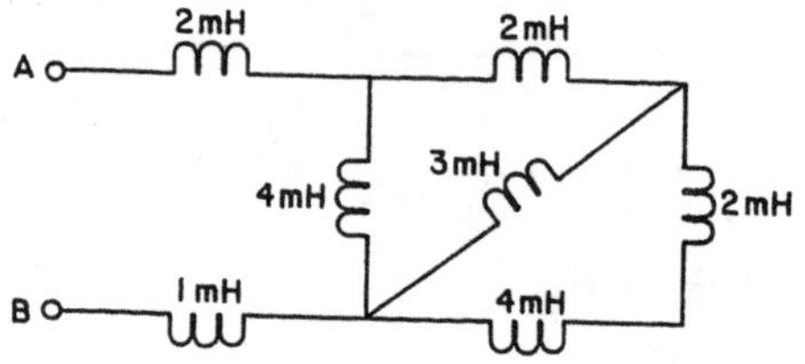

Solution: Remembering that all inductor values are in mH, if we start at the right-most end of the network and combine the elements we obtain

$$L_T = \{([(2+4)||3]+2)||4\}+2+1 = 5mH$$

10. Find the equivalent inductance at the terminals A-B in the following network.

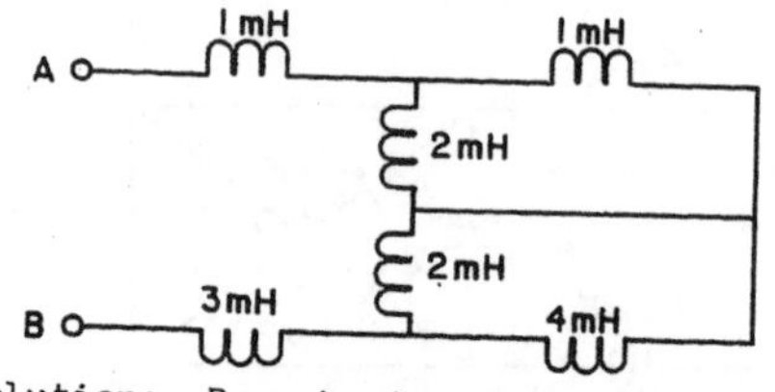

Solution: Remembering that all inductor values are in mH, the equivalent inductance is

$$L_{eq} = (1||2)+(2||4)+1+3 = 6mH$$

11. Find the equivalent inductance for the network below.

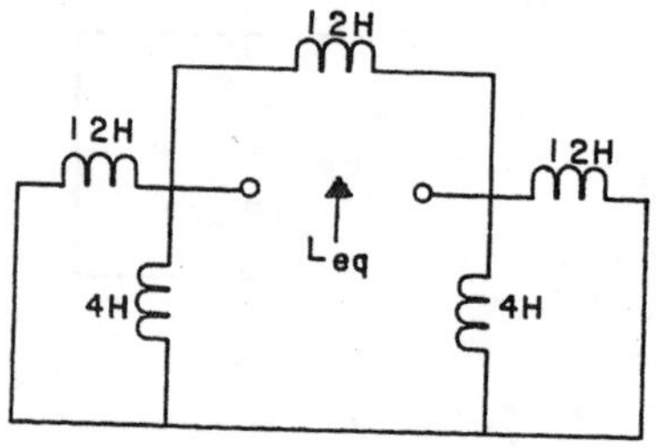

Solution: The equivalent inductance is

$$L_{eq} = [(4||12)+(4||12)]||12 = 4H$$

12. The current i(t) in a 10mH inductor is shown below. Determine the voltage across the inductor and plot it as a function of time.

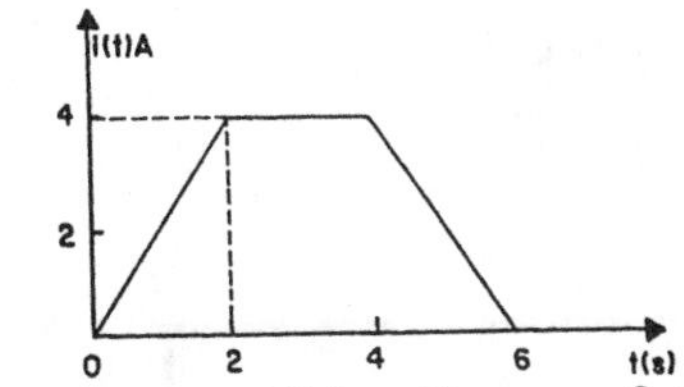

Solution: $i(t) = 2t \qquad 0 \leq t \leq 2s$

$= 4 \qquad 2 \leq t \leq 4s$

$= -2t+12 \qquad 4 \leq t \leq 6s$

$= 0 \qquad t > 6s$

Using $v(t) = L\dfrac{di(t)}{dt}$ yields

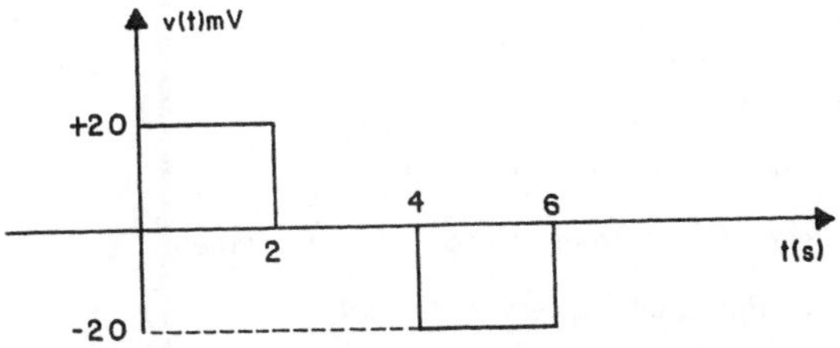

13. The current i(t) in a 4-Henry inductor is shown below. Determine the waveform for the inductor voltage.

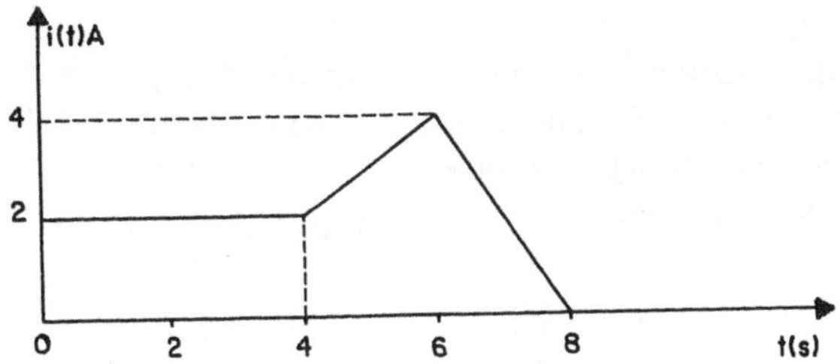

Solution: $i(t) = 2 \qquad 0 \leq t \leq 4s$

$= t-2 \qquad 4 \leq t \leq 6s$

$= -2t+16 \qquad 6 \leq t \leq 8s$

$= 0 \qquad t > 8s$

Using $v(t) = L\dfrac{di(t)}{dt}$ we obtain

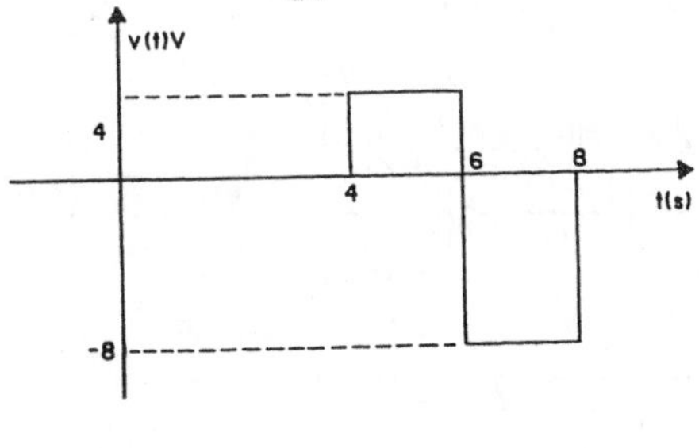

14. The current in a 20-mH inductor is given by the waveform in the following figure. Derive the waveform for the inductor voltage if the equation for the current in the interval $0 < t \leq 2s$ is $i(t) = 2.5t^2 A$.

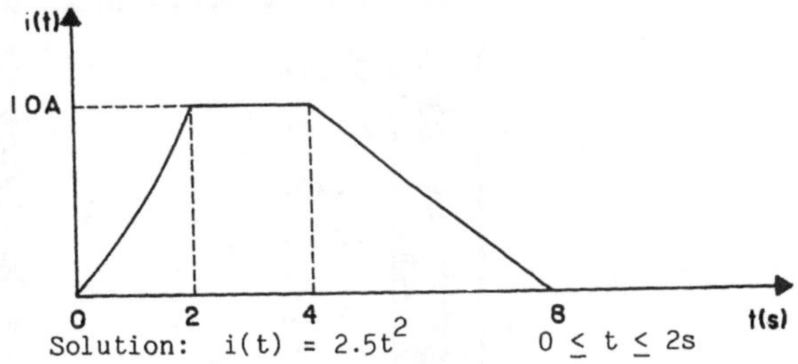

Solution: $i(t) = 2.5t^2 \qquad 0 \leq t \leq 2s$

$= 10 \qquad 2 \leq t \leq 4s$

$= -\frac{10}{4}t + 20 \qquad 4 \leq t \leq 8s$

$= 0 \qquad t > 8s$

using $v(t) = L\dfrac{di(t)}{dt}$ we obtain

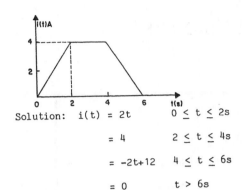

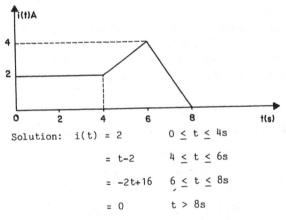

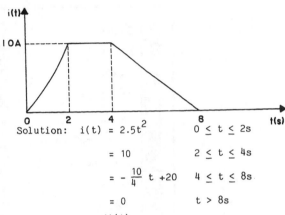

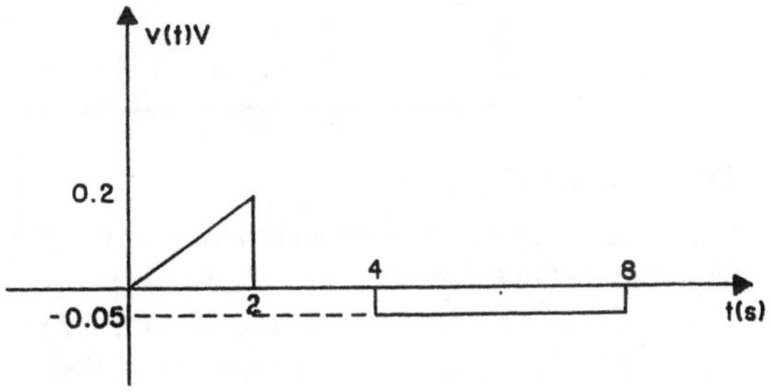

Chapter 6 FIRST-ORDER TRANSIENT CIRCUITS

TRANSIENT SIMULATIONS IN PSPICE

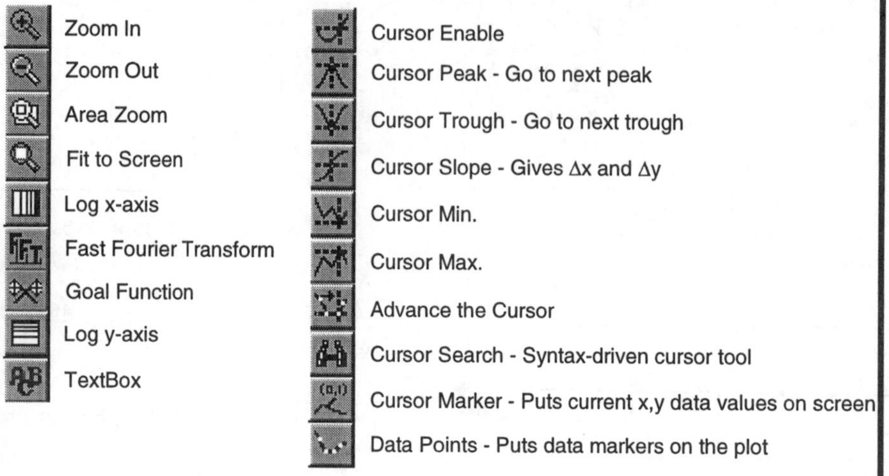

Zoom In
Zoom Out
Area Zoom
Fit to Screen
Log x-axis
Fast Fourier Transform
Goal Function
Log y-axis
TextBox
Cursor Enable
Cursor Peak - Go to next peak
Cursor Trough - Go to next trough
Cursor Slope - Gives Δx and Δy
Cursor Min.
Cursor Max.
Advance the Cursor
Cursor Search - Syntax-driven cursor tool
Cursor Marker - Puts current x,y data values on screen
Data Points - Puts data markers on the plot

Figure 6.1. PROBE hot buttons for plot and cursor manipulation.

THE CURSOR BUTTONS

Before jumping into transient simulations, we should ask the question, "How are we going to get accurate data from the PROBE plots?" The best way is to use the cursors in PROBE's Tools/Cursor menu. Using cursors allows us to extract exact plot values. You'll find it very convenient to use the cursor hot buttons shown in Figure 6.1. Now, we'll proceed with some first order transient simulations.

PSPICE SIMULATIONS

Simulation One - Marking Data Values

Let's simulate the *Schematics* circuit in Figure 6.2a. to find the following characteristics of the output voltage, V_O: the value before switching and the time at which V_O is half its peak value. We will use the VIEWPOINT part at V_O to determine the dc value of the output voltage before switching occurs. The transient nature of V_O will be plotted in PROBE.

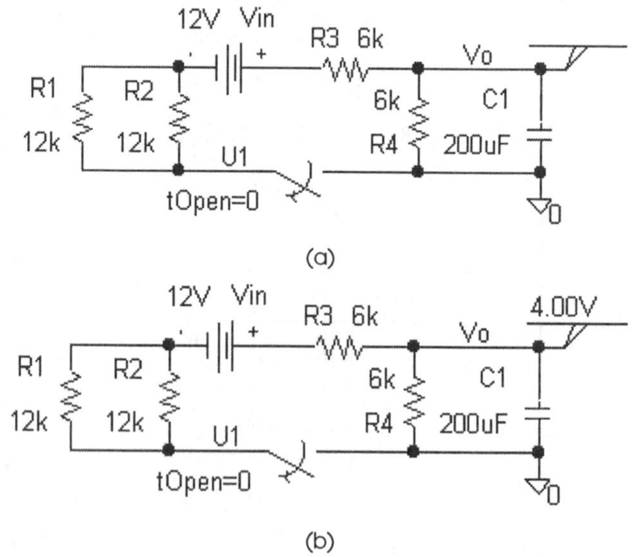

Figure 6.2. PSpice *Schematics* circuit for the first transient simulation, (a) before simulation and (b) after simulation.

An important part of transient simulations is determining how long the simulation should last. As a rule of thumb, estimate the time constant using the smallest resistor in the circuit for the time constant equivalent resistance. Multiply the time constant estimate by 5 and use that number as the Final Time in the Transient setup. For example, in this circuit, we estimated the time constant resistance to be about 6 kΩ which yields a time constant of 1.2 seconds. So, the transient **Final Time** is set to 6 seconds. From the dc simulation shown in Figure 6.2b, $V_O = 4$ V before switching occurs. Transient data is taken from the PROBE plot in Figure 6.3. The maximum value of V_O is 4V. Using the cursor, we find the time at $V_O = 2$V is 830 ms and mark it on the plot using the Cursor Marker hot button. This simulation can be reviewed by running the Visual Tutor, TRANSIENT.EXE.

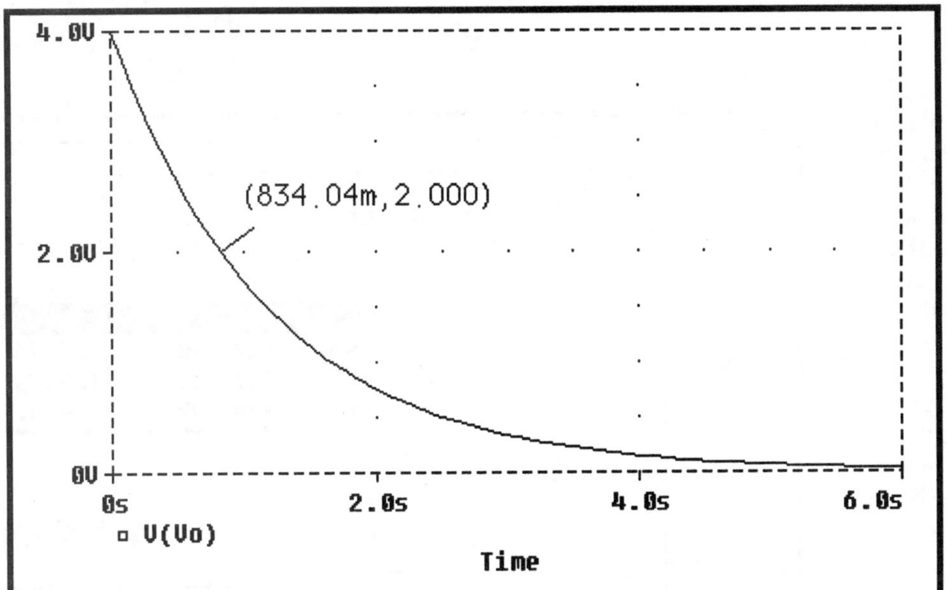

Figure 6.3. Transient results for first PSpice simulation.

Simulation Two - Finding Time Constants

The time constant is by far the most important characteristic of a first-order transient circuit. In PSpice, it can be easily gotten from the PROBE plot using the cursors. To illustrate the procedure, let's simulate the first-order inductive circuit in Figure 6.4 to find the voltage across the current source, I_{in}.

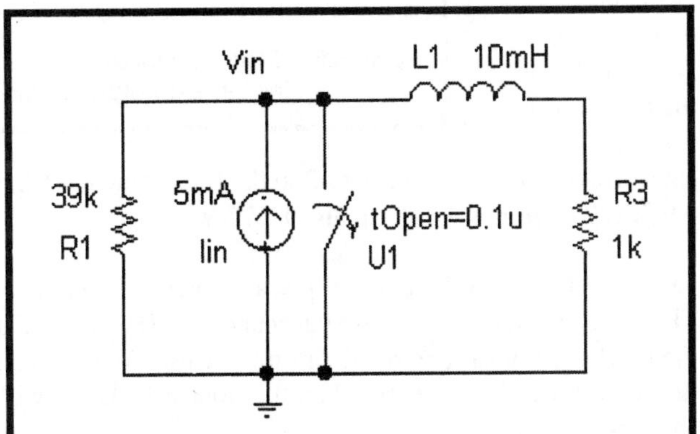

Figure 6.4. PSpice Schematic for the second simulation. Note that the switch time is offset by 0.1 μs.

A easy way to view signal waveforms including the time before switching occurs is to set the switch time to a non-zero value. In Figure 6.4, the switch moves at 0.1 μs. THEREFORE, FOR THIS SIMULATION, OUR TIME SCALE IS SHIFTED FORWARD BY 0.1 μs. Next we must decide on a suitable time duration for the simulation. After some hit-and-miss, we settled on a 2.5 μs simulation. The results are shown in Figure 6.5.

Extracting the time constant begins with the first-order general solution for dc excitation

$$v(t) = K_1 + K_2 e^{-t/\tau} \tag{6.1}$$

where $v(t)$ could be also be a current. Next, we find the slope of the signal at $t = 0$

$$\left.\frac{dv(t)}{dt}\right|_{t=0} = \frac{-K_2}{\tau} \tag{6.2}$$

Recall that K_2 is just the initial value ($t = 0$) minus the final value ($t \rightarrow \infty$) of the signal $v(t)$. Solving for the time constant yields

$$\tau = \frac{-\Delta V}{slope} \tag{6.3}$$

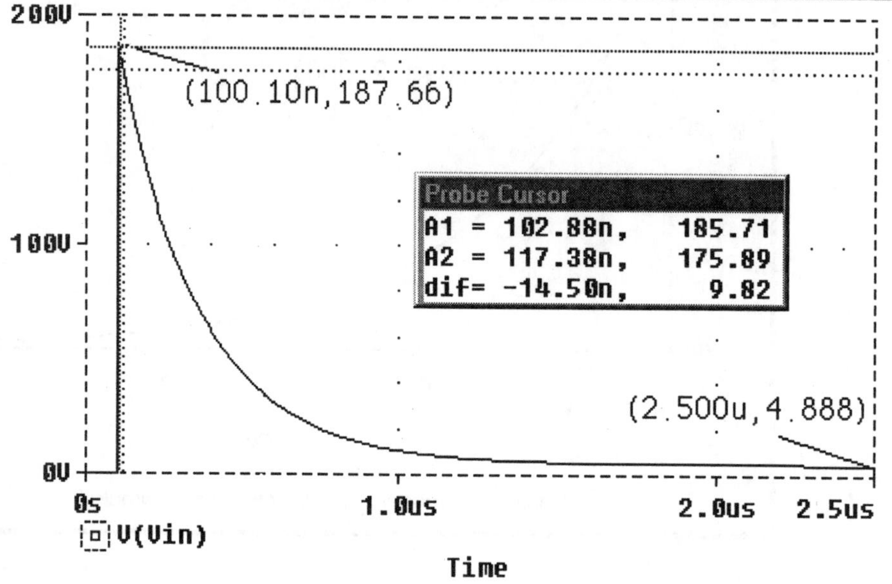

Figure 6.5. PROBE plot for second PSpice simulation. Cursor information needed to find the time constant is included.

where ΔV is the total change in $v(t)$. The slope can be gotten from the PROBE plot of $v(t)$. First, activate the cursors using either the Tools/Cursor menu or the Cursor hot button in Figure 6.1. Next, to determine ΔV, use the left and right arrow keys to move cursor 1 to the point just after switching, t = 0.1 µs in this case. Next, hold down the Shift key and use the left and right arrow keys to move cursor 2 to the final value of $v(t)$. From these two data points, marked in Figure 6.5, ΔV is 187.66 minus 4.888 equals 182.77 V.

Now, we'll find the slope of $v(t)$ just after switching has occurred. Move cursor 1 to a point just after switching. Then, move cursor 2 to a point past cursor 1. The cursor data box, inset in Figure 6.5, will change, containing the new cursor positions and their differences. To get the slope, take the ratio of the differences. Obviously, if the cursors are close to each other, the slope calculation will be more accurate. From the cursor data in, the slope is

$$slope = \frac{9.82}{-14.50\,x10^{-9}} = -6.77\,x10^{8}$$

Armed with the slope at $t = 0$, ΔV and (6.3), the time constant is found to be

$$\tau = \frac{-\Delta V}{slope} = \frac{182.77}{6.77\,x10^{6}} = 0.27\ \mu s$$

which is within 8% of the actual value of 0.25 µs.

Simulation 3 - A Design Approach - The Performance Analysis Feature

An engineer hastily designed the circuit in Figure 6.6 to have a maximum output voltage between 18 and 22 V. Everything looked good on paper. PSpice simulations produced the output seen in Figure 6.7 - looks good. But, when the design was committed to manufacture, almost none of the circuits met the output voltage specification, even when low tolerance resistors ($\leq 1\%$) were used. A co-op student heard about the trouble and decided to use her PSpice expertise to solve the problem. She did so by using the Performance Analysis feature in PROBE. With in an hour the circuit was saved and at minimal cost. Let's so how she did it.

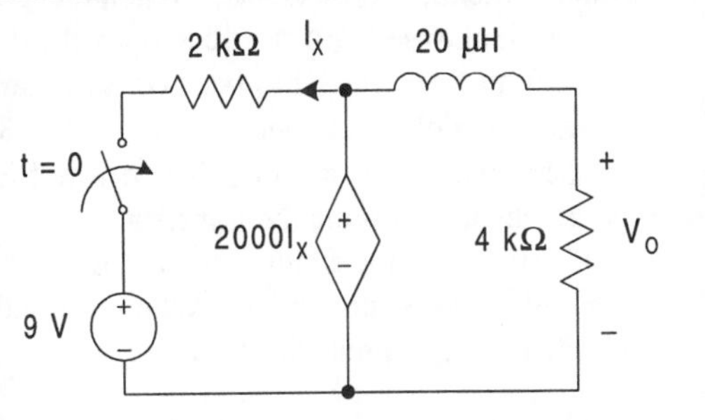

Figure 6.6. A seemingly simple circuit.

The Performance Analysis feature is used to optimize designs. That is, to find component values that optimize a circuit's performance. Obviously, there are two questions that must be answered before we can setup PSpice to use this feature: which component value will we vary and what characteristic of the circuit do we want to optimize. In this example, we will vary the resistor R_1 and investigate the maximum value of the output voltage.

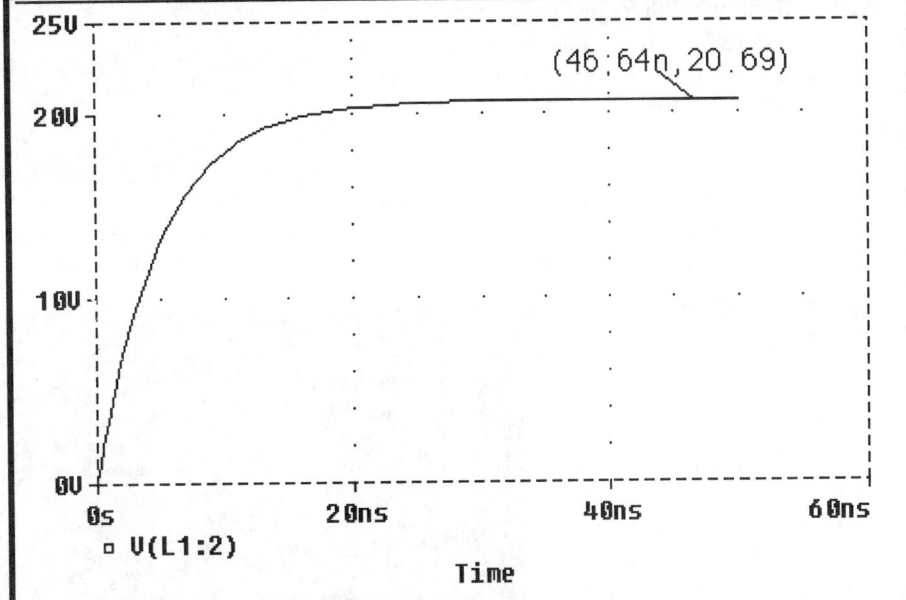

Figure 6.7. PSpice simulation for the circuit in Figure 6.6.

The procedure for using Performance Analysis is a bit involved, so we will describe the procedure as we work through this example. First, we must create a variable name and relate it to R_1. This is done using the PARAM part which goes right on the schematic as seen in Figure 6.8. In a single PARAM part, we can create three variables in the NAME fields. Their values go in the VALUE fields. Figure 6.8 shows that we have already created a variable called Rvariable and given it a value of 2 kΩ.

Next, we relate our R_1 to the variable we just named Rvariable. To do this, change the value of R1 in the schematic to {Rvariable} as shown in Figure 6.8. The braces tell PSpice that the value of R1 is a function of the PARAM variable - in particular, R1 = Rvariable. Note that the expression inside the braces can be just about any function of the three PARAM part variables.

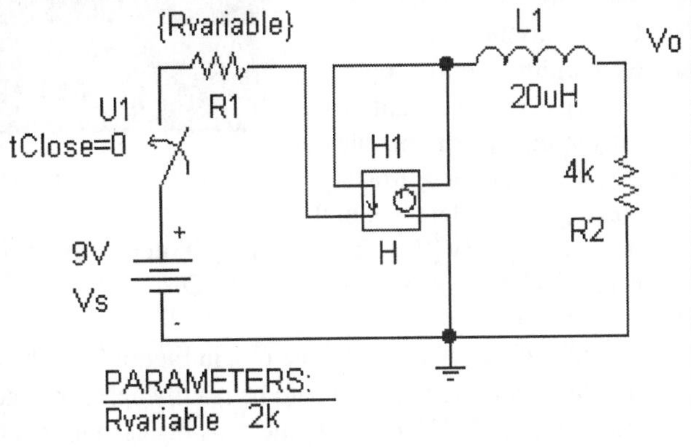

Figure 6.8. PSpice schematic for the circuit in Figure 6.6 including the PARAM part for Rvariable.

Next, we must specify a range for R1. From the **Analysis** menu, choose **Setup**. When the **Setup** box opens, choose **Parametric** to open the dialog box in Figure 6.9 which has been edited to sweep Rvariable, a global parameter, from 1 kΩ to 3kΩ in 50 Ω steps. After finishing the **Parametric** box, go ahead and setup the transient specifications for the simulation. In this example, the **Final Time** is set to 50 ns. When finished with the setup, simulate the circuit.

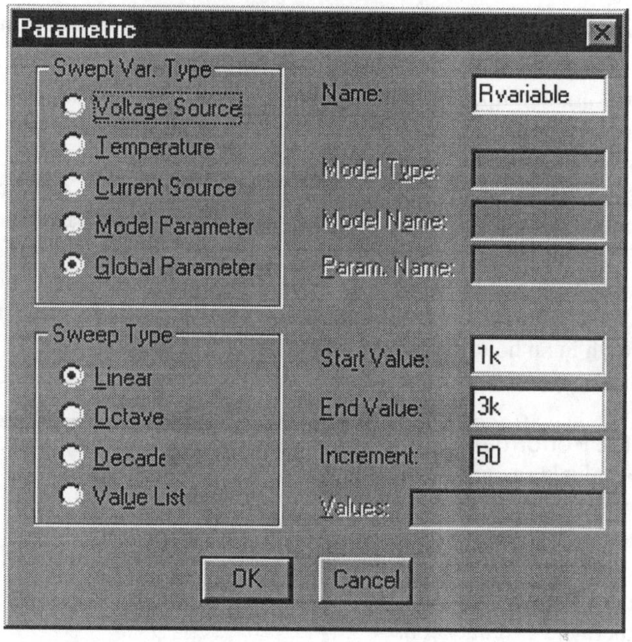

Figure 6.9. Parametric dialog box set to sweep Rvariable from 1kΩ to 3kΩ in steps of 50Ω.

After simulation, the window in Figure 6.10 will appear listing all the individual simulations, **called Sections**, PSpice just performed. Since we swept Rvariable from 1000 to 3000 in steps of 50, there 41 sections available for viewing. For this example, we want all of them, so, just select **All**.

As usual, the PROBE window will open. From the **Traces** menu, select **Performance Analysis**. The information in Figure 6.11 will appear. Select **OK** and a new PROBE screen appears with Rvariable on the *x*-axis!

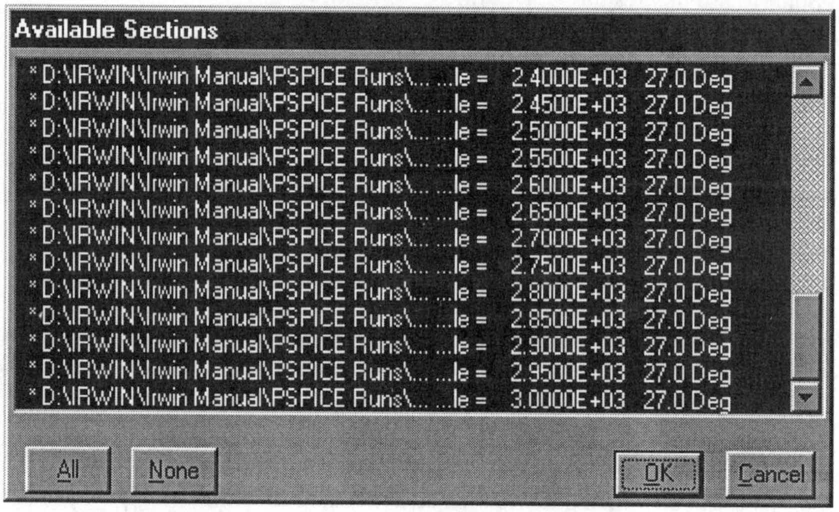

Figure 6.10. The Available Sections dialog box.

Finally, we must choose the circuit characteristic, PSpice calls them **Goal Functions,** that we wish to investigate. Selecting **Add** from the **Traces** menu will open the **Add Traces** box in Figure 6.12. Notice that the right half of the box contains the **Goal Functions** (i.e. bandwidth, risetime, pulse width). Since we want to look at the maximum value of Vo, we will choose Max(1). Next, go to the left side of the box and select V(Vo). The **resulting Trace Expression** at the bottom of the box should match that in Figure 6.12. Finish the process by selecting **OK**.

All this effort pays off with the plot in Figure 6.13 showing the maximum value of V_O versus Rvariable. The problem with the design should be readily apparent - at the specified value of R_1 (2kΩ), V_O is very sensitive to Rvariable (R_1). Even though the resistors used in manufacture were within their tolerance spec., the variation was large enough to cause a large number of failures.

How did our co-op solve the dilemma? Notice that at R_1 = 1.1kΩ the output voltage is 20V and is much less sensitive to changes in R_1. The co-op simply change the value of R_1 from 2kΩ to 1.1kΩ $\pm$ 5%. The step-by-step process for using Performance Analysis is summarized in Table 6.1 and demonstrated visually in the Visual Tutor, TRANSIENT_PA.EXE.

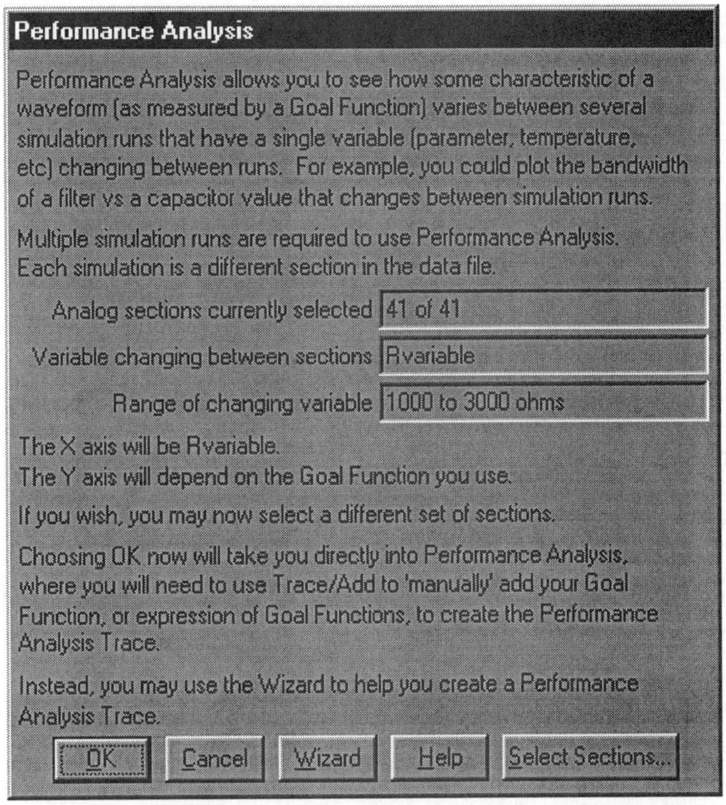

Figure 6.11. Information about the Performance Analysis setup for this simulation.

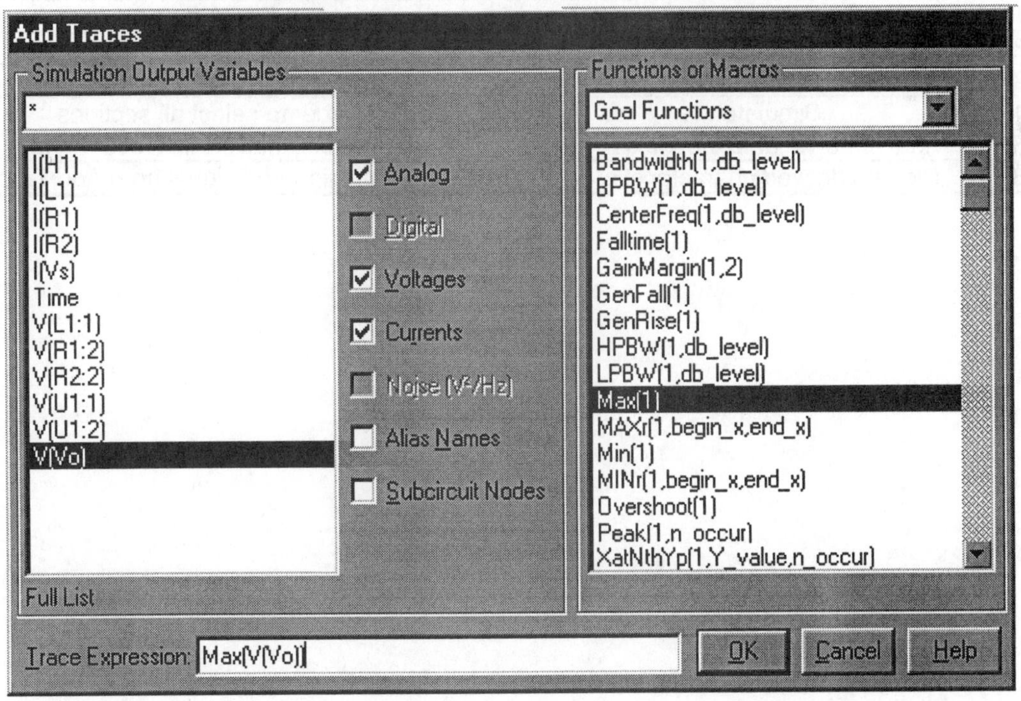

Figure 6.12. The Add Traces box used to enter Goal Functions during a Performance Analysis simulation. In this case we will plot the maximum value of Vo.

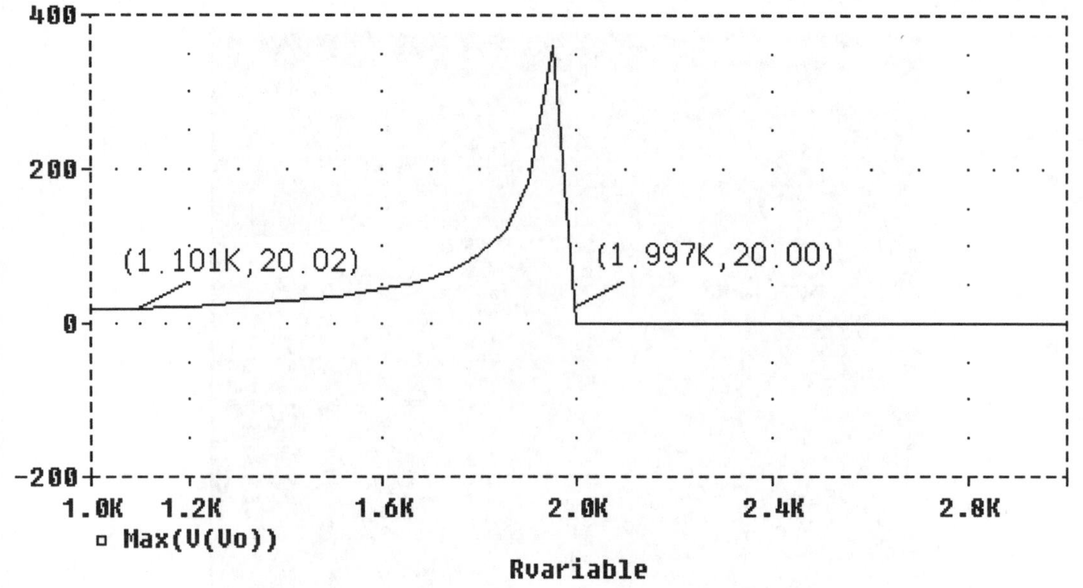

Figure 6.13. The final results of the Performance Analysis simulation - maximum output voltage versus R_1.

Table 6.1

Procedure for Performance Analysis Simulations

Step	Activity	PSpice Part/Operation
1	Create variables	Use a PARAM part
2	Relate variables to components	Use braces { }
3	Specify variable range	Parametric in the Setup box
4	Simulate	Don't forget to select all sections
5	Make the variable the x-axis	Performance Analysis in Trace menu
6	Plot the desired characteristic	Choose Goal Function and signal from Add Trace box

EWB SIMULATIONS

Let's use EWB to find time at which $v_O(t)$ is 10 V for the circuit in Figure 6.14. In the requisite EWB diagram, shown in Figure 6.15, we have delayed the switching time by 1 second. Thus, we must subtract 1 from any cursor time-base information we use. From Figure 6.17, $v_O(t)$ is 10V at a simulation time of right at 2 s, which corresponds to $t = 1.0$ s. The node numbers in Figure 6.15 were generated automatically in EWB and must be specified when performing the transient analysis. For example, in this simulation, we have to ask for the voltage at node 4 to produce the plot in Figure 6.17.

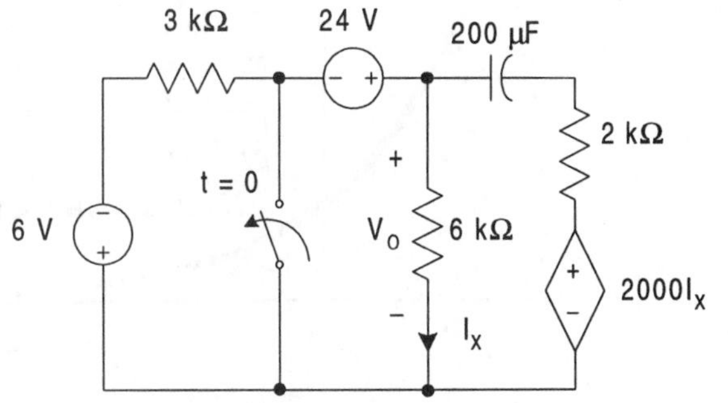

Figure 6.14. The circuit for the first EWB simulation.

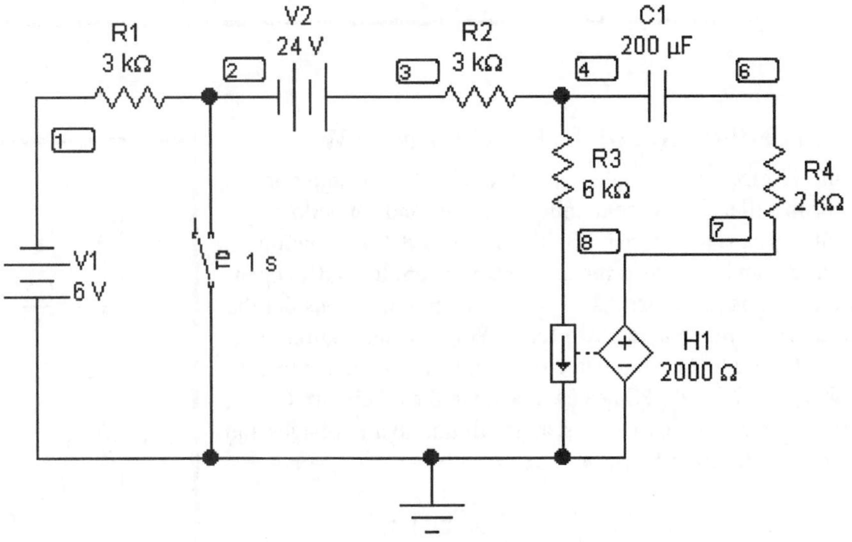

Figure 6.15. EWB diagram for the first EWB simulation. The node numbers were generated automatically in EWB.

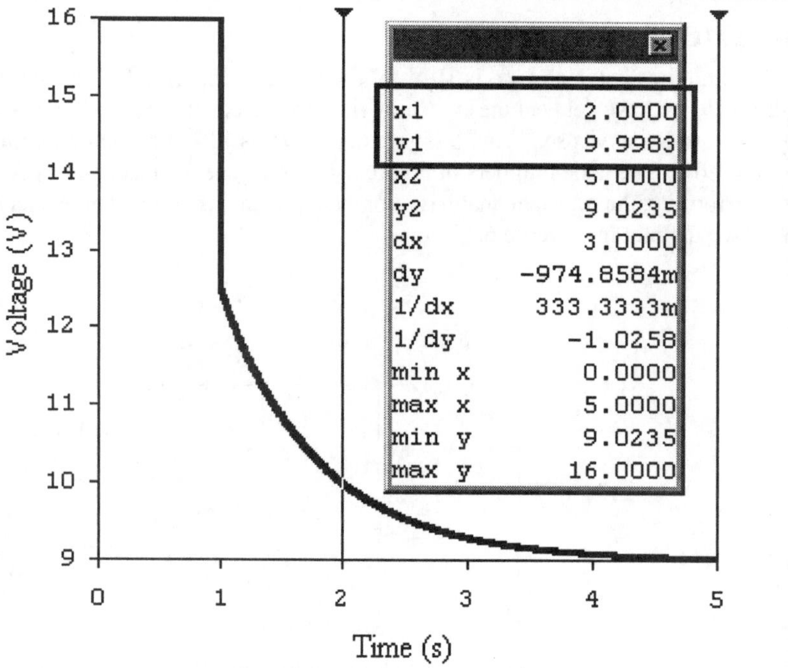

Figure 6.17. Results for the first EWB simulation. As seen in the box, the output is 10V after two seconds of simulation.

CONVERGENCE PROBLEMS IN PSPICE AND EWB

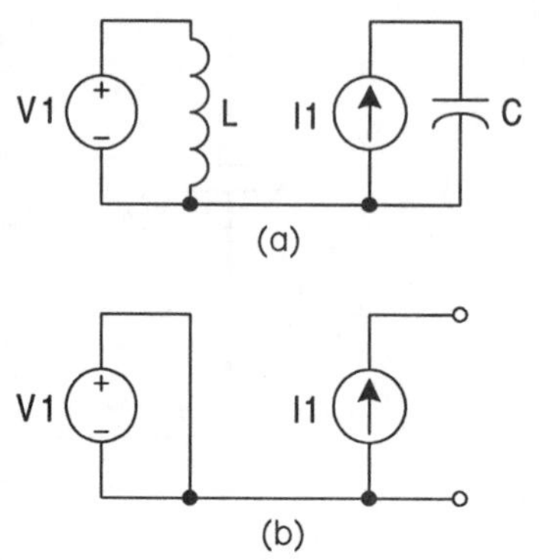

Figure 6.16. A problematic circuit for PSpice and EWB.

If you work with PSpice long enough, you will discover two simulation errors that occur in some circuits containing inductors and capacitors. Whenever a simulation is performed, PSpice first does a dc simulation for all node voltages and branch currents. Capacitors are treated as open-circuits and inductors as short-circuits. There's the trouble. Consider the circuit in Figure 6.16a, which is redrawn for a PSpice dc simulation in Figure 6.16b. In Figure 6.16b, the source current, I_S, has no place to go and node number 1 is isolated. PSpice cannot solve the dc circuit, that is, the simulation cannot converge to a consistent solution to all voltages and currents, and you get the error message

```
Node $N_0002 is floating
```

The inductor in Figure 6.16b is also irksome since PSpice cannot determine the branch current. Again, convergence fails and the error message reads

```
Voltage source and/inductor loop involving V_V1
```

How do we fix these errors? In the case of the capacitor, place a very large resistor in parallel with the capacitor. Now, during dc simulation, node 2 is not floating. Just be sure to make the resistor value is much larger than the other resistances in the circuit. As for the inductor, place a very small resistor in series with the inductor. The resistor value should be much less than the other resistors in the circuit.

The authors of EWB have included a means to avoid the capacitor "floating node" problem. A resistance, called Rshunt, is placed across all components. You can access this feature in the Analysis/Analysis Options/Global menu as shown in Figure 6.18. Rshunt be set to any value (in Ohms) or it can be disabled. Unfortunately, Rshunt does not solve inductor-based convergence errors.

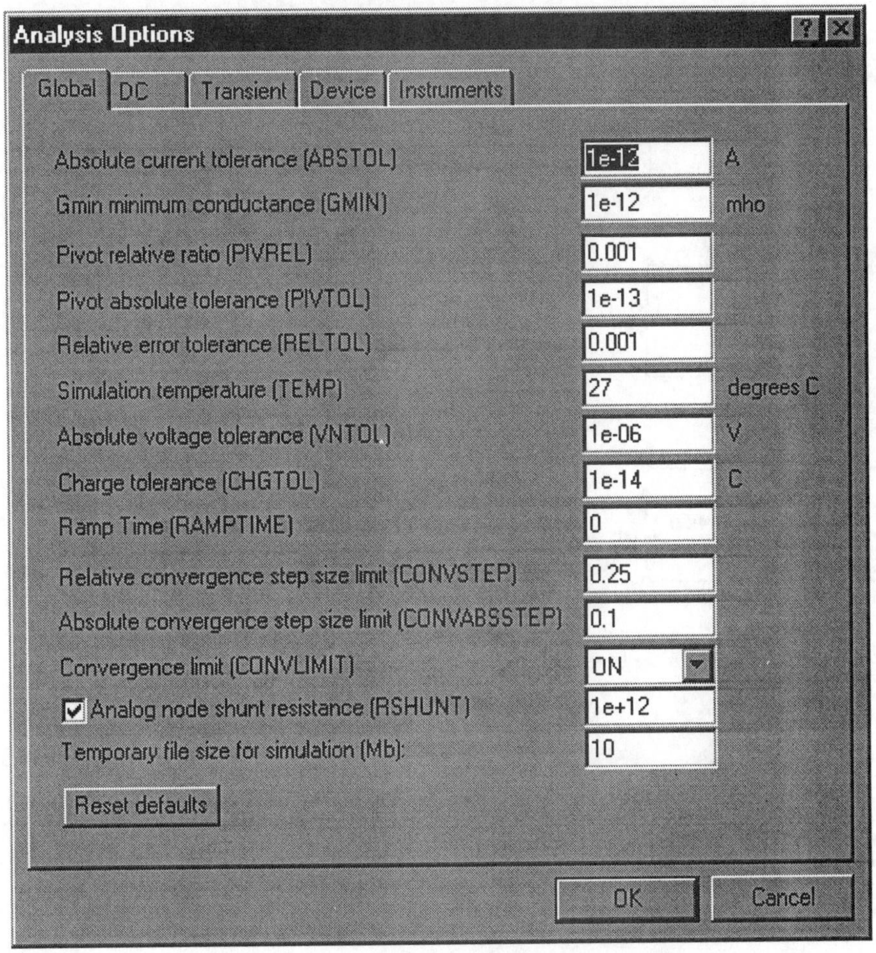

Figure 6.18. Accessing the Rshunt feature in EWB.

NOTES

PROBLEM SOLVING EXAMPLES

1. Find $v_o(t)$ for $t>0$ in the following network

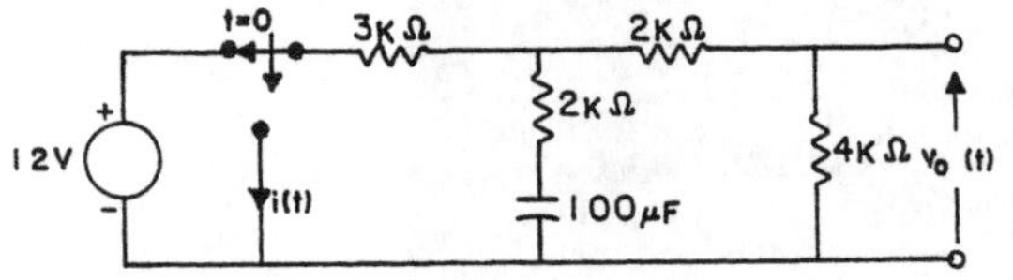

Solution:

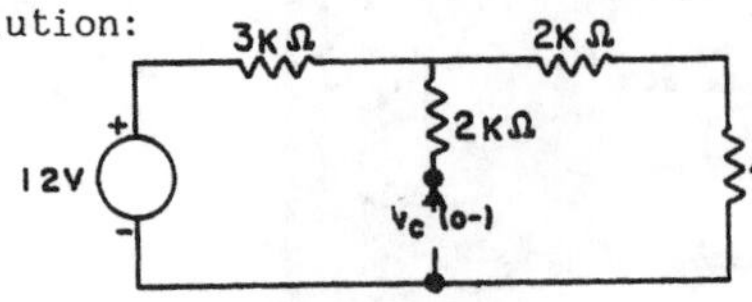

$v_C(0-) = (12)(\frac{6}{9}) = 8V$

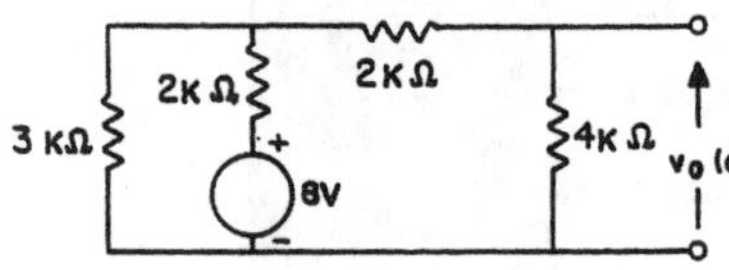

$v_o(0) = (\frac{8}{2K+2K})(\frac{3K}{9K})(4K) = \frac{8}{3}$ V

$v_o(\infty) = 0$

$T_c = R_{TH}C = (4K)(100\mu F) = 0.4s$

Hence $v_o(t) = \frac{8}{3} e^{-t/0.4} V$

2. Find i(t) for $t>0$ in the network in Problem 1.

Solution: Using the data from the previous problem

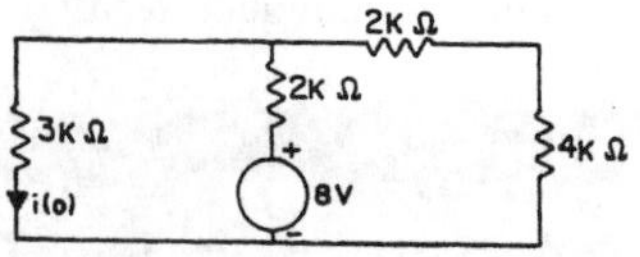

$i(0) = (\frac{8}{2K+2K})(\frac{6K}{9K}) = \frac{4}{3}$ ma

$i(\infty) = 0$ and $T_c = 0.4s$ hence

$i(t) = \frac{4}{3} e^{-t/0.4}$ ma.

3. Find $v_o(t)$ for $t>0$ in the network below.

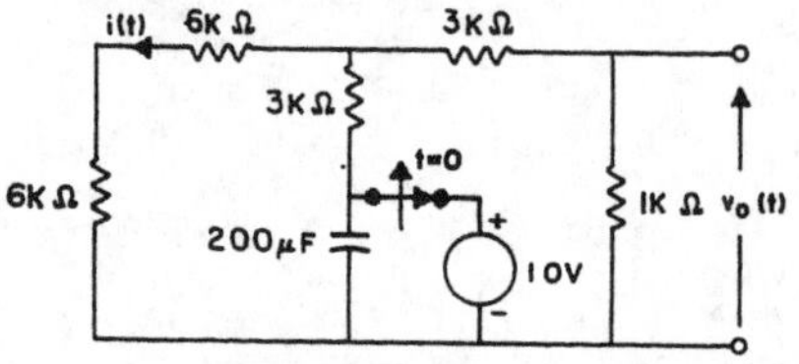

Solution: $v_C(0-) = v_C(0+) = 10V$

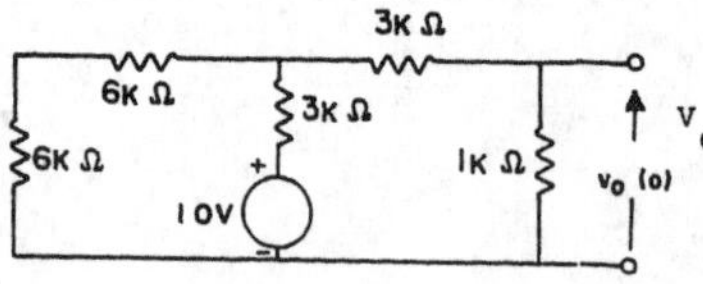

$v_o(0) = (\frac{10}{3K+3K})(\frac{12K}{16K})(1K)$

$= \frac{5}{4}$ V

$v_o(\infty) = 0$

$T_c = R_{TH}C = (3K+12K||4K)200\mu F = 1.2s$

Hence $v_o(t) = \frac{5}{4} e^{-t/1.2}$ V.

4. Find i(t) for $t>0$ in the network in Problem 3.

Solution: Using the data from the previous problem

$v_C(0-) = v_C(0+) = 10V$

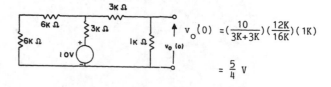

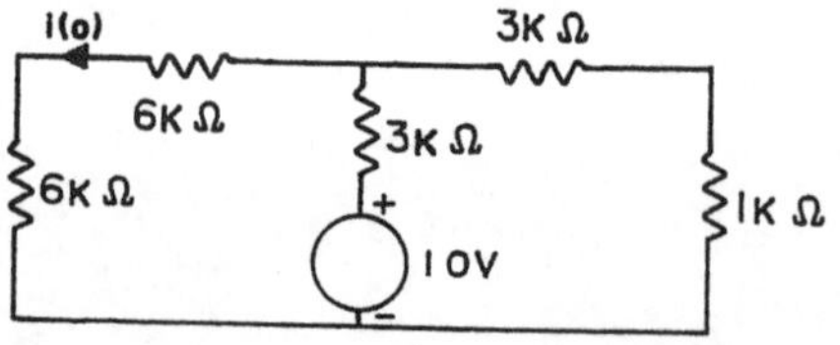

$i(0) = (\frac{10}{3K+3K})(\frac{4K}{12K+4K}) = \frac{5}{12}$ ma

$i(\infty) = 0$ and $T_c = 1.2s$

Hence $i(t) = \frac{5}{12} e^{-t/1.2}$ ma.

5. Find $v_o(t)$ for t>0 in the network below

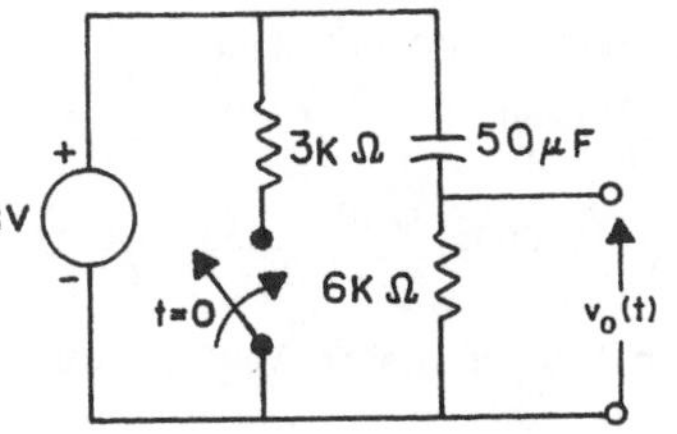

Solution: $v_C(0-) = v_C(0+) = 18V$

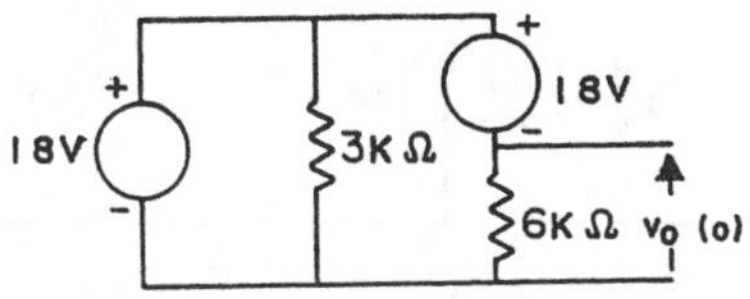

$v_o(0) = 0$

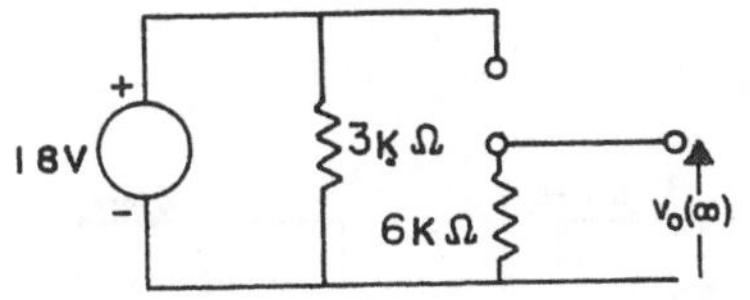

$v_o(\infty) = 0$

$R_{TH} = 6K\Omega$ and $T_c = (6K)(50\mu F) = 0.3s$

$v_o(t) = 0$

6. Find $v_o(t)$ for t>0 in the following circuit

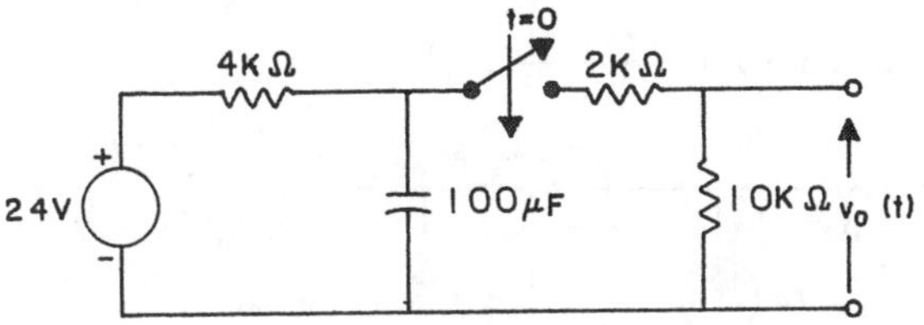

Solution: $v_C(0-) = v_C(0+) = 24V$

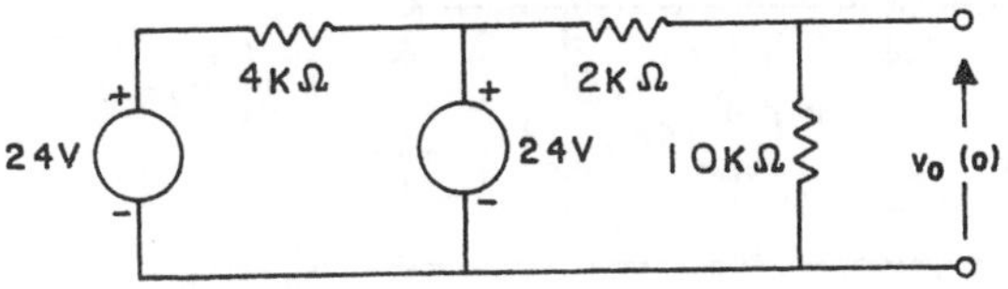

$v_o(0) = (\frac{24}{12K})(10K) = 20V$

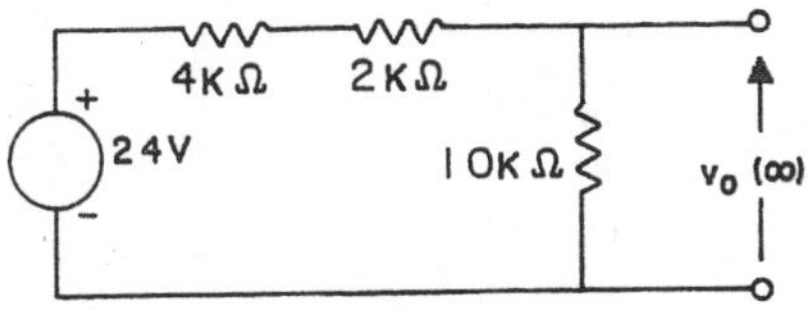

$V_o(\infty) = (\frac{24}{16K})(10K) = 15V$

$R_{TH} = 4K||12K = 3K$ and $T_c = (3K)(100\mu F) = 0.3s$

Hence $v_o(t) = 15+(20-15)e^{-t/0.3}V$.

7. Find $i_o(t)$ for t>0 in the network below

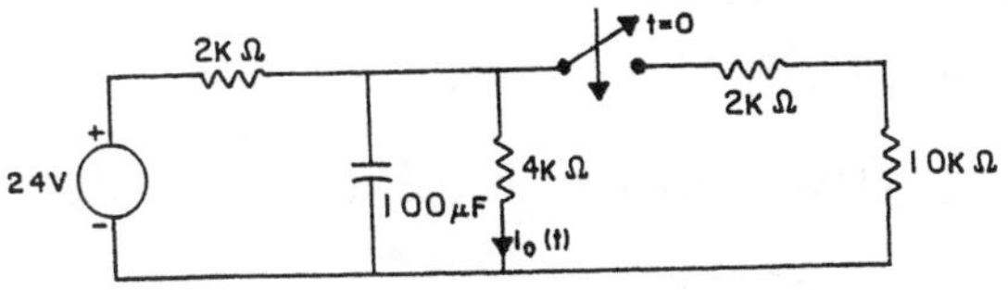

Solution: $v_C(0-) = v_C(0+) = (\frac{24}{6K})(4K) = 16V$

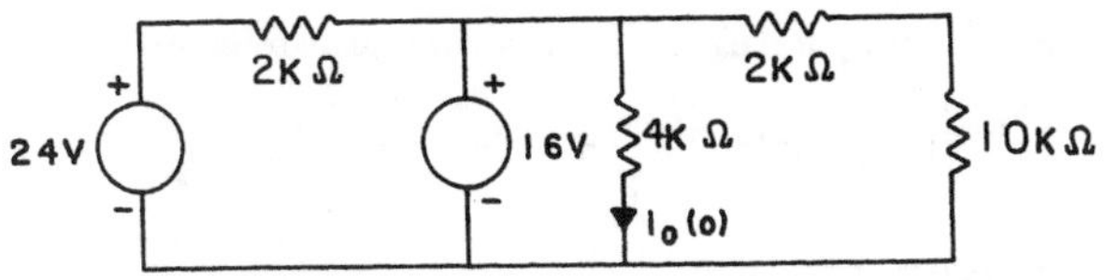

$i_o(0) = \frac{16}{4K} = 4$ ma

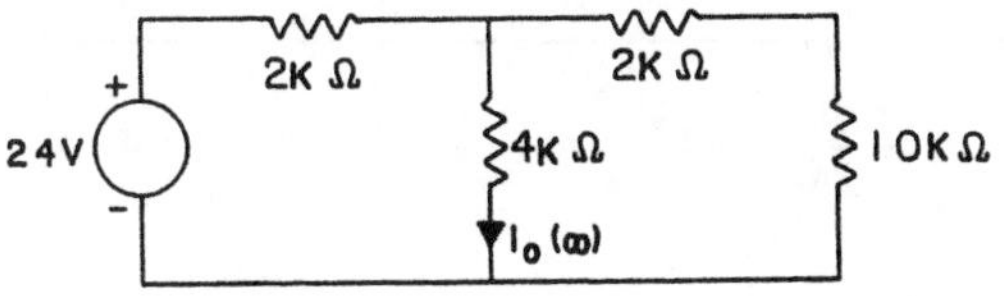

$i_o(\infty) = \frac{24}{2K+(4K||12K)}(\frac{12K}{16K}) = \frac{18}{5}$ ma

$R_{TH} = 2K||3K = \frac{6}{5}$ K and $T_c = (\frac{6}{5}$ K$)(100\mu F) = 0.12s$

Hence $i_o(t) = \frac{18}{5} +(4- \frac{18}{5})e^{-t/0.12}$ ma.

8. Find $v_o(t)$ for t>0 in the following network.

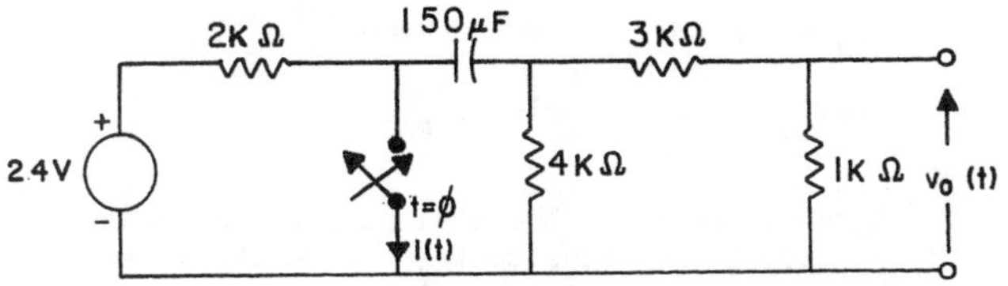

Solution:

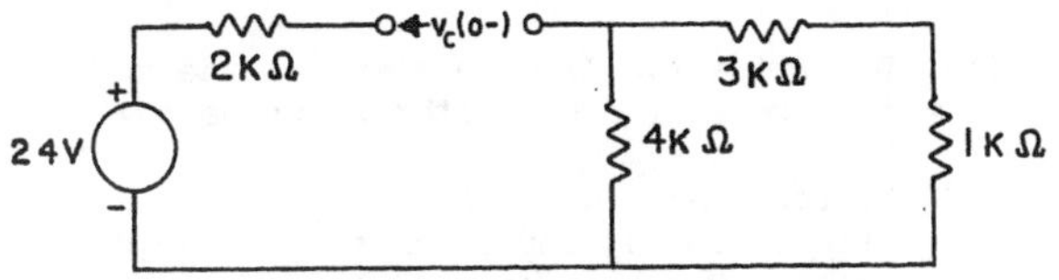

$v_C(0-) = v_C(0+) = 24V$

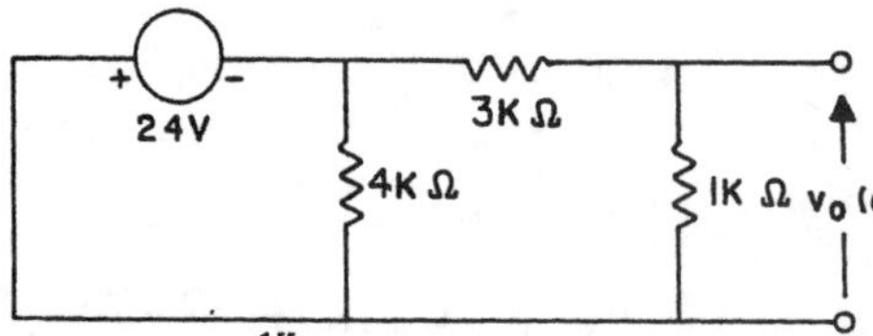

$v_o(0) = -24(\frac{1K}{4K}) = -6V$

$v_o(\infty) = 0$

$R_{TH} = 4K||4K = 2K\Omega$ and $T_c = (2K)(150\mu F) = 0.3s$

$v_o(t) = -6e^{-t/0.3}$ V

9. Find i(t) for t>0 in the network in the previous problem.

Solution: Using data from the previous problem

$v_C(0-) = v_C(0+) = 24V.$

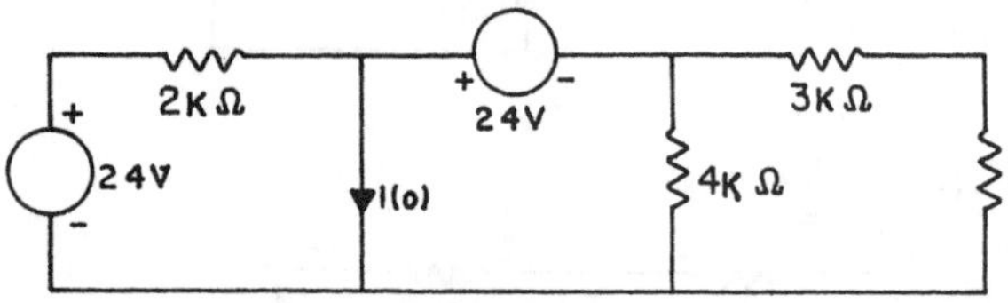

$i(0) = \frac{24}{2K} + \frac{24}{4K||4K} = 24ma$

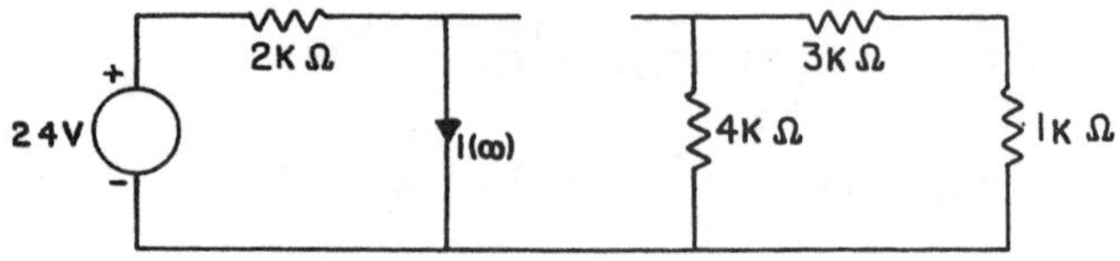

$i(\infty) = \frac{24}{2K} = 12ma$

$R_{TH} = 2K\Omega$ and $T_c = 0.3s$. Hence

$i(t) = 12+(24-12)e^{-t/0.3}$ ma.

10. Although we model the capacitor as an ideal device, in practice there is normally a very large leakage resistance in parallel which provides a conduction path between the capacitor plates.

(a) Determine the time constant of the model of a realistic capacitor shown below.

(b) Suppose this capacitor is placed in a television set and is charged to 10,000 volts. How much time will you have to wait, after the T.V. is turned off, until the capacitor has discharged to less than 10 volts?

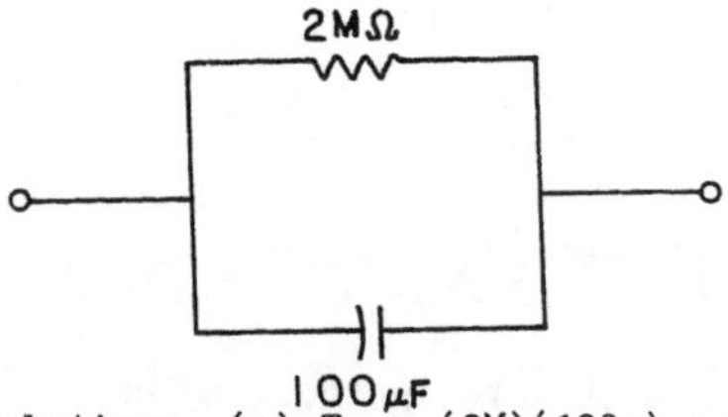

Solution: (a) $T_c = (2M)(100\mu) = 200s.$

(b) $v(\infty) = 0$, $v(0) = 10,000$ Hence

$v(t) = 10,000e^{-t/200}$

If $v(t) = 10 = 10,000e^{-t/200}$ then $t \simeq 23$ minutes.

11. Find $v_o(t)$ for t>0 in the following network.

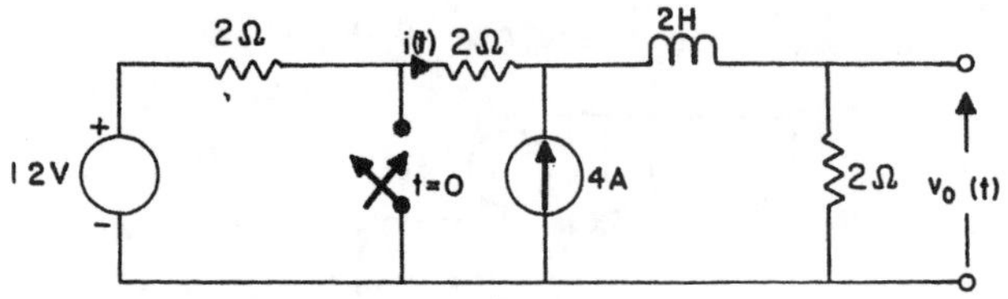

Solution:

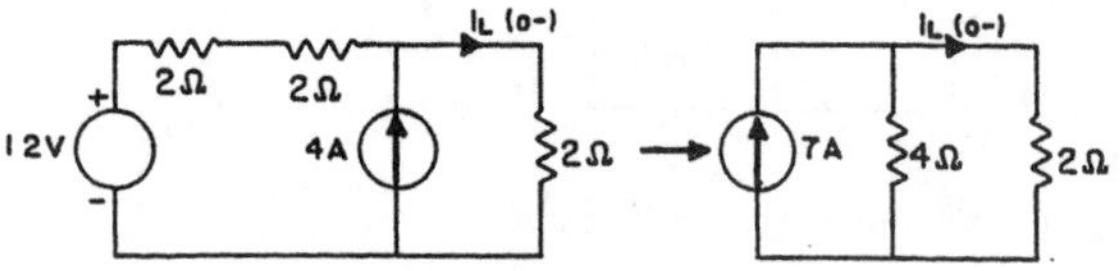

Therefore by using source transformation we see that $i_L(0-) = \frac{(7)(4)}{6} = \frac{14}{3}$ A.

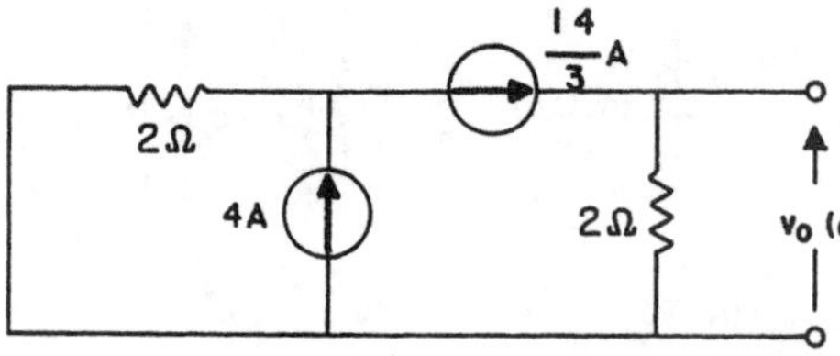

$v_o(0) = \frac{28}{3}$ V

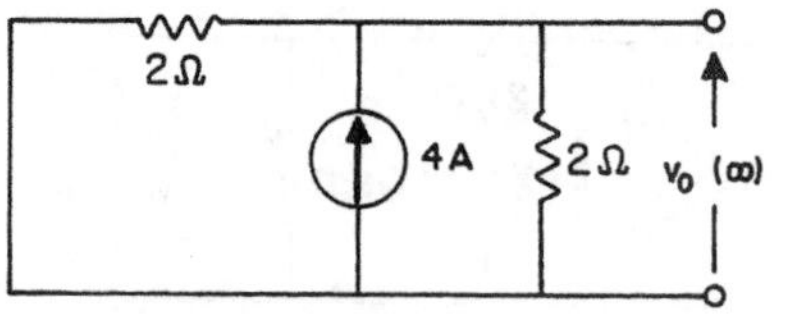

$v_o(\infty) = 4V$

$R_{TH} = 4\Omega$ and $T_c = \frac{L}{R_{TH}} = \frac{1}{2}$ s.

Hence

$v_o(t) = 4+(\frac{28}{3} - 4)e^{-2t}$ V.

12. Find i(t) for t>0 in the network in Problem 11.

Solution: Using data from the previous problem

$i_L(0-) = i_L(0+) = \frac{14}{3}$ A.

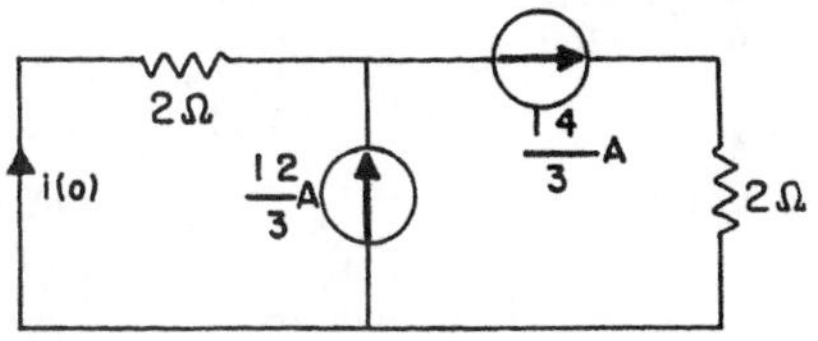

$i(0) = \frac{2}{3}$ A

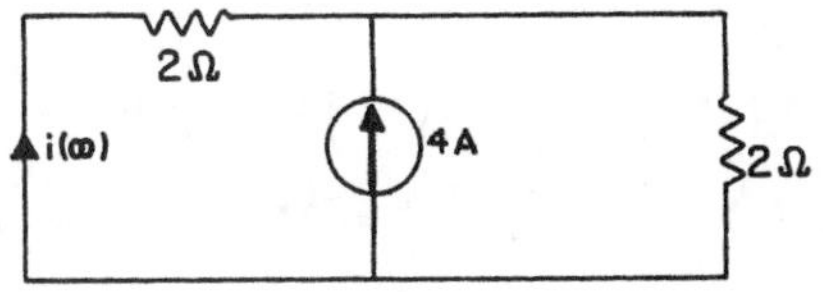

$i(\infty) = -2$A

$R_{TH} = 4\Omega$ and $T_c = \frac{1}{2}$ s.

Hence $i(t) = -2+(\frac{2}{3}+2)e^{-2t}$A.

13. Find $v_o(t)$, t>0, in the following circuit.

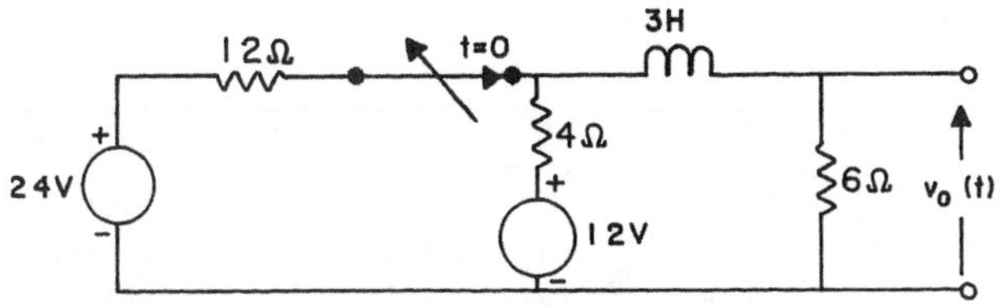

Solution:

Using superposition

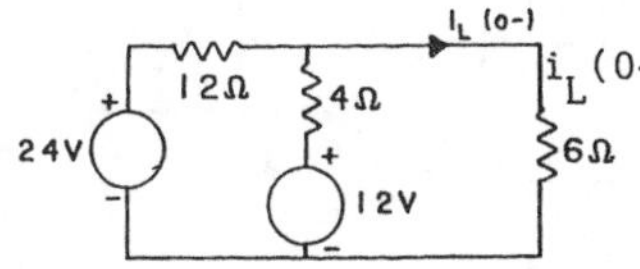

$$i_L(0-) = \frac{24}{12+\frac{24}{10}}(\frac{4}{10}) + \frac{12}{4+\frac{72}{18}}(\frac{12}{18})$$

$$= \frac{5}{3} \text{ A}$$

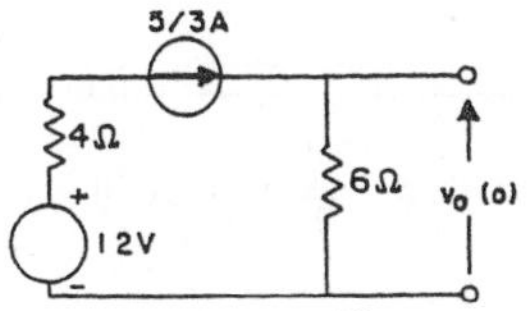

$v_o(0) = (\frac{5}{3})(6) = 10$V

$v_o(\infty) = \frac{12}{4+6}(6) = 7.2$V

$R_{TH} = 10\Omega$ and $T_c = 0.3$s.

Hence $v_o(t) = 7.2+(10-7.2)e^{-t/0.3}$V.

14. Find $v_o(t)$, t>0 in the network below.

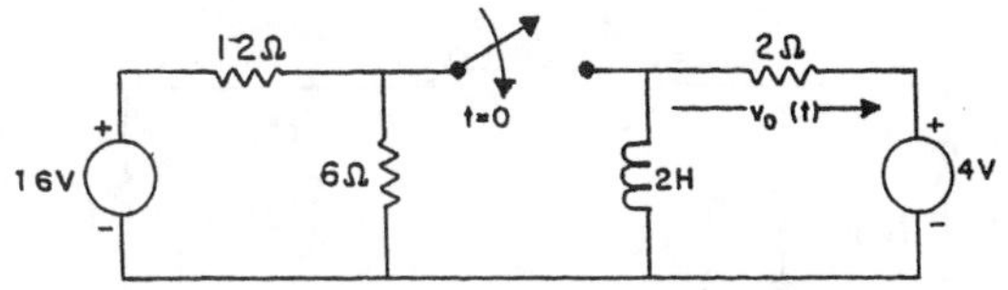

Solution: $i_L(0-) = i_L(0+) = \frac{4}{2} = 2$A

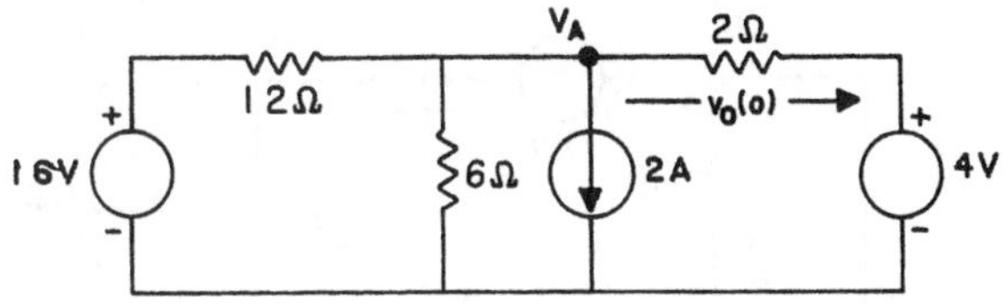

Using a nodal equation we can find V_A.

$$\frac{V_A-16}{12} + \frac{V_A}{6} + 2 + \frac{V_A-4}{2} = 0 \text{ and hence } V_A = \frac{16}{9} \text{ V}.$$

Then $v_o(0) = 4-V_A = \frac{20}{9}$ V. At $t = \infty$ the inductor looks like a short circuit and the 2Ω resistor is therefore in parallel with the 4V source. Hence $v_o(\infty) = 4$V, $R_{TH} = 2||6||12 = \frac{4}{3}\ \Omega$ and $T_c = \frac{L}{R_{TH}} = \frac{3}{2}$ s. Therefore $v_o(t) = 4+(\frac{20}{9}-4)e^{-2t/3}$ V.

15. Find v(t) for t>0 for the network below.

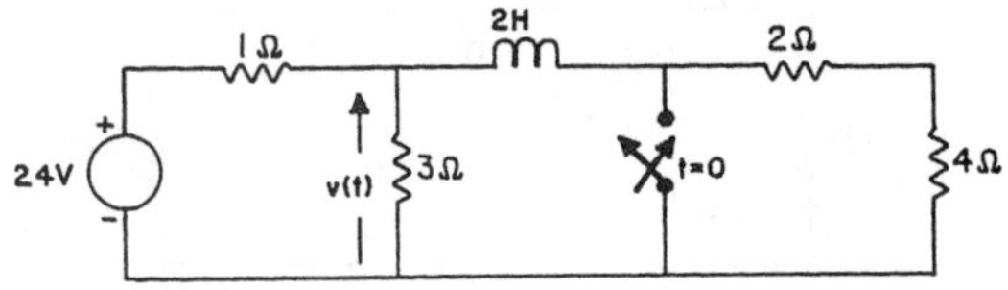

Solution: $i_L(0-) = i_L(0+) = \frac{24}{1+3||6}(\frac{3}{3+6}) = \frac{8}{3}$ A.

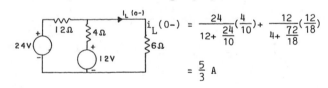

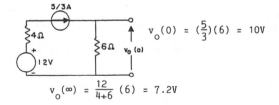

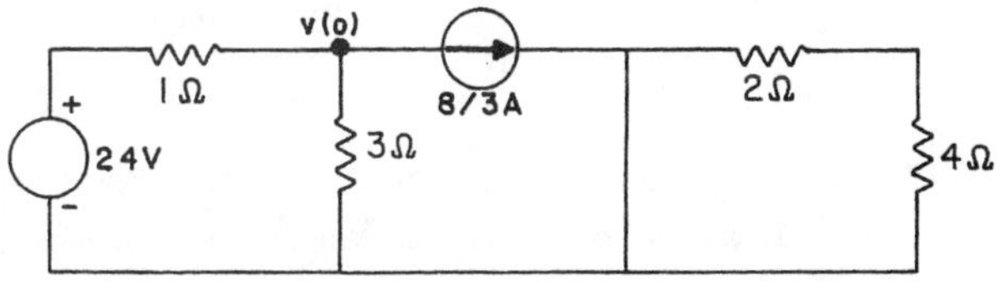

Using a nodal equation

$$\frac{v_o(0)-24}{1} + \frac{v_o(0)}{3} + \frac{8}{3} = 0 \text{ and hence } v_o(0) = 16\text{V}.$$

At $t = \infty$ the inductor and switch cause $v_o(\infty) = 0$, $R_{TH} = 1||3 = \frac{3}{4}\ \Omega$ and $T_c = \frac{L}{R_{TH}} = \frac{8}{3}$ s. Hence $v_o(t) = 16e^{-\frac{8}{3}t}$ V.

16. Find $i_o(t)$ for t>0 for the following network.

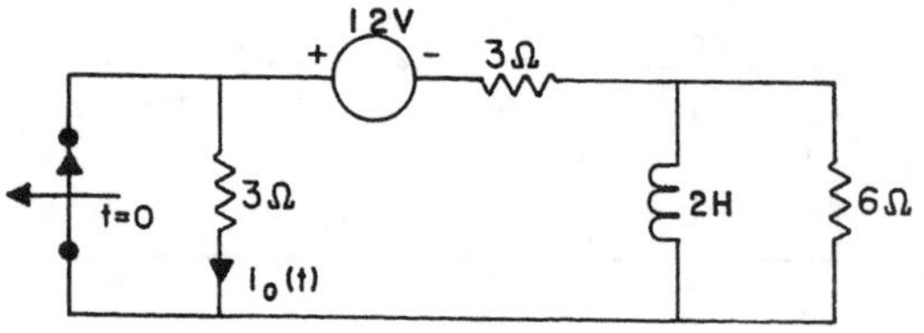

Solution:

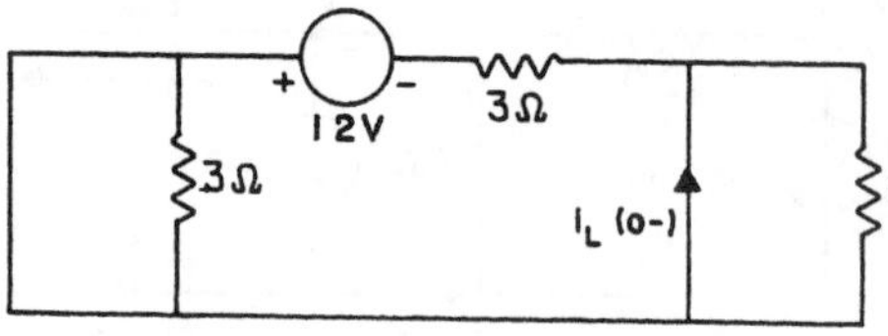

$i_L(0-) = \frac{12}{3} = 4A$

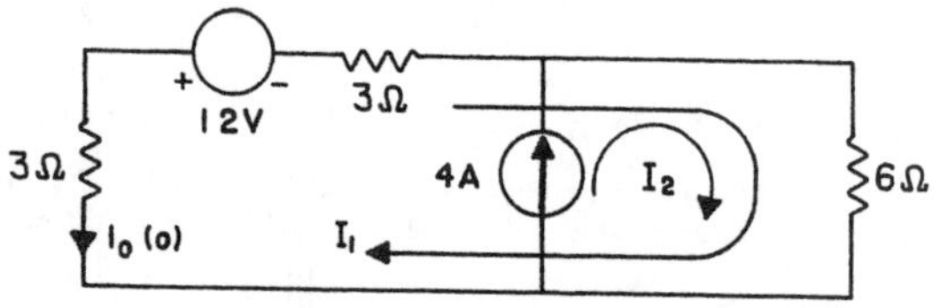

The loop equations are $12I_1+6I_2 = -12$ and $I_2 = 4$. Hence $I_1 = -3A$ and therefore $i_o(0) = 3A$.

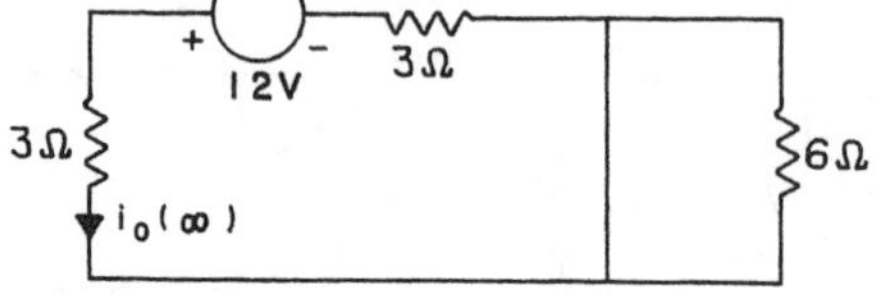

$i_o(\infty) = \frac{12}{6} = 2A$

$R_{TH} = 6||6 = 3\Omega$ and $T_c = \frac{L}{R_{TH}} = \frac{2}{3}$ s. Hence

$i_o(t) = 2+(3-2)e^{-3t/2}A$.

17. Determine the expression for $v_o(t)$, t>0, in the following circuit if the input i(t) is shown below.

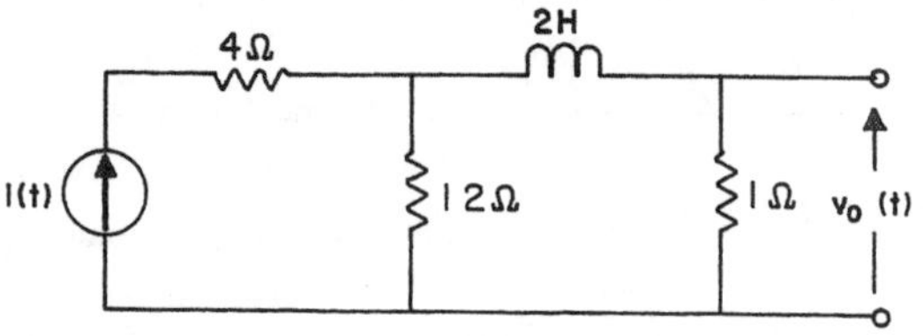

(a)

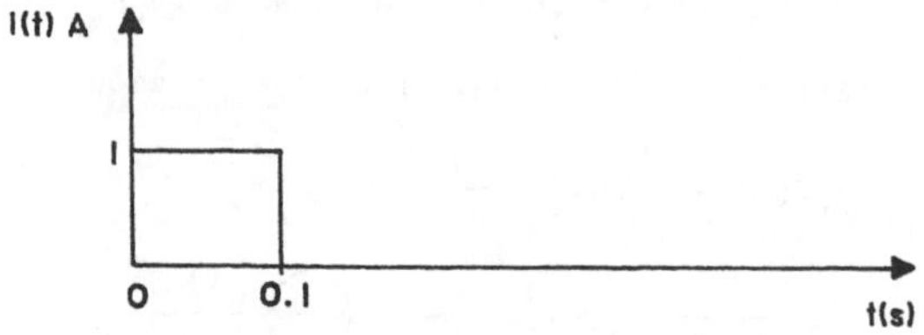

(b)

Solution: $i_1(0-) = i_L(0+) = 0$ and $v_o(0+) = 0$.

For a step function $v_o(\infty) = (1)(\frac{12}{13})(1) = \frac{12}{13}$ V. $R_{TH} = 4\Omega$ and $T_c = \frac{L}{R_{TH}} = \frac{1}{2}$ s. Hence

$v_o(t) = \frac{12}{13}(1-e^{-2t})\ 0 \le t \le 0.1$.

At $t = 0.1$, $v_o(0.1) = \frac{12}{13}(1-e^{-0.2})$. Therefore $v_o(t) = \frac{12}{13}(1-e^{-0.2})e^{-(t-0.1)2}$, $t > 0.1$. Hence

$v_o(t) = \frac{12}{13}(1-e^{-2t})[u(t)-u(t-0.1)]+ \frac{12}{13}(1-e^{-0.2})e^{-2(t-0.1)}u(t-0.1)V$.

A plot of this function looks like

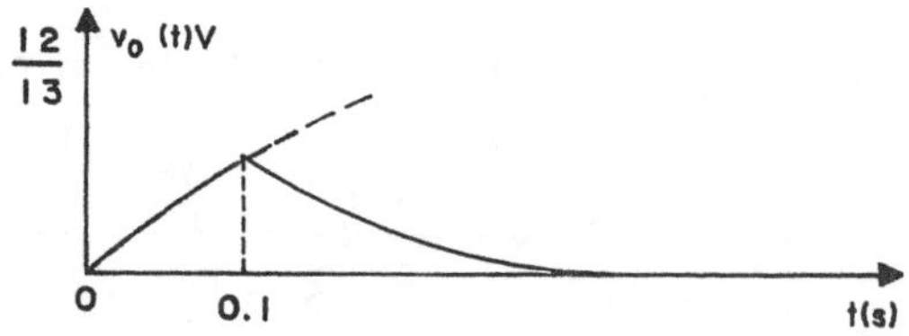

18. Show that the following circuit is a differential integrator; that is, the output is the integral of the difference between the two input voltages.

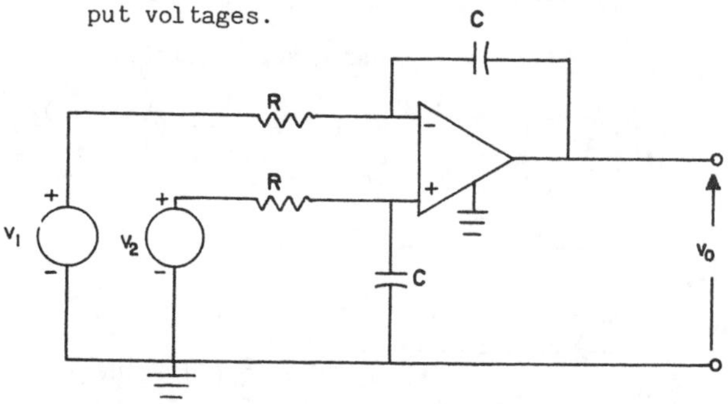

Solution: The KCL equations at the input terminals are

$$\frac{v_1-v_-}{R} + C\frac{d(v_o-v_-)}{dt} = i_-$$

$$\frac{v_2-v_+}{R} = C\frac{dv_+}{dt} + i_+$$

But $i_+ = i_- = 0$ and $v_- = v_+$ therefore

Chapter 7 SECOND-ORDER TRANSIENT CIRCUITS

At this point, we are learning about 2nd order systems and their responses to stimulus. While the text focuses on electrical circuits, we should recognize that the discussions in BECA regarding under-, critically and over-damped responses apply to any system characterized by the 2nd order differential equation

$$A\frac{d^2x(t)}{dt^2}+B\frac{dx(t)}{dt}+Cx(t)=f(t) \tag{7.1}$$

As an example, consider the automobile shock absorber sketched in Figure 7.1a. A reasonable model for the absorber is given in Figure 7.1b which is a 2nd order system that can be written in the form of (7.1) where $f(t)$ is the force applied to the absorber. Therefore, we expect shock absorbers, and the car that they're on, to respond to bumps and potholes in accordance with one of the 2nd order responses - overdamped, critically damped or underdamped. These scenarios are shown graphically in Figure 7.1c.

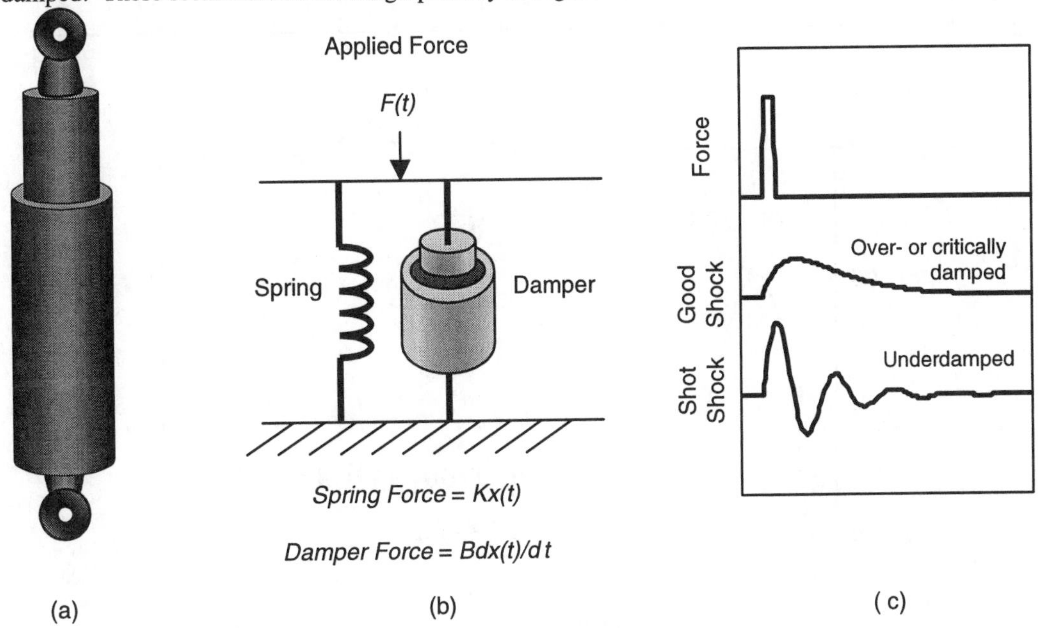

Figure 7.1. An example of a 2nd order system - a shock absorber.

PSPICE SIMULATIONS

SIMULATION ONE

For our first PSpice simulation, we will investigate the underdamped circuit in Figure 7.2. In particular, we want to find the frequency of oscillation , ω_d , the damping factor, ζ , the natural frequency , ω_n ,and the pole locations, s_1 and s_2. The requisite *Schematics* diagram and PROBE plot of the output voltage are shown in Figure 7.3a and Figure 7.3b respectively. To extract the required information, we first find the *x, y* data points at two peaks as seen in Figure 7.3b. The points are also listed in Table 7.1 for easy reference. Next, we develop equations for requested data.

Table 7.1

Listing of marked data points in Figure 7.2b.

t_1, V_1	1.783 ms, 1.017 V
t_2, V_2	8.026 ms, 44.09 mV
t_2 - t_1,	6.245 ms
V_2/V_1	23.11

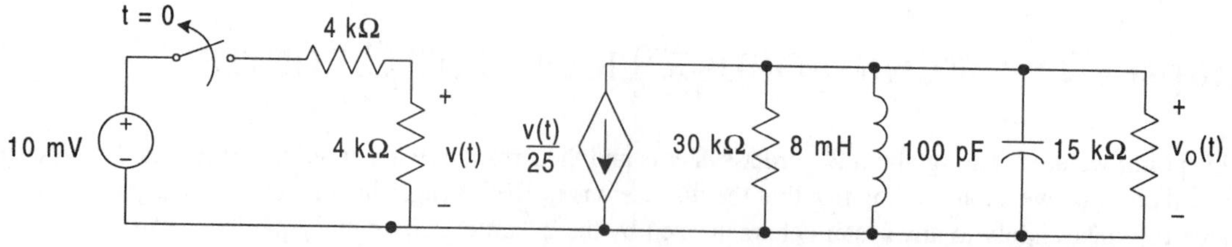

Figure 7.2. The circuit diagram for the first PSice simulation.

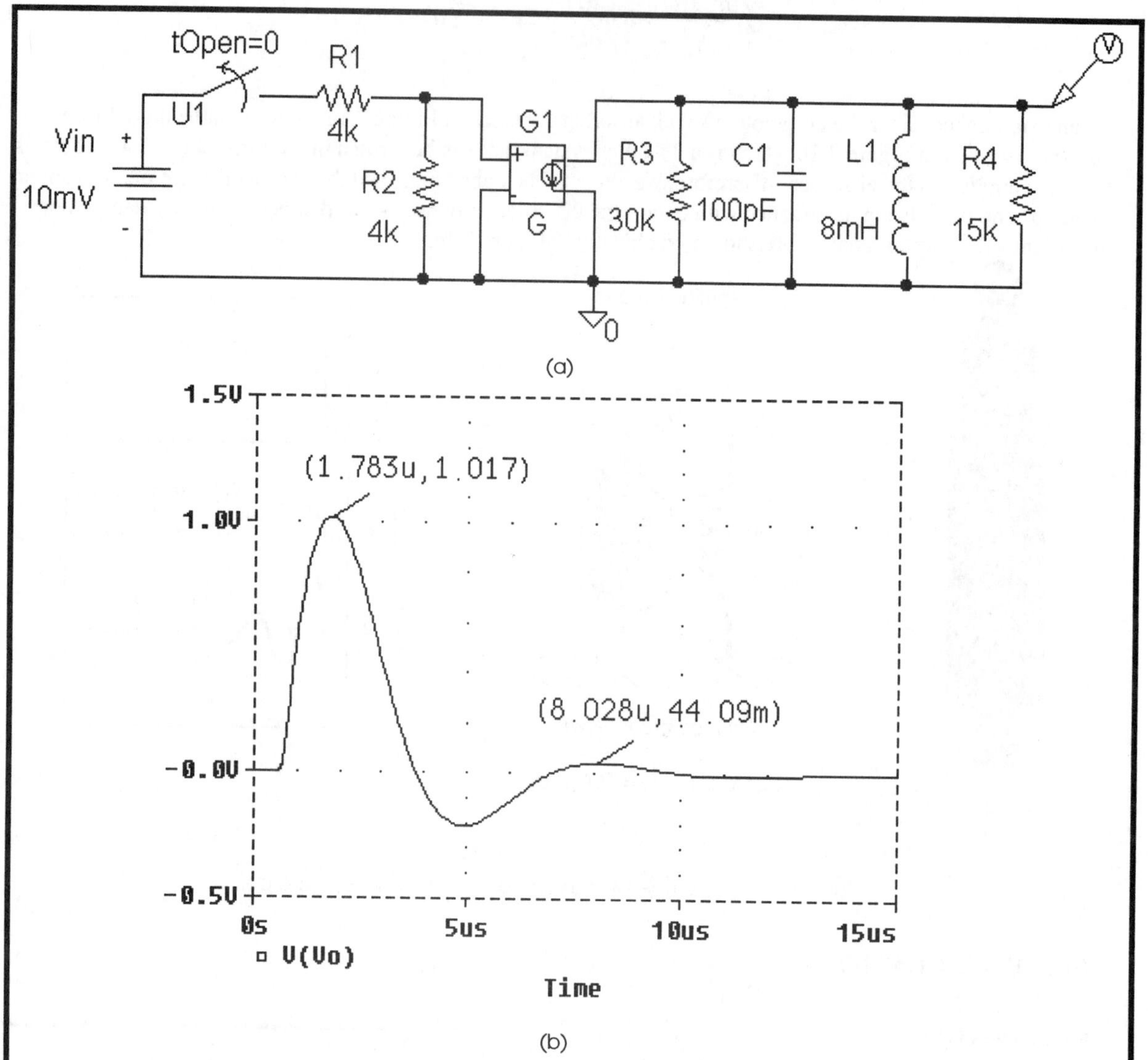

Figure 7.3. The (a) *Schematics* diagram and (b) PROBE results for the first PSpice simulation.

The frequency of oscillation is related to the period, T, which is just t_2 - t_1, by the relationship

$$\omega_d = \frac{2\pi}{T} = \frac{2\pi}{t_2 - t_1} = \frac{2\pi}{6.245\, x 10^{-6}} = 1.0 \text{ Mrad/s}$$

To extract damping info, let's first look at the general underdamped response equation.

$$v(t) = e^{-\sigma t}\left[A_1 \cos(\omega_d t) + A_2 \sin(\omega_d t)\right] \tag{7.2}$$

Note that at both labeled data points in Figure 7.3b, the bracketed trigonometric part of (7.2) has the same value. This allows us to write

$$\frac{v(t_1)}{v(t_2)} = \frac{e^{-\sigma t_1}}{e^{-\sigma t_2}} = e^{\sigma(t_2 - t_1)}$$

Solving for σ and substituting values from Table 7.1 yields,

$$\sigma = \frac{\ln\left(\dfrac{v(t_1)}{v(t_2)}\right)}{t_2 - t_1} = \frac{\ln(23.11)}{6.245\,x\,10^{-6}} = 5.03\,x\,10^5$$

We now have the pole frequencies! They are

$$s_{1,2} = -\sigma \pm j\omega_d = -0.50 \pm j1.0 \text{ MRad/s}$$

Turning to ζ and ω_n , we relate to σ and ω_d as

$$\sigma = \zeta\omega_n \quad \text{and} \quad \omega_d = \omega_n\sqrt{1-\zeta^2}$$

Solving for ζ and ω_n involves a simple quadratic equation. The results are $\zeta = 0.45$ and $\omega_n = 1.12$ Mrad/s.

SIMULATION TWO

In the simple 2nd order circuit in Figure 7.4, the inductor current should rise from 0 A to 1 A when the switch closes. We have two design restrictions however: first, we want the current to reach 1 A as fast as possible, and second, the overshoot beyond 1 A must not exceed 10 % - 0.1 A. We will vary the inductor value to optimize the performance.

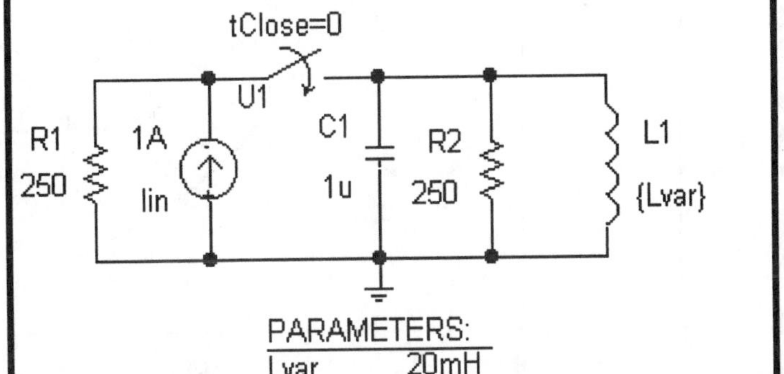

Figure 7.4. *Schematics* diagram for the second PSpice simulation.

Since we want to investigate waveform characteristics as a component value is changing, we must conduct a Performance Analysis simulation. The procedure for this kind of simulation is discussed on pages 55 -58 of this guide. Particular to this simulation, using the PARAM part in Figure 7.4, we create a variable called Lvar. Next, we set L1 = Lvar. Third, in the simulation SETUP, we set the Parametric fields as shown in Figure 7.5. Then, we simulate. After selecting all sections (see Figure 6.10) PROBE opens and we select Performance Analysis from the Add Traces menu. Next, we add the two traces - Overshoot(I(L1) and Risetime(I(L1)). The result is the plot in Figure 7.6 which indicates that as the inductor gets larger, the overshoot decreases and the risetime increases (the circuit gets slower). Using the cursors, it is found that the percent overshoot is satisfied when Lvar ≥ 22 mH. Also, the fastest risetime (321 ms) occurs at Lvar = 22 mH. Therefore, a 22 mH inductor should be used to push the circuit to the edge.

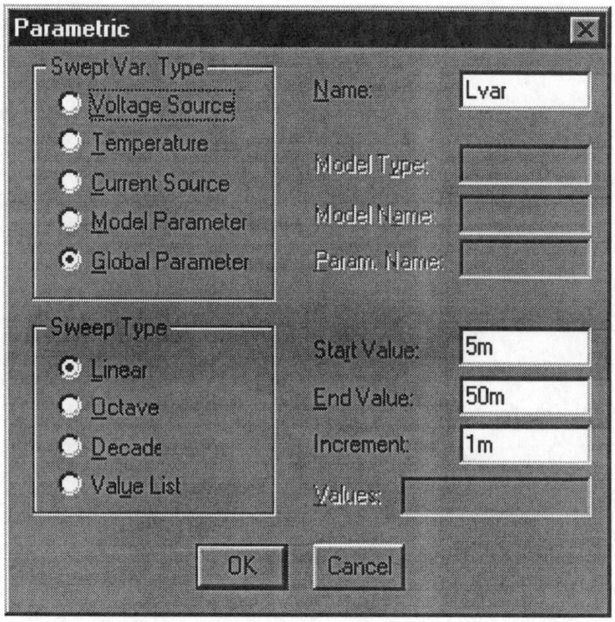

Figure 7.5. The Parametric settings used in the second PSpice simulation.

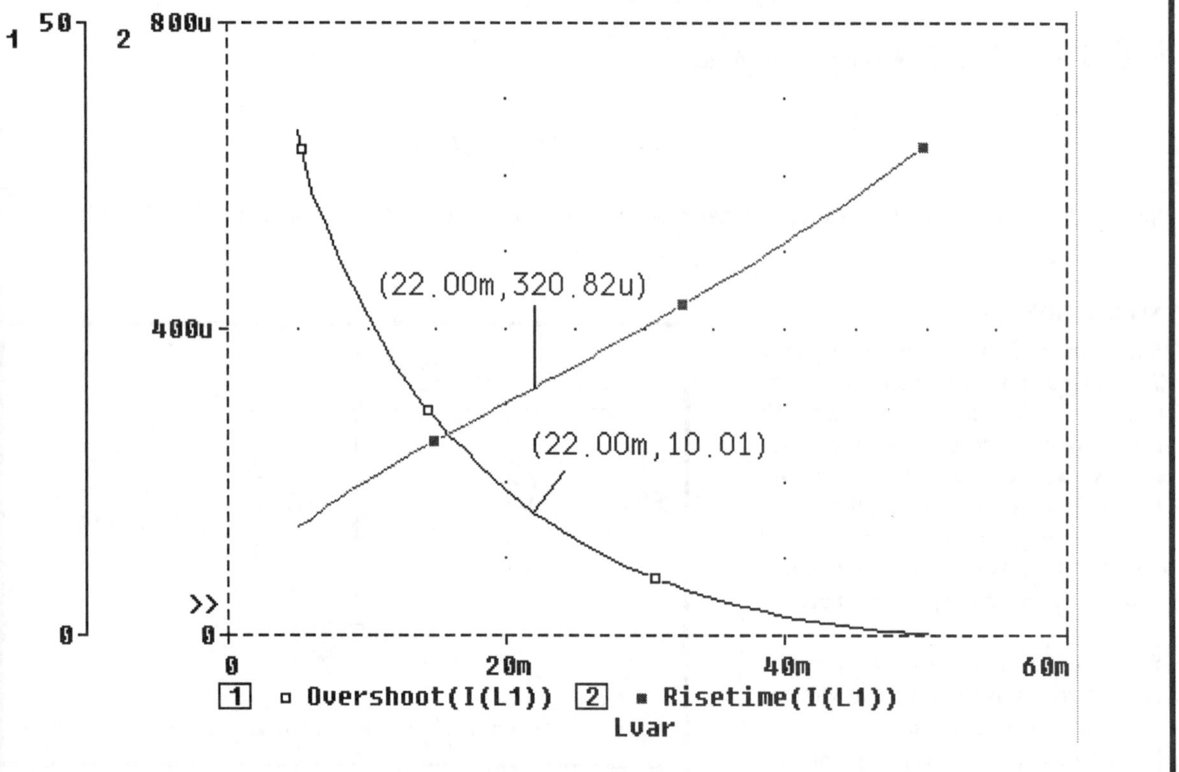

Figure 7.6. Performance Analysis results for 2nd PSpice simulation indicating an optimum inductor value of 22 mH.

EWB SIMULATION

Referring back to our discussion about inductors in Chapter 5, we found that inductors have some resistance. Of course, this is due to resistance of the wire used to make the coil. Also, from the crosstalk simulations on page 46, we know that wires laid side by side share a parasitic capacitance. Since the wraps of wire in the inductor lay side by side as well atop one another, we should suspect that there is a small amount of capacitance in every inductor - and there is.

Let's look at the effect of parasitic capacitance in the inductor. We will simulate the simple, 1st order circuit in Figure 7.8a for two different inductor models: an ideal inductor, Figure 7.8b, and the RLC model in Figure 7.8c. To make the comparison meaningful, we'll use a Panasonic ELJ-FC100JF, 10 μH inductor. Panasonic specs the resistance and capacitance at 6.33 Ω and 2.47 pF respectively. The EWB circuits and simulation results are shown in Figure 7.7 and Figure 7.9. Notice the voltage spike the instant of switching. It's as large as the final value! Luckily, its width is only about 5 ns. The obvious question then is this, "Under what circumstances can we use this inductor?" The real question is, can the rest of the circuitry respond to a 5 ns pulse? If not, then this inductor is appropriate. If not, buy a better inductor.

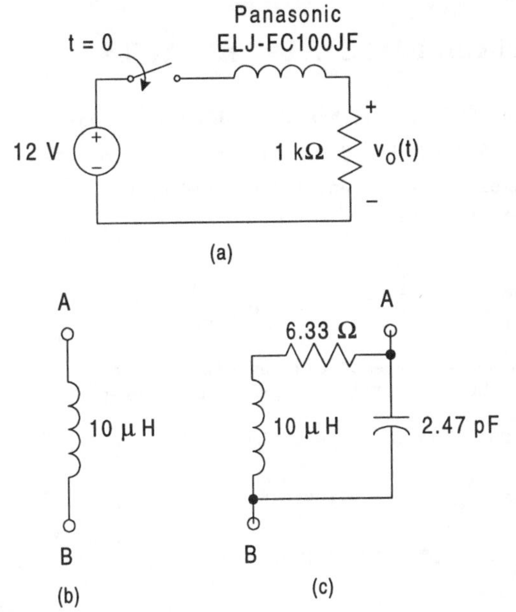

Figure 7.8. Circuitry for the EWB inductor simulation.

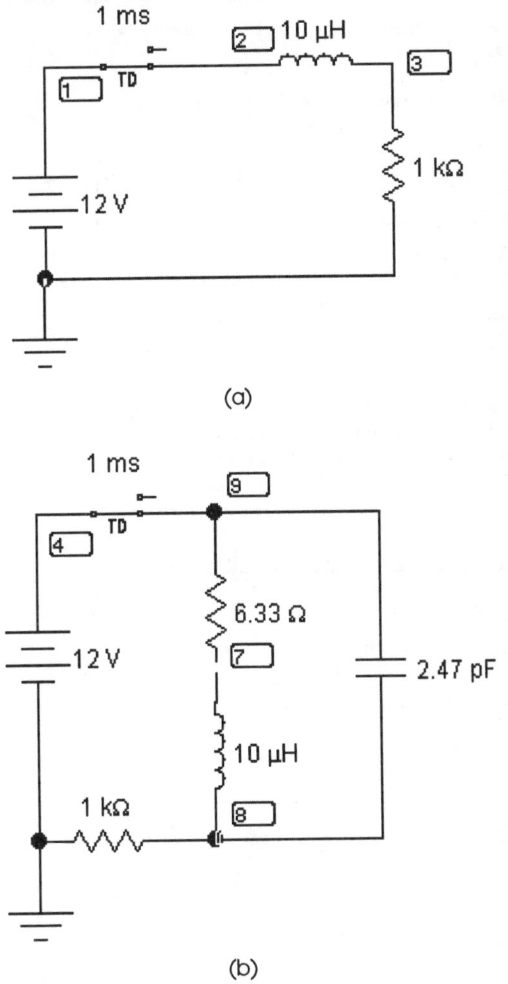

Figure 7.7. EWB circuits for the inductor simulations.

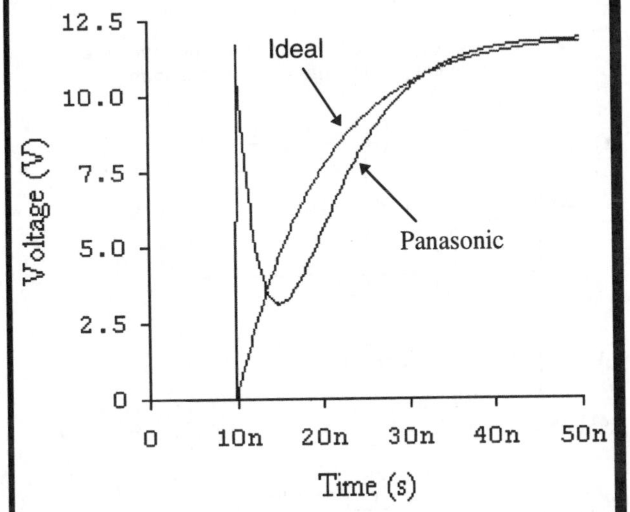

Figure 7.9. EWB results showing the transient effect of parasitic resistance and capacitance in an inductor.

PROBLEM SOLVING EXAMPLES

1. A series RLC circuit has the following parameters: $R = 4\Omega$, $L = \frac{1}{6}$ H, and $C = \frac{1}{16}$ F. Compute the damping coefficient and the undamped natural frequency.

   Solution: $\alpha = \frac{R}{2L} = 12$ and $\omega_o = \frac{1}{\sqrt{LC}} = 4\sqrt{6}$ r/s .

2. Determine the type of damping exhibited by the series RLC circuit in the previous Problem.

   Solution: The characteristic equation is

   $s^2 + \frac{R}{L} s + \frac{1}{LC} = 0$ or $s^2 + 24s + 96 = 0$

   $s_1, s_2 = -5.07, -18.93$ therefore the circuit is overdamped.

3. A series RLC circuit contains a resistor $R = 2\Omega$ and a capacitor $C = \frac{1}{8}$ F. Select a value of the inductor so that the circuit is critically damped.

   Solution: The characteristic equation is

   $s^2 + \frac{R}{L} s + \frac{1}{LC} = 0$ or $s^2 + \frac{2}{L} s + \frac{8}{L} = 0$ then $s_1, s_2 =$

   $\frac{-\frac{2}{L} \pm \sqrt{\left(\frac{2}{L}\right)^2 - \frac{32}{L}}}{2}$. The radical will be zero if

   $\frac{4}{L^2} - \frac{32}{L} = 0$ or $L = \frac{1}{8}$ H.

4. For the circuit shown below, determine values of the capacitor which will produce a current response that is (a) underdamped, (b) overdamped and (c) critically damped.

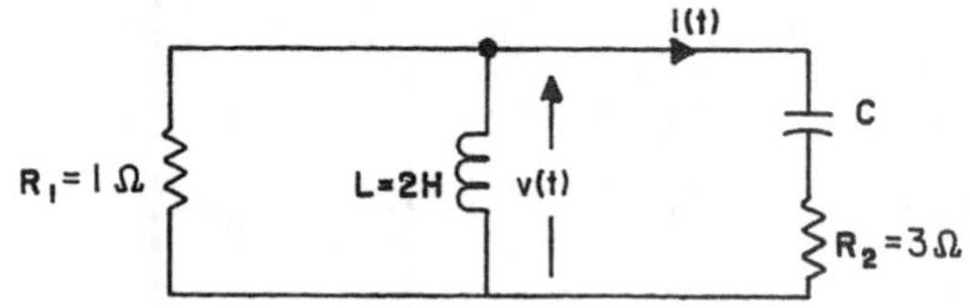

Solution: The network equations are

$$\frac{v(t)}{R_1} + \frac{1}{L} \int v(t)dt + i(t) = 0 \text{ and } v(t) = \frac{1}{C} \int i(t)dt + R_2 i(t).$$

Taking the derivative of the first equation and then substituting the second equation above into that resultant equation yields

$$\frac{1}{R_1 C} i(t) + \frac{R_2}{R_1} \frac{di(t)}{dt} + \frac{1}{LC} \int i(t)dt + \frac{R_2}{L} i(t) + \frac{di(t)}{dt} = 0.$$

Taking the derivative of this expression yields

$$\frac{d^2 i(t)}{dt^2} + \left[\frac{L + R_1 R_2 C}{LC(R_1 + R_2)}\right] \frac{di(t)}{dt} + \frac{R_1}{LC(R_1 + R_2)} i(t) = 0.$$

The characteristic equation is then $s^2 + \frac{2+3C}{8C} s + \frac{1}{8C} = 0$.

Therefore

(a) $\left(\frac{2+3C}{8C}\right)^2 - \frac{1}{2C} < 0$ and $\frac{2}{9} < C < 2$

(b) $\left(\frac{2+3C}{8C}\right)^2 - \frac{1}{2C} = 0$ and $C = \frac{2}{9}$, $C = 2$

(c) $\left(\frac{2+3C}{8C}\right)^2 - \frac{1}{2C} > 0$ and $C < \frac{2}{9}$, $C > 2$

5. Find the damping coefficient and the undamped natural frequency for the circuits shown below.

(a)

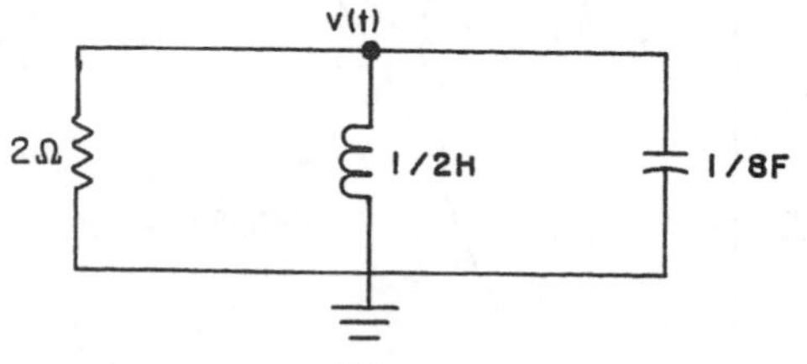

(b)

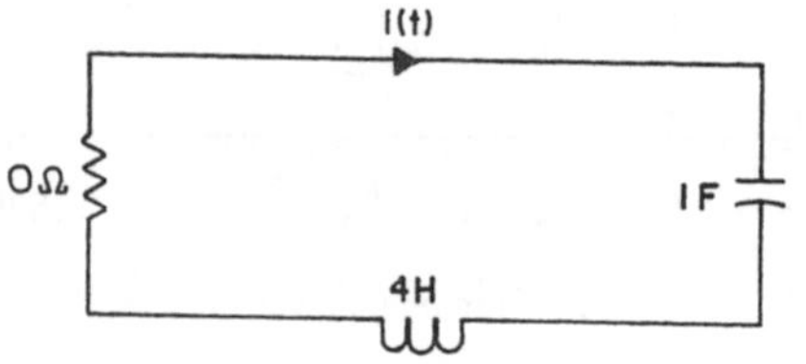

Solution: (a) $\frac{v(t)}{R} + \frac{1}{L} \int v(t)dt + C \frac{dv(t)}{dt} = 0$ and

$\frac{d^2 v(t)}{dt^2} + \frac{1}{RC} \frac{dv(t)}{dt} + \frac{1}{LC} v(t) = 0$. Hence

$\omega_o = \frac{1}{\sqrt{LC}} = 4 r/s$ and $\alpha = \frac{1}{2RC} = 2.0$.

(b) $Ri(t)+L\frac{di(t)}{dt} + \frac{1}{C}\int i(t)dt = 0$ and

$\frac{d^2i(t)}{dt^2} + \frac{R}{L}\frac{di(t)}{dt} + \frac{1}{C}i(t) = 0$, therefore $\omega_o =$

$\frac{1}{\sqrt{LC}} = \frac{1}{2}$ r/s and $\alpha = \frac{R}{2L} = \frac{5}{4}$.

6. Given the following overdamped circuit below, determine the voltage v(t) if the initial conditions on the storage elements are $i_L(0) = 2A$ and $v_C(0) = 10V$.

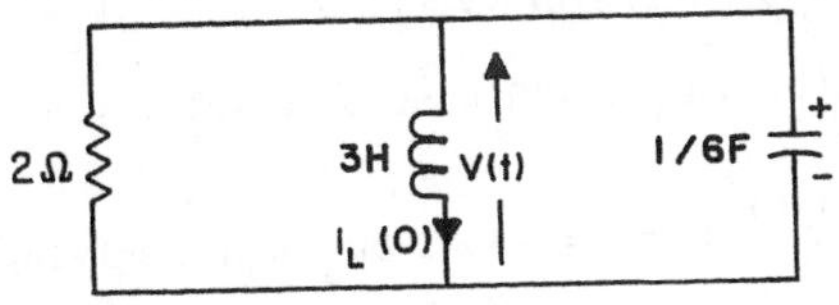

Solution: The network equation is

$$\frac{d^2v(t)}{dt^2} + \frac{1}{RC}\frac{dv(t)}{dt} + \frac{1}{LC}v(t) = 0$$

$s^2+3s+2 = 0$ and hence $s = -2$, $s = -1$.

Therefore

$$v(t) = k_1e^{-2t}+k_2e^{-t}$$

and

$$v(0) = 10 = k_1+k_2 \qquad (1)$$

Also

$$C\frac{dv(t)}{dt} + \frac{v(t)}{R} + i_L(t) = 0$$

$$\frac{dv(0)}{dt} = -3(10)-6(2) = -42$$

In addition

$$\frac{dv(t)}{dt} = -2k_1e^{-2t}-k_2e^{-t} .$$

Hence

$$-42 = -2k_1-k_2 \qquad (2)$$

Solving for k_1 and k_2 using equations (1) and (2) yields

$$v(t) = 32e^{-2t}-22e^{-t}V .$$

7. Given the overdamped circuit below, determine the current i(t) if the initial conditions on the storage elements are $i_L(0) = 4A$ and $v_C(0) = 6V$.

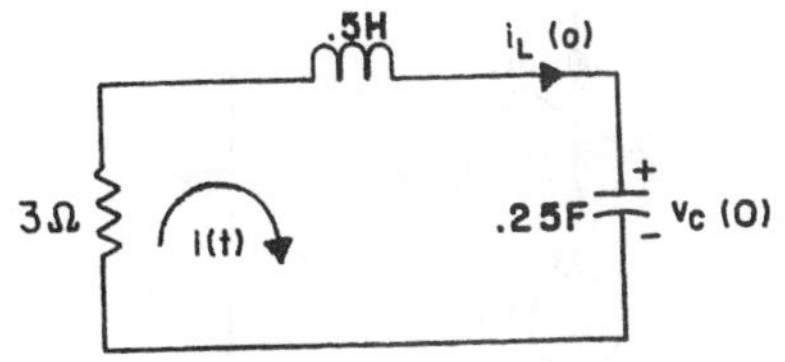

Solution: The network equation is

$\frac{d^2i(t)}{dt^2} + \frac{R}{L}\frac{di(t)}{dt} + \frac{1}{LC}i(t) = 0$ or $s^2+6s+8 = 0$.

Hence $s_1 = -2$ and $s_2 = -4$ and thus $i(t) = k_1e^{-2t}+k_2e^{-4t}$. $i_L(0) = i(t{=}0) = 4 = k_1+k_2$. In addition $Ri(t)+L\frac{di(t)}{dt} +v_C(t) = 0$. Therefore

$\frac{di(t)}{dt} = -\frac{R}{L}i(t)-\frac{v_C(t)}{L}$ and $\frac{di(0)}{dt} = -36$. Hence

$-36 = -2k_1-4k_2$ and solving the two equations for k_1 and k_2 yields $i(t) = -10e^{-2t}+14e^{-4t}A$.

8. In the following circuit switch action occurs at t = 0. Find the equation for the voltage v(t) for t > 0.

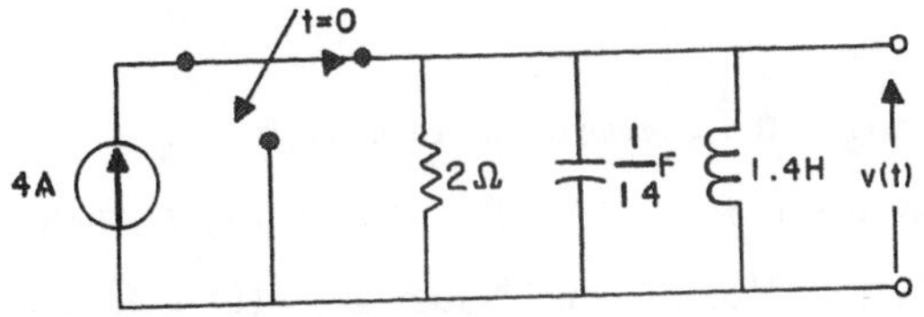

Solution: The characteristic equation for the network is $s^2+7s+10 = 0$ and therefore $s_1 = -2$ and $s_2 = -5$. Hence $v(t) = k_1e^{-2t}+k_2e^{-5t}$. An analysis of the circuit shows that $i_L(0) = 4$ and $v_C(0) = 0$. From the above equation $v(0) = k_1+k_2$ $= 0$ and $\frac{dv(0)}{dt} = -\frac{1}{RC}v(0) - \frac{i_L(0)}{C} = -56$, however $\frac{dv(0)}{dt} = -2k_1-5k_2$. Therefore $-2k_1-5k_2 = -56$.

Solving the equations for k_1 and k_2 yields $k_1 = -\frac{56}{3}$ and $k_2 = \frac{56}{3}$. Therefore $v(t) = -\frac{56}{3}(e^{-2t}-e^{-5t})$ V.

9. In the following circuit the switch opens at t = 0. Determine the equation for i(t), t>0.

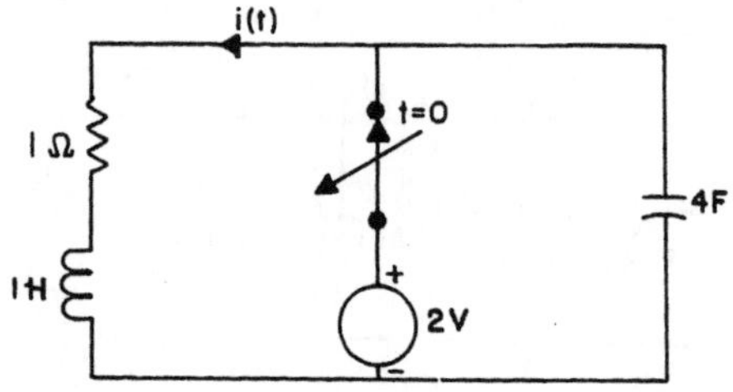

Solution: The characteristic equation is

$s^2+\frac{R}{L}s+\frac{1}{LC} = 0$ or $s^2+s+.25 = 0$ and $s_1 = s_2 = 0.5$.

Then $i(t) = k_1e^{-t/2}+k_2te^{-t/2}$. The initial conditions are $i(0) = 2$ and $v_C(0) = 2$. Note $i(0) = 2 = k_1$. Also $\frac{di(0)}{dt} = 0 = -\frac{k_1}{2}+k_2 = -\frac{R}{L}i(0) - \frac{v(0)}{L} = -4 = -\frac{k_1}{2}+k_2$. Hence $k_2 = -3$ and therefore

$i(t) = 2e^{-t/2}-3te^{-t/2}$A.

10. Given the circuit shown below, find an expression for v(t), t>0.

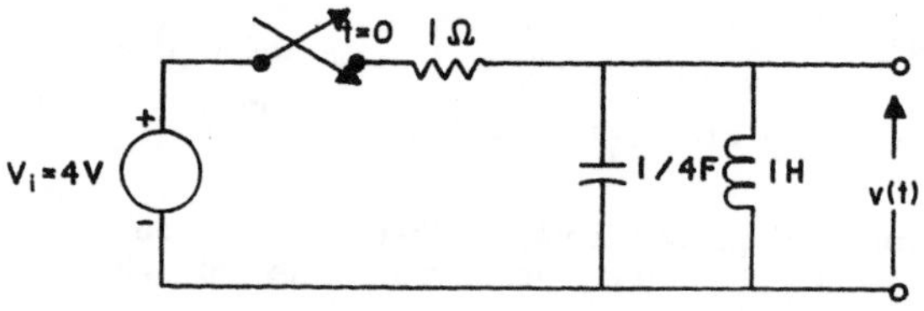

Solution: The characteristic equation is

$s^2+8s+20 = 0$ or $s_1 = -4+j2$ and $s_2 = -4-j2$. Then

$v(t) = k_1e^{-4t}\cos 2t+k_2e^{-4t}\sin 2t$ and $v(0) = 10 = k_1$.

$\frac{dv(t)}{dt} = -2k_1e^{-4t}\sin 2t-4k_1e^{-4t}\cos 2t+2k_2e^{-4t}\cos 2t$

$-4k_2e^{-4t}\sin 2t$. Hence $\frac{dv(0)}{dt} = -4k_1+2k_2 = -\frac{v_C(0)}{RC}$

$-\frac{i(0)}{C} = -120$. Solving for k_1 and k_2 yields

$k_1 = 10$, $k_2 = -40$.

$v(t) = 10e^{-4t}\cos 2t-40e^{-4t}\sin 2t$ V

11. Determine the voltage across the capacitor for t>0 if $R_o = 2\Omega$ in the following figure.

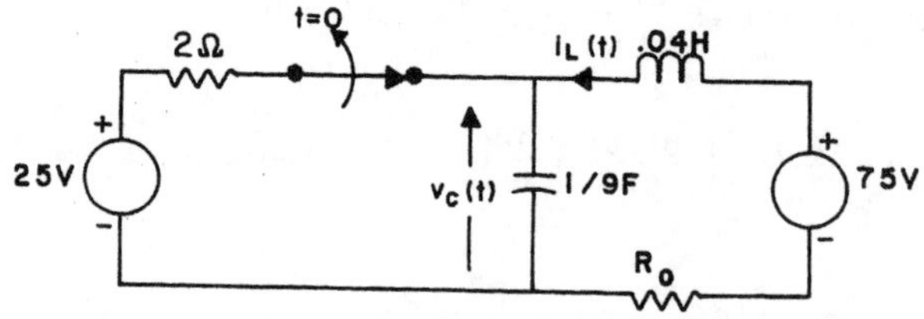

Solution: The characteristic equation is

$s^2+50s+225 = 0$ or $s_1 = -45$ and $s_2 = -5$.

$i_L(0) = \frac{75-25}{2+2} = \frac{25}{2}$. $v_C(\infty) = 75$ and $v_C(0) = 25+(\frac{25}{2})2 = 50$. Therefore $v_C(t) = k_1e^{-45t}+k_2e^{-5t}+k_3$ and $v_C(0) = k_1+k_2+75 = 50$ or $k_1+k_2 = -25$.

$\frac{dv_C(0)}{dt} = \frac{i(0)}{C} = \frac{225}{2} = -45k_1-5k_2$. Hence solving for the constants yields $k_1 = \frac{5}{16}$, $k_2 = -\frac{405}{16}$ and $k_3 = 75$. Therefore $v_C(t) = \frac{5}{16}e^{-45t}-\frac{405}{16}e^{-5t}+75$ V.

Chapter 8 AC STEADY-STATE ANALYSIS

This chapter of the BECA text is an introduction to a very important area of electrical engineering - *ac steady-state analysis*. To better understand the significance of this chapter, let's look at each word in the phrase "ac steady-state". The term "ac" means the source voltages and/or currents are sinusoidal functions of time given by the general expression

$$v(t) = V_M \cos(\omega t + \theta) \tag{8.1}$$

This expression is one of the most important in electrical engineering power utility voltages are sinusoidal. So, every appliance, every industrial motor, every lighting fixture, every anything that plugs in an electrical outlet, was intended and designed to operate from an "ac" supply.

"Steady-state" means that the ac circuit has been on long enough for all transients to have settled out. As an example, consider what happens when we start a clothes dryer. Initially, the dryer motor needs a lot of current to generate the torque needed to start turning the drum. As motor speed increases, the required torque and current drop. When the motor speed becomes constant, the current, while still a sinusoid, has reached a fixed magnitude. This is what we mean by "steady-state". The ac steady-state analysis techniques you are learning here accurately predict circuit performance only after the start-up transients have settled out. They cannot predict transient performance at all. One final remark, since "ac" implies that the power company is involved, and since the utilities tightly regulate the sinusoid frequency, we perform our analyzes at the power utility frequency - 60 Hz in the U.S. and 50 Hz in Europe.

PHASORS AND COMPLEX NUMBERS

Phasors - a word that strikes fear into the hearts of Klingons and sophomores alike. The idea of abandoning the cosine function in (8.1) in favor of a magnitude and a phase can be disconcerting. When we add complex impedances, which also have magnitude and phase, it can get downright confusing. There is, however, a fundamental difference between a phasor and a general complex number - *phasors correspond directly to a time domain sinusoid function.* Since complex numbers such as impedances are ratios of voltages and currents and do not represent time domain sinewaves, they are not phasors.

How can a phasor (magnitude and phase) correspond to the function in (8.1)? If all circuit elements in a network, resistors, inductors, etc., are linear, then all currents and voltages in the network are sinusoidal at the same frequency. Therefore, recording the frequency of each voltage and current is redundant. Referring to (8.1), the only remaining characteristics that are unique to a particular voltage or current are the magnitude and phase, which constitute the phasor representation. So, a phasor can be viewed as a shorthand notation for all of the unique characteristics of a particular signal in an ac steady-state circuit.

Representing resistors, inductors and capacitors in ac steady-state circuitry as complex numbers is quite ingenious. It replaces the derivatives and integrals in the capacitor and inductor I-V relationships with the complex impedances listed in Table 8.1. This approach has two benefits. First, and most obviously, ac steady-state circuits can be described by algebraic equations rather than differential equations. The second benefit is more subtle. Looking back at the circuits chapters 6 and 7, we see that 1st order circuit have one capacitor/inductor and 2nd order circuits have two. If follows that a circuit with 5 capacitors/inductors would be described by a 5th order differential equation - enjoy. But in the frequency domain, the number of capacitors and inductors has little effect on the complexity of the analysis.

Table 8.1

Time-domain/Frequency Domain Conversions

Sinusoid ↔ Phasor
$A\cos(\omega t \pm \theta) \leftrightarrow A\angle{\pm\theta}$
$A\sin(\omega t \pm \theta) \leftrightarrow A\angle{\pm\theta - 90^\circ}$
i(t) - v(t) ↔ I -V
$i(t) = C(dv(t)/dt) \leftrightarrow \mathbf{I} = j\omega C\mathbf{V}$
$v(t) = L(di(t)/dt) \leftrightarrow \mathbf{V} = j\omega L\mathbf{I}$

SINGLE FREQUENCY AC SIMULATIONS IN PSPICE

PSpice performs ac steady-state analysis using complex impedances and phasor representations for voltages and currents. We will use the VAC and IAC voltage and current source parts, shown in Figure 8.1, for our simulations. Note that we can choose both magnitude and phase for VAC whereas only the magnitude of IAC editable - the phase is zero. Therefore, simulations of circuits containing current sources may require us to shift the phases of the simulation results.

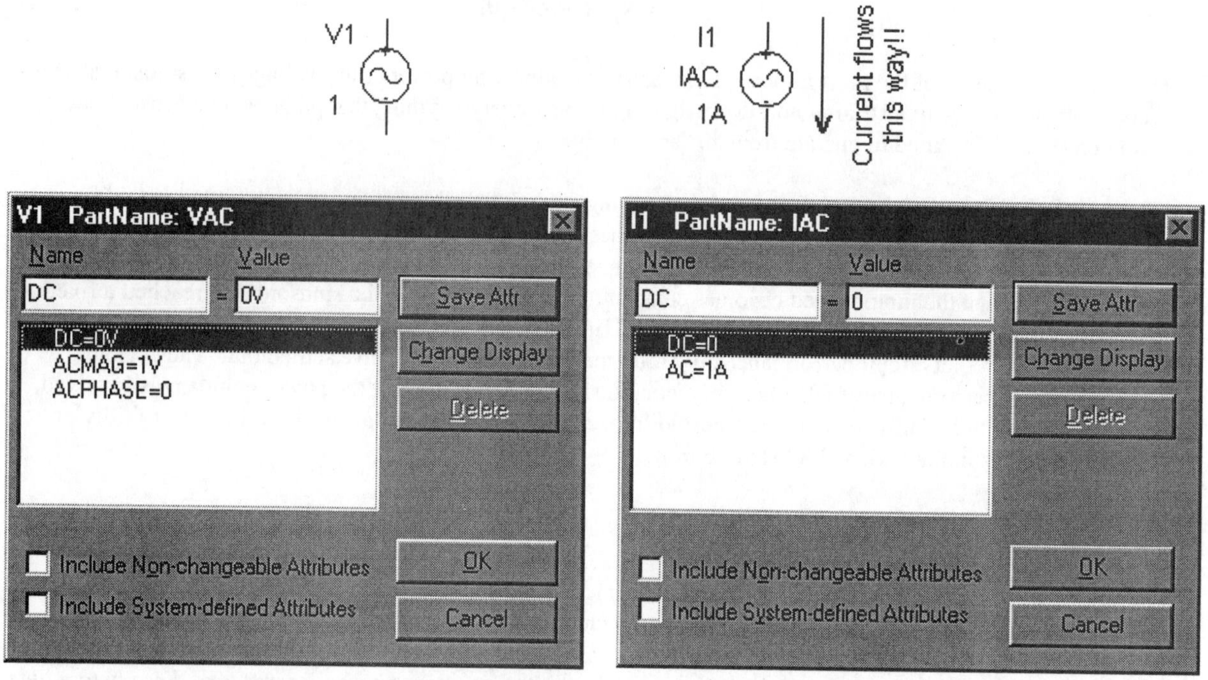

Figure 8.1. The VAC and IAC parts and their attributes boxes. Note that the IAC current flows out it the positively marker terminal. This is consistent with the passive sign convention.

What we call ac steady-state analysis, PSpice calls an AC Sweep. This kind of analysis is requested by selecting Setup in the Analysis menu. When the Setup dialog box appears, select AC Sweep to open the AC Sweep box shown in Figure 8.2. From the dialog box, it is apparent that PSpice was originally written to vary the frequency over a specified range. This will come in handy in later chapters, but right now we want to simulate at a single frequency. The AC Sweep dialog box fields in Figure 8.2 specify an ac simulation at only one frequency - 60 Hz.

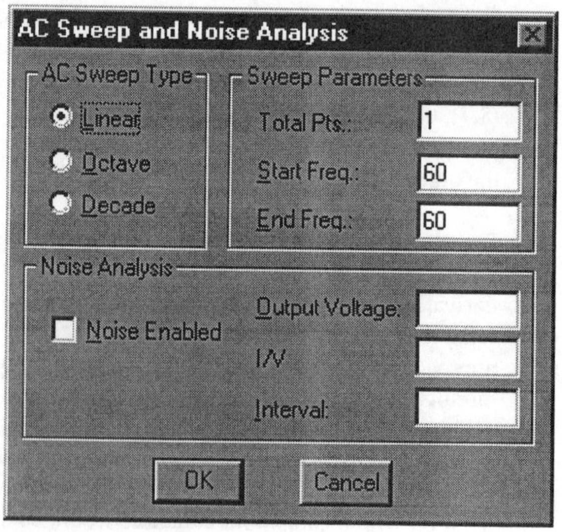

Figure 8.2. The AC Sweep attribute box set for a simulation at 60 Hz.

THE VPRINT1 AND IPRINT PARTS

Results of AC Sweep simulations are normally plotted in PROBE where frequency is the *x*-axis variable. Since we are simulating at a single frequency, making such a plot is not sensible. Instead, we should print the results into a file. This is done using two parts in the SPECIAL library, the VPRINT1 and IPRINT parts shown in Figure 8.3. Obviously, VPRINT1 parts function like voltmeters, measuring the voltage at one node with respect to the ground node. (There is VPRINT2 part as well that measures voltage between two non-ground nodes.) Similarly, IPRINT parts act as ammeters and must be in series-connected. These two parts tell PSpice to print the simulation results to an output file. Exactly what gets printed is specified in the parts attribute box. Figure 8.3 shows a VPRINT1 attribute box with the fields set to request ac simulation results for magnitude and phase. Attributes for each VPRINT and IPRINT part in a *Schematics* circuit must edited.

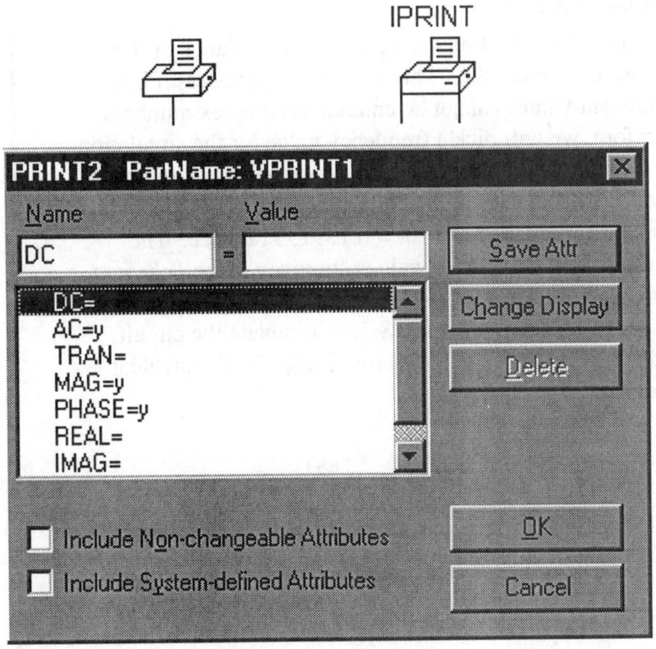

Figure 8.3. The VPRINT1 and IPRINT parts. The VPRINT1 attribute box is set to gather magnitude and phase data from an AC Sweep simulation.

THE OUTPUT FILE

Having set the AC Sweep, VPRINT and IPRINT attributes, we simulate the circuit (if PROBE opens, close it down). The results go to output file which can be opened by selecting Examine Output from the Analysis menu. The output file is the original Spice file format, predating PROBE and *Schematics*. It contains a lot of information about the circuit besides the requested VPRINT/IPRINT data that we will not go into. You'll find the ac data will be at the bottom of the file. Fortunately, the output file is just a text file that can be edited using programs like Notepad or the PSpice Text Editor. This allows you eliminate unwanted info and edit the file to fit your tastes or your professor's requests.

PSPICE SIMULATIONS

SIMULATION ONE

We'll keep the first example simple, simulating the circuit in Figure 8.5 to determine $\mathbf{V_O}$ and $\mathbf{I_S}$. The required *Schematics* diagram including VPRINT1 and IPRINT parts is shown in Figure 8.4. The attributes of the AC Sweep, the VPRINT1 and IPRINT parts are given in Figure 8.2 and Figure 8.3 respectively. From the output file, the requested phasors are given below. The step-by-step procedures used in this simulation are demonstrated in Visual Tutor, ACTUTOR.EXE.

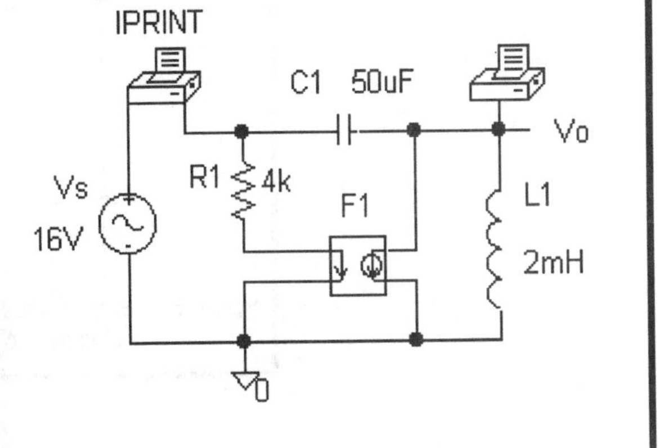

Figure 8.4. *Schematics* diagram for the first PSpice simulation.

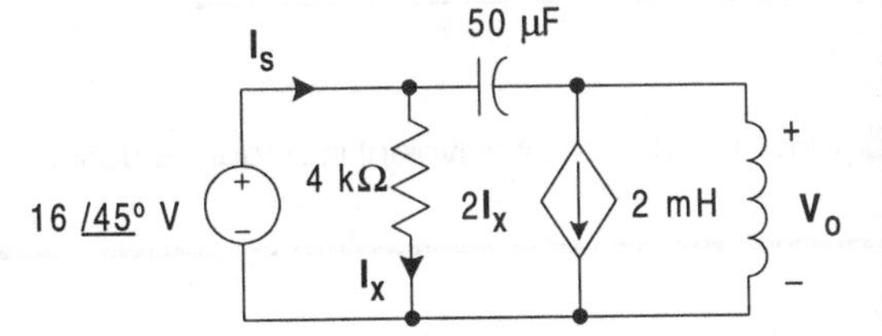

Figure 8.5. Circuit for the first PSpice simulation.

FREQ	Vo-magnitude	Vo-phase	Is-magnitude	Is-phase
60.0 Hz	0.2308 V	-133.5 degrees	0.3060 A	134.3 degrees

SIMULATION TWO

Let's use PSpice to find the equivalent impedance of the network in Figure 8.6. This is a two step process. First, component values cannot be entered as complex numbers Therefore, we will pick a frequency value for the simulation and calculate the inductor and capacitor values necessary to yield the impedances in the circuit diagram. A convenient simulation frequency of $1/2\pi = 0.159155$ rad./sec. The corresponding component values are given in the *Schematics* diagram in Figure 8.7. The second step is including a current source, $\mathbf{I_{TEST}}$, seen in Figure 8.7, to stimulate the circuit, producing a voltage, $\mathbf{V_{TEST}}$, across $\mathbf{I_{TEST}}$. The equivalent impedance is

$$\mathbf{Z_{eq}} = \frac{\mathbf{V_{TEST}}}{\mathbf{I_{TEST}}}$$

With foresight, we set $\mathbf{I_{TEST}} = 1 \underline{/0^\circ}$, then $\mathbf{Z_{eq}} = \mathbf{V_{TEST}}$. The output file results for $\mathbf{V_{TEST}}$ are given below indicating an equivalent impedance of $3.8 + j0.6\ \Omega$.

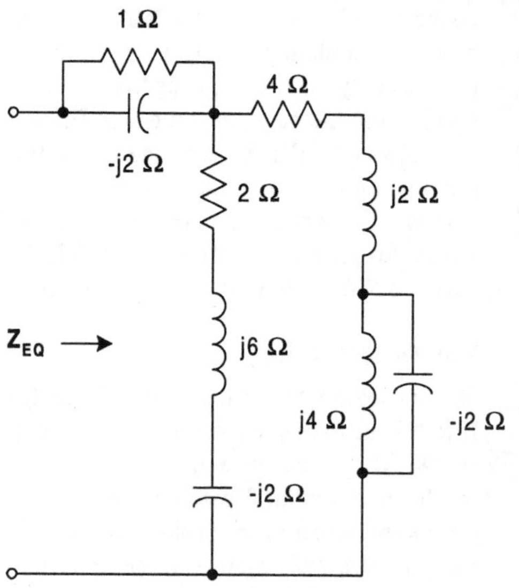

Figure 8.6. Circuit for the second PSpice simulation.

```
FREQ (r/s)    0.159
Vtest
   mag.     3.847 V
   phase    8.973 degs
   real     3.800 V
   imag.    0.600 V
```

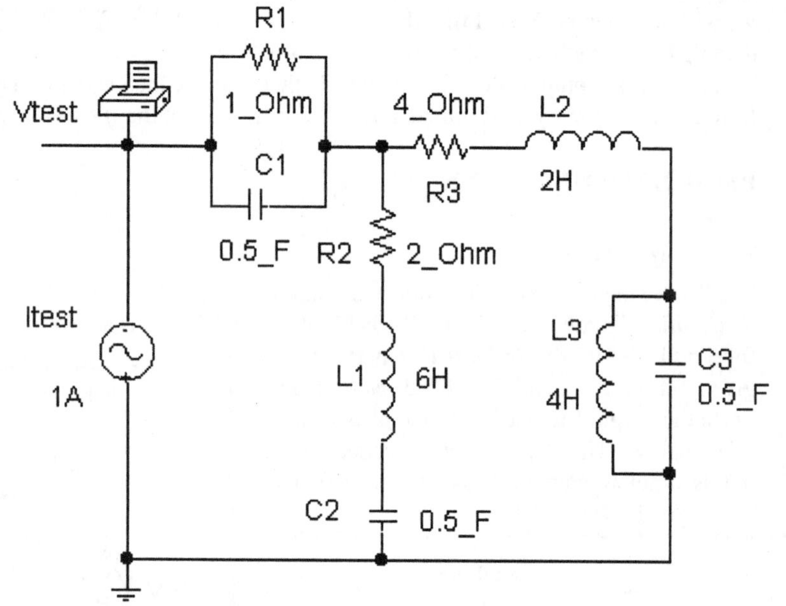

Figure 8.7. *Schematics* diagram for finding the equivalent impedance defined in Figure 8.6

EWB SIMULATIONS

Electronics Workbench, at least the Student Edition, can neither perform single frequency simulations nor can its ac simulation results be printed to a file. Instead, magnitude and phase must be plotted over a frequency range so data can be extracted using cursors. Additionally, currents cannot be plotted directly as they can in PROBE. Current-controlled voltage sources (gain = 1V/A) must be used to create voltages proportional to the currents of interest.

We will examine these features/limitations by simulating the circuit in Figure 8.8 to find the currents **I** and $\mathbf{I_C}$. The requisite EWB schematic diagram is shown in Figure 8.9 where the dependent sources produce voltages at nodes 6 and 7 that are proportional to the currents **I** and $\mathbf{I_C}$ respectively. Since the gains of the dependent sources are 1 V/A, the voltages and 6 and 7 are numerically equal to the currents. The voltage source specifications bear some inspection as well. In EWB, an ac voltage source is a time-domain sinewave, not a phasor. So, the source specifications in Figure 8.9 are not the values that will be used in the ac simulation. To set the source's phasor specification, double click on the source and go to its Analysis Setup tab. This opens the dialog box in Figure 8.10 which has been properly edited for this simulation. Also, since the circuit operates at about 795 Hz (5000/2π rad/s.), the frequency range for the simulation was chosen to be 790 Hz to 800 Hz.

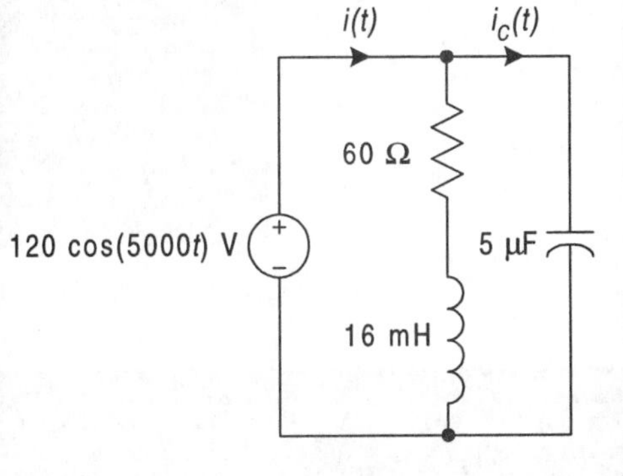

Figure 8.8. Time-domain circuit for EWB simulation.

The simulation results are shown in Figure 8.11, where, from the cursor data, we find

$$\mathbf{I} = 2.163\,\underline{/70.6^\circ}\text{ A}$$

$$\mathbf{I_C} = 3.0\,\underline{/90^\circ}\text{ A}$$

or, in the time domain

$$i(t) = 2.163\cos(\omega t + 70.6^\circ)\text{ A}$$

$$i(t) = 3.0\cos(\omega t + 90^\circ)\text{ A}$$

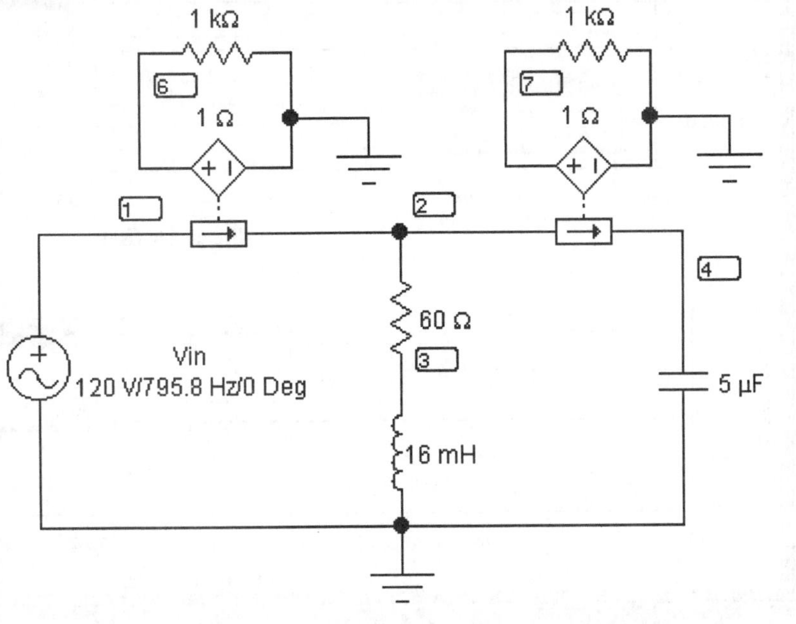

Figure 8.9. The required EWB diagram.

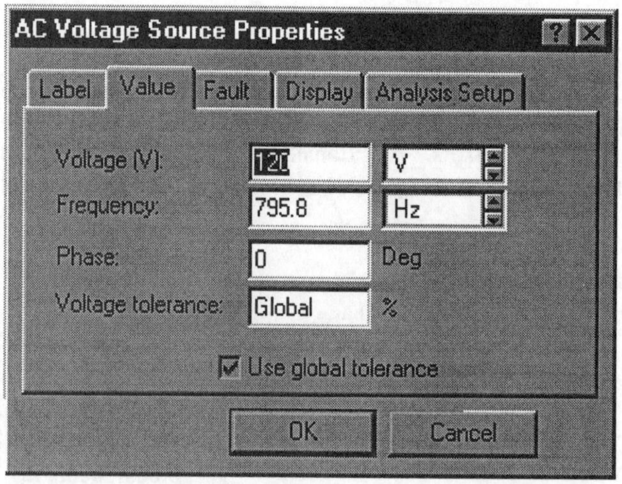

Figure 8.10. The EWB dialog box edited for the circuit in Figure 8.8.

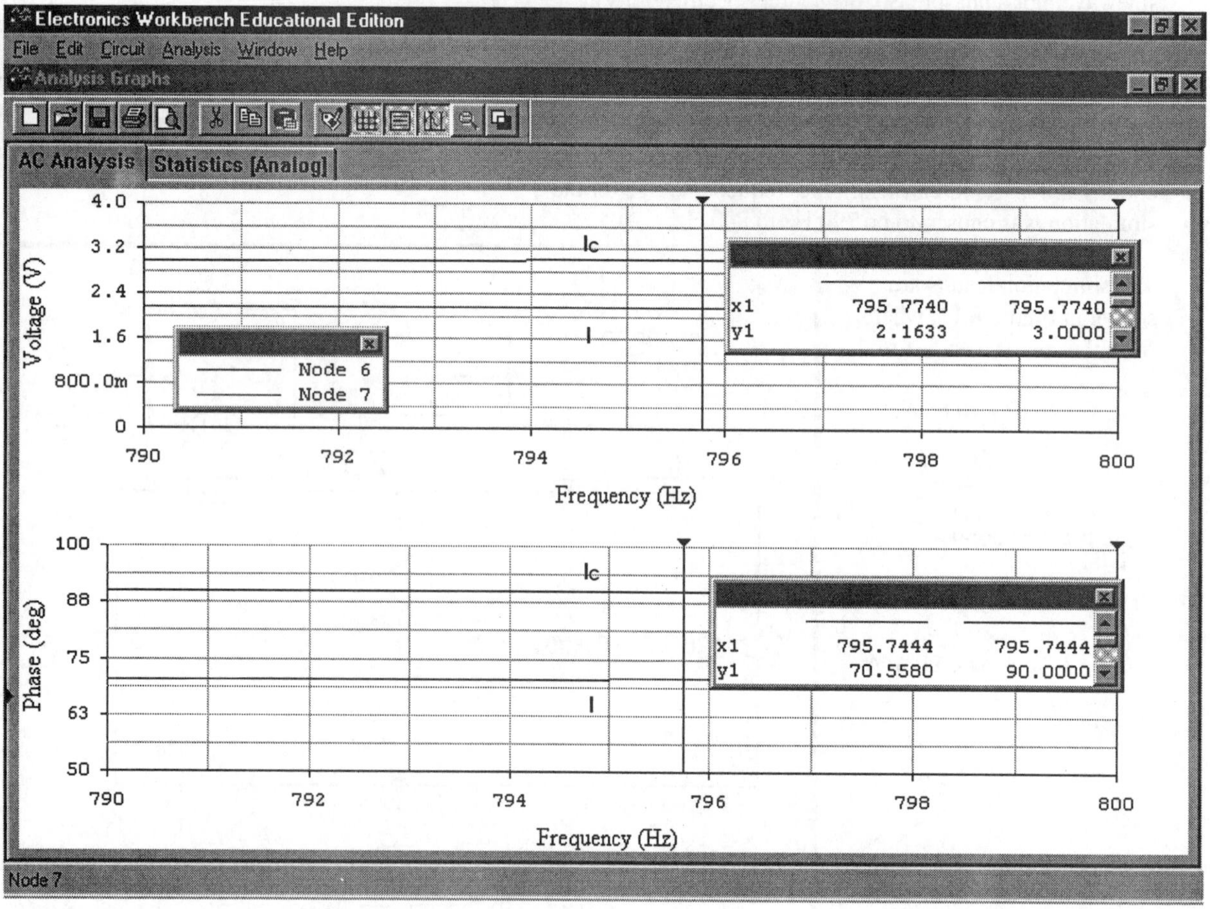

Figure 8.11. The much-edited EWB simulation results for the circuit in Figure 8.7. The cursor windows list the phasor data for I and Ic.

MATLAB SOLUTIONS

SOLUTION ONE

This MATLAB file finds $\mathbf{Z_{EQ}}$ at a user-specified frequency as well as plots $\mathbf{Z_{EQ}}$ across a user-specified frequency range for the circuit in Figure 8.12. We choose 377 r/s for the single frequency calculation and 10 to 10,000 r/s for the plot range. The resulting plots are shown in Figure 8.13.

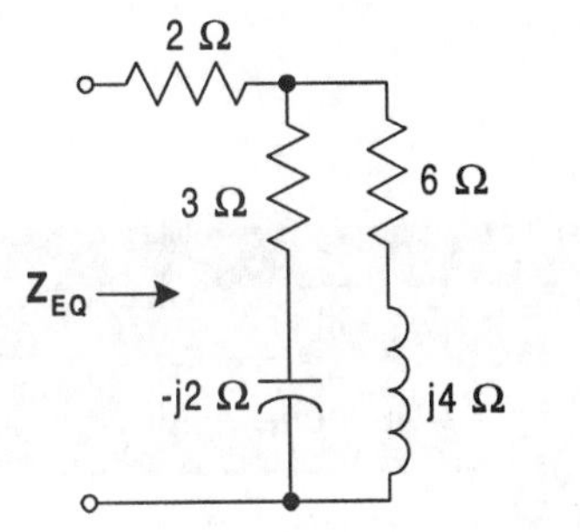

Figure 8.12. Circuit for first MALTAB solution. Find Z_{EQ}.

```
% MATLAB solution to the circuit in Figure 8.12
% Equivalent Impedance
% define the impedances. Note that MATLAB uses
% either "i" or "j" as the imaginary operator.
Z1 = 2;
Z2 = 3;
Z3 = 6;
Z4 = -2j;
Z5 = 4j;
Zeq = Z1 + 1/( 1 / (Z3+ Z5) + 1 /( Z2 + Z4)    );
fprintf ('The equivalent impedance is Eeq = %g %+gj \n', real(Zeq), imag(Zeq) );
fprintf ('Magnitude Zeq = %g Phase Zeq = %g (deg) \n',abs(Zeq),phase(Zeq)*180/pi);
% Suppose the impedances were determined at a particular radian frequency,
% determine C and L
w_comp = input ('Enter Radian Frequency to Determine Component Values ->');
L = Z5/(j*w_comp);
C = -j/(Z4*w_comp);
fprintf ('At w = %g rad/sec L = %g Henries C = %g Farads\n',w_comp,L,C);
% Perform a frequency sweep over a user specified range
fprintf ('\nPerforming a frequency sweep\n');
w_low = input ('Enter Lowest Radian Frequency ->');
w_high = input ('Enter Highest Radian Frequency ->');
num_points = 1000;
w_step = (w_high - w_low)/num_points;
w(1) = w_low;
for k=2:num_points
       w(k) = w(k-1) + w_step;
end;

% Z1, Z2, and Z3 do not change with frequency only Z4 and Z5 need to be recalculated
for k=1:num_points
       Z4 = -j/(w(k)*C);
       Z5 = j*w(k)*L;
       Zeq(k) = Z1 + 1/( 1 / (Z3+ Z5) + 1 /( Z2 + Z4)    );
end;

mag = abs(Zeq);
ph = phase(Zeq)*180/pi;
subplot(2,2,1), plot(w,mag);
xlabel ('Frequency in Radians');
ylabel ('Mag of Zeq (Ohms)');
title ('Magnitude - Linear Scale');
grid on;
subplot(2,2,2), plot(w,ph);
xlabel ('Frequency in Radians');
ylabel ('Phase of Zeq (deg)');
title ('Phase - Linear Scale');
grid on;
subplot(2,2,3), semilogx(w,mag);
xlabel ('Frequency in Radians (log scale)');
ylabel ('Mag of Zeq (Ohms)');
title ('Magnitude - Semilog Scale');
grid on;
subplot(2,2,4), semilogx(w,ph);
xlabel ('Frequency in Radians (log scale)');
ylabel ('Phase of Zeq (deg)');
title ('Phase - Semilog Scale');
grid on;
```

The single frequency solution is

```
The equivalent impedance is Zeq = 4.75294 - 0.611765j
             Magnitude is Zeq = 4.79215  Phase Zeq = -7.33438 (deg)
```

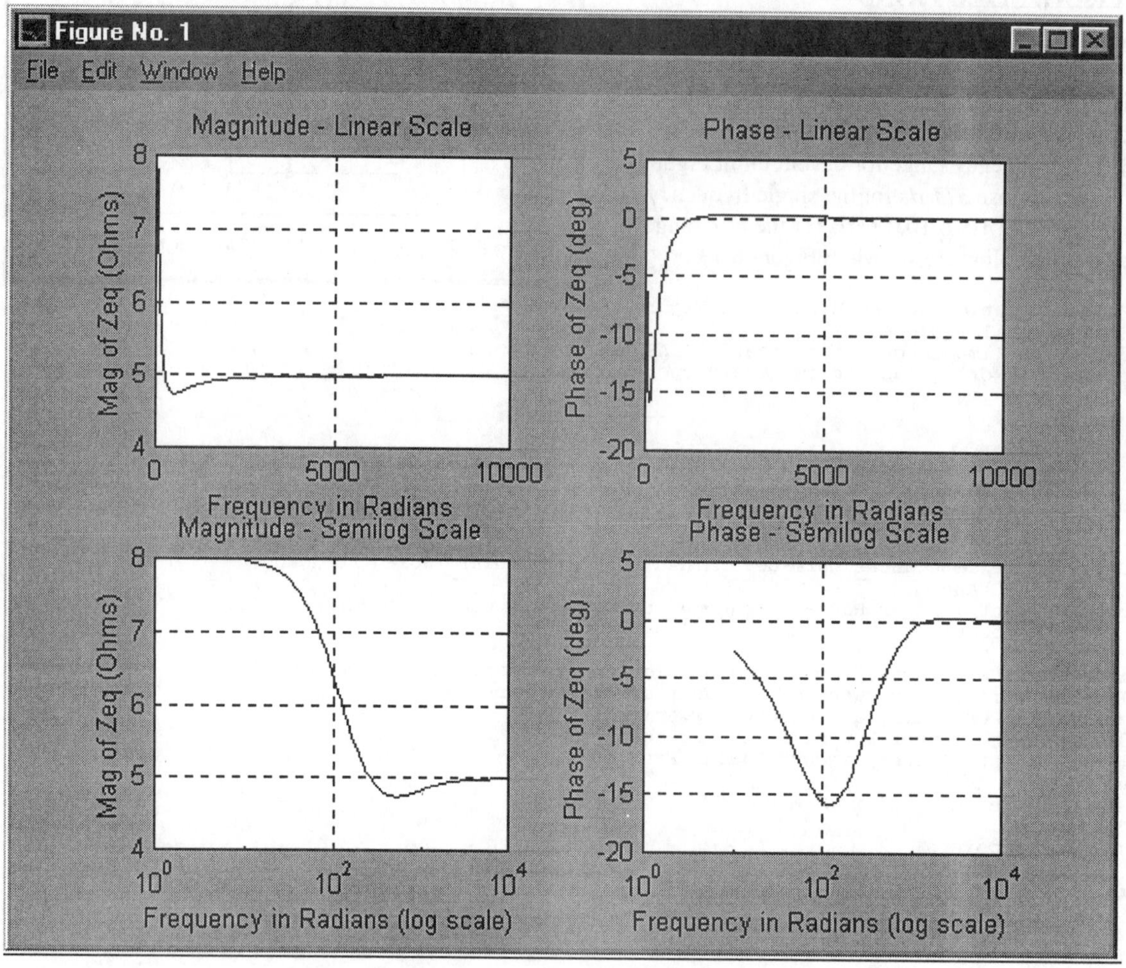

Figure 8.13. MATLAB plots for ZEQ in MATLAB Solution One.

SOLUTION TWO

```
% MATLAB solution to the circuit in Figure
8.14
% Steady state AC Analysis

% assign component values
C1 = 15.83E-06;
L1 = 10E-03;
R1 = 1E+03;
L2 = 442E-03;

% Describe the voltage source as a cosine
f = input ('Frequency of Cosine in Hz ->');
w= 2*pi*f;
amp = input ('Amplitude of Cosine ->');
VS = (amp +j*0);

% calculate component impedances at given
freq.
ZC1 = -j*(1/(w*C1));
ZL1 = j*w*L1;
ZR1 = R1;
ZL2 = j*w*L2;

% calculate the equivalent impedance of the top right block
Zeq = 1/(1/ZC1 + 1/ZL1 + 1/ZR1);

% loop current.  Also total current
I1 = VS /(Zeq + ZL2 - ZL2*1.0E-04*Zeq);

V0 = I1*Zeq;
V1 = VS - V0;

% currents through all the devices
IL2 = V1 / ZL2;
IC1 = V0 / ZC1;
IL1 = V0 / ZL1;
IR1 = V0 / R1;

% Power supplied by the power supply
S = VS * conj(I1);

fprintf ('\n');
fprintf ('The voltage accross the resistor (R1) is %g %+gj volts.\n',real(V0), imag(V0));
fprintf ('Magnitude = %g Phase = %g (deg) \n\n',abs(V0),phase(V0)*180/pi);

fprintf ('The voltage accross the inductor (L2) is %g %+gj volts.\n',real(V1), imag(V1));
fprintf ('Magnitude = %g Phase = %g (deg) \n\n',abs(V1),phase(V1)*180/pi);

fprintf ('The current through the inductor (L2) is %g %+gj milliamps.\n',1000*real(IL2),
1000*imag(IL2));
fprintf ('Magnitude = %g Phase = %g (deg) \n\n',1000*abs(IL2),phase(IL2)*180/pi);

fprintf ('The current through the inductor (L1) is %g %+gj milliamps.\n',1000*real(IL1),
1000*imag(IL1));
fprintf ('Magnitude = %g Phase = %g (deg) \n\n',1000*abs(IL1),phase(IL1)*180/pi);

fprintf ('The current through the capacitor (C1) is %g %+gj milliamps.\n',1000*real(IC1),
1000*imag(IC1));
fprintf ('Magnitude = %g Phase = %g (deg) \n\n',1000*abs(IC1),phase(IC1)*180/pi);

fprintf ('The current through the resistor (R1) is %g %+gj milliamps.\n',1000*real(IR1),
1000*imag(IR1));
fprintf ('Magnitude = %g Phase = %g (deg) \n\n',1000*abs(IR1),phase(IR1)*180/pi);

fprintf ('The complex power supplied by the power supply %g %+gj\n',real(S), imag(S));
fprintf ('Magnitude = %g Phase = %g (deg) \n\n',abs(S),phase(S)*180/pi);
```

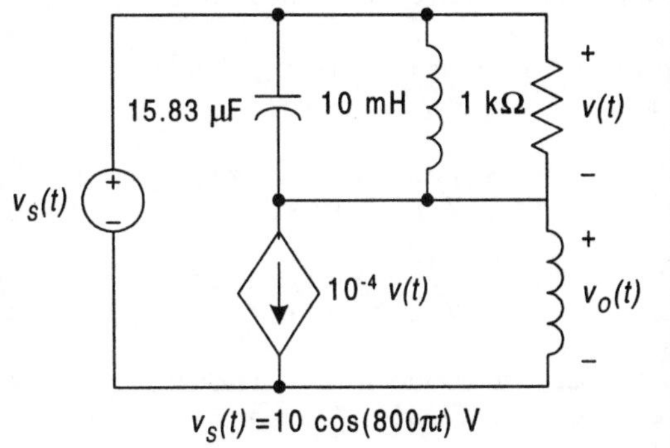

Figure 8.14. Circuit for second MATLAB solution. Use phasors to find v(t).

The solution to MATLAB Solution Two is

```
Frequency of Cosine in Hz ->60
Amplitude of Cosine ->120

The voltage across the resistor (R1) is 2.71456 -0.00920911j volts.
Magnitude = 2.71457 Phase = -0.194375 (deg)

The voltage across the inductor (L2) is 117.285 +0.00920911j volts.
Magnitude = 117.285 Phase = 0.0044988 (deg)

The current through the inductor (L2) is 0.0552668 -703.867j milliamps.
Magnitude = 703.867 Phase = -89.9955 (deg)

The current through the inductor (L1) is -2.44279 -720.059j milliamps.
Magnitude = 720.063 Phase = -90.1944 (deg)

The current through the capacitor (C1) is 0.0549579 +16.1999j milliamps.
Magnitude = 16.1999 Phase = 89.8056 (deg)

The current through the resistor (R1) is 2.71456 -0.00920911j milliamps.
Magnitude = 2.71457 Phase = -0.194375 (deg)

The complex power supplied by the power supply 0.0392067 +84.4642j
Magnitude = 84.4642 Phase = 89.9734 (deg)
```

NOTES

PROBLEM SOLVING EXAMPLES

1. Write the following functions as cosine functions with positive amplitudes.
(a) $A \sin(\omega t+30^\circ)$
(b) $-B \sin(\omega t+45^\circ)$
(c) $-C \cos(\omega t+60^\circ)$

Solution:
(a) $A \sin(\omega t+30^\circ) = A \cos(\omega t-60^\circ)$
(b) $-B \sin(\omega t+45^\circ) = B \sin(\omega t+225^\circ) = B \cos(\omega t+135^\circ)$
(c) $-C \cos(\omega t+60^\circ) = C \cos(\omega t+240^\circ)$

2. Three node voltages in a circuit are found to be

$v_1(t) = 6 \cos(\omega t-20^\circ)V$

$v_2(t) = 4 \sin(\omega t+40^\circ)V$

$v_3(t) = -12 \cos(\omega t+30^\circ)V$

Determine the phase angles by which $v_2(t)$ lags $v_1(t)$ and $v_3(t)$ lags $v_1(t)$.

Solution: $v_2 = 4 \sin(\omega t+40^\circ) = 4 \cos(\omega t-50^\circ)$.

Hence v_2 lags v_1 by $-20^\circ-(-50^\circ) = 30^\circ$

$v_3 = -12 \cos(\omega t+30^\circ) = 12 \cos(\omega t+210^\circ)$.

Therefore v_3 lags v_1 by $-20^\circ-(210^\circ) = -230^\circ$.

3. Given the function $12 \sin \omega t+6 \cos \omega t$, express it in the form $A \cos(\omega t-\theta)$.

Solution: $A \cos(\omega t-\theta) = A \cos \omega t \cos \theta + A \sin \omega t \sin \theta$.
$A \cos \theta = 6$ and $A \sin \theta = 12$. $\tan \theta = \frac{12}{6}$ and

$\theta = 63.43^\circ$. $A \sin 63.43^\circ = 12$ and hence $A = 13.42$. Therefore $12 \sin \omega t + 6 \cos \omega t = 13.42 \cos(\omega t-63.43^\circ)$.

4. Express the following currents in phasor notation.
(a) $i_1(t) = 6 \sin(377t+85^\circ)A$
(b) $i_2(t) = 10 \cos(377t-120^\circ)A$

Solution: (a) $I_1 = 6\angle -5^\circ A$ (b) $I_2 = 10\angle -120^\circ A$

5. Convert the following sinusoidal expressions to phasor notation and exponential notation.
(a) $5 \cos(\omega t)$ (c) $3 \sin(\omega t)$
(b) $7 \cos(\omega t+30^\circ)$ (d) $4 \sin(\omega t-45^\circ)$

Solution: (a) $5\angle 0^\circ$, $5e^{j\omega t}$ (b) $7\angle 30^\circ$, $7e^{j(\omega t+30^\circ)}$ (c) $3 \cos(\omega t-90^\circ)$, $3\angle -90^\circ$, $3e^{j(\omega t-90^\circ)}$, (d) $4 \cos(\omega t-45^\circ-90^\circ)$, $4\angle -135^\circ$, $4e^{j(\omega t-135^\circ)}$

6. Convert the following phasors to an exponential form and a sinusoidal form.
(a) $3\angle 0^\circ$ (c) $-8\angle -180^\circ$
(b) $-2\angle 35^\circ$ (d) $4\angle -360^\circ$

Solution: (a) $3e^{j\omega t}$, $3 \cos \omega t$, (b) $2\angle -145^\circ$, $2e^{j(\omega t-145^\circ)}$, $2 \cos(\omega t-145^\circ)$ (c) $8\angle 0^\circ$, $8e^{j\omega t}$, $8 \cos \omega t$, (d) $4\angle 0^\circ$, $4e^{j\omega t}$, $4 \cos \omega t$

7. Add the following phasors.
(a) $4\angle 20^\circ+7\angle -80^\circ$ (c) $3\angle -160^\circ+5\angle 90^\circ$
(b) $5\angle -45^\circ+8\angle 175^\circ$ (d) $4\angle -25^\circ+2\angle 100^\circ$

Solution:
(a) $(3.759+j1.368)+(1.216-j6.894) = 4.974-j5.526 = 7.435\angle -48^\circ$
(b) $-4.434-j2.838 = 5.265\angle -147.4^\circ$
(c) $-2.819+j3.974 = 4.872\angle 125.4^\circ$
(d) $3.278+j0.279 = 3.29\angle 4.9^\circ$

8. Determine the following impedances in polar form:
(a) a .5H inductor at 377 rad/sec.
(b) a 3H inductor at 60 Hz.
(c) a 100μF capacitor at 377 rad/sec.
(d) a 10Ω resistor at 50 Hz.

Solution:
(a) $Z = j\omega L = j188.5 = 188.5\angle 90^\circ$
(b) $Z = j\omega L = j1131 = 1131\angle 90^\circ$
(c) $Z = \frac{1}{j\omega C} = -j26.5 = 26.5\angle -90^\circ$
(d) $Z = 10\angle 0^\circ$

9. Determine the equivalent impedance Z_{eq} in the circuit shown below.

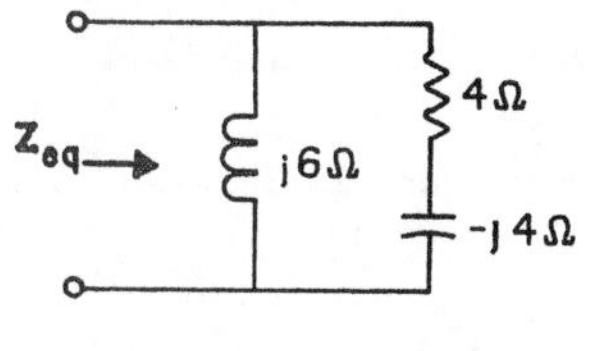

Solution:

$$Z_{eq} = \frac{(4-j4)(j6)}{4-j4+j6} = \frac{24+j24}{4+j2} = \frac{33.94\angle 45^\circ}{4.47\angle 26.57^\circ}$$

$$= 7.59\angle 18.43^\circ \Omega$$

10. Determine the equivalent impedance of the following network.

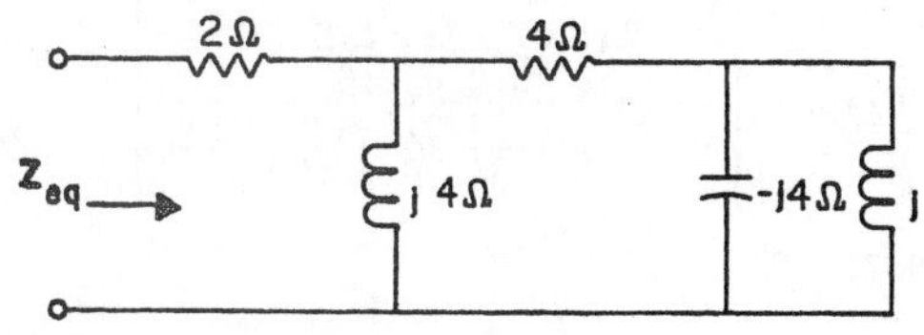

Solution:

$$\mathbf{Z}_{eq} = 2+\frac{j4[4+\frac{(-j4)(j2)}{-j4+j2}]}{j4+4+\frac{(-j4)(j2)}{-j4+j2}}$$

$$= 2+\frac{j4(4+j4)}{j4+4+j4}$$

$$= 2+\frac{-16+j16}{4+j8}$$

$$= 2.8+j2.4\Omega$$

11. Calculate the equivalent admittance of the circuit shown below.

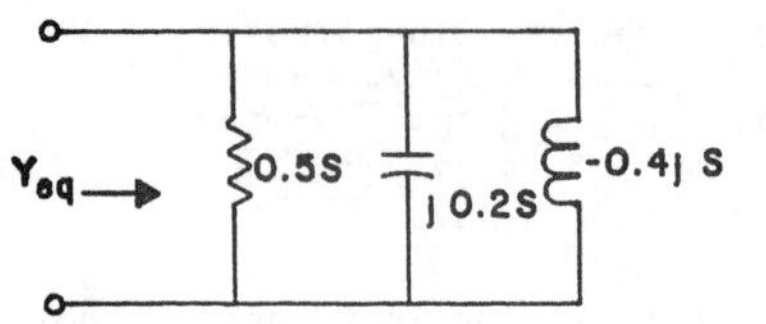

Solution: $\mathbf{Y}_{eq}$ = .05+.2j-.4j = 0.5-j0.2 S

12. Determine the magnitude and phase of the current flowing through the capacitor in the following circuit.

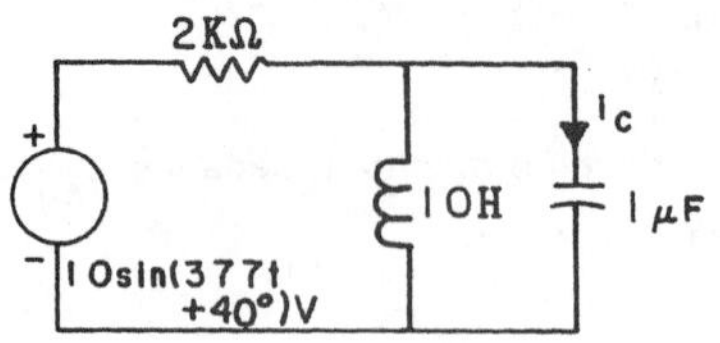

Solution: The network is redrawn in the following form

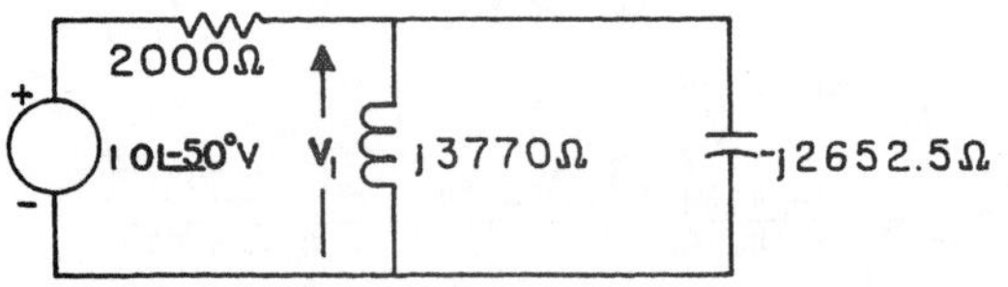

$$\mathbf{V}_1 = 10\underline{/-50°}\left[\frac{\frac{(j3770)(-j2652.5)}{j3770-j2652.5}}{2000+\frac{(j3770)(-j2652.5)}{j3770-j2652.5}}\right]$$

$$= 9.76\underline{/-62.6°}\text{V}$$

Then

$$I_C = \frac{9.76\underline{/-62.6°}}{2652.5\underline{/-90°}} = 3.68\underline{/27.4°}\text{ma}$$

13. Find the magnitude and phase of the voltage across the inductor in the following circuit.

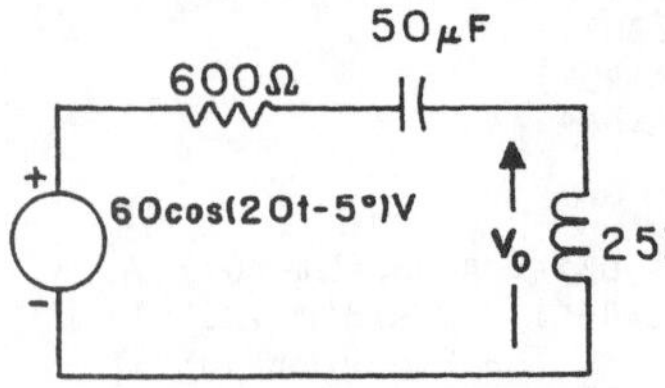

Solution: The network is redrawn as follows.

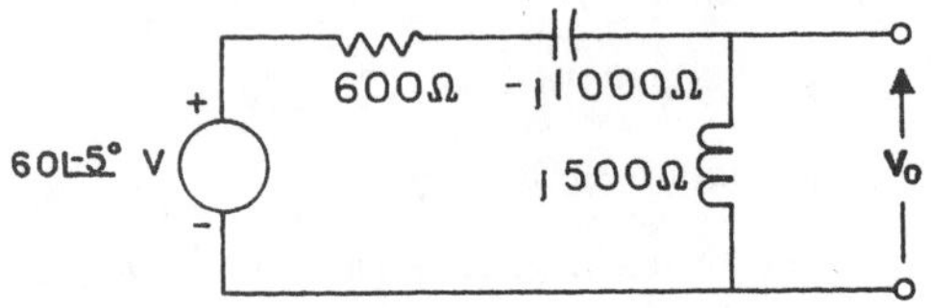

$$\mathbf{V}_o = 60\underline{/-5°}\left(\frac{500\underline{/+90°}}{600-j500}\right) = 38.4\underline{/124.8°}\text{V}$$

14. Determine the magnitude and phase of the current flowing through the capacitor in the network below.

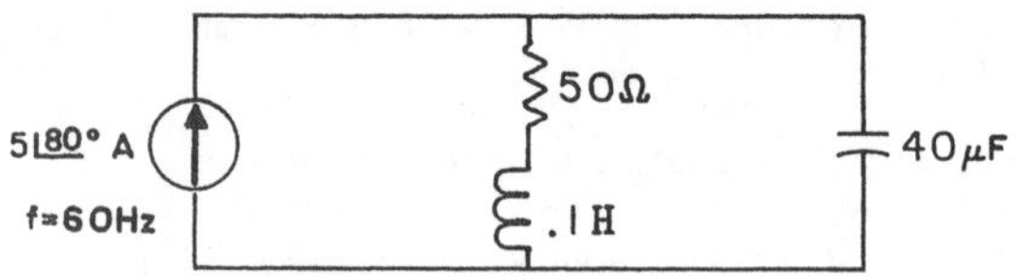

Solution: Redrawing the network we obtain

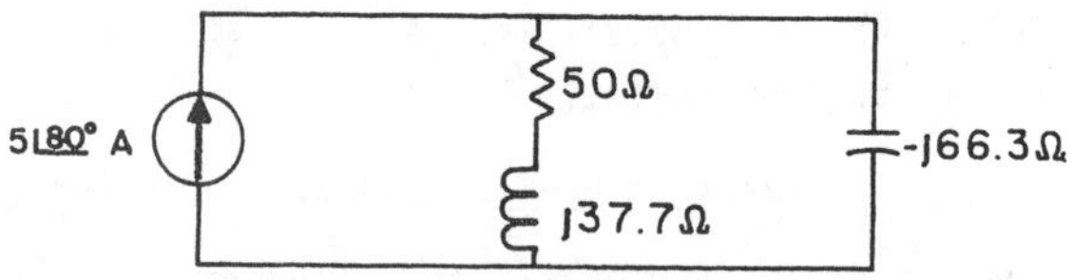

$$I_C = 5\underline{/80°}\left(\frac{50+j37.7}{50+j37.7-j66.3}\right)$$

$$= 5.44\underline{/146.8°}\text{A}$$

15. Determine the voltage $\mathbf{V}_o$ in the following network if the impedance $\mathbf{Z}$ contains
(a) a 0.5H inductor
(b) a 10μF capacitor

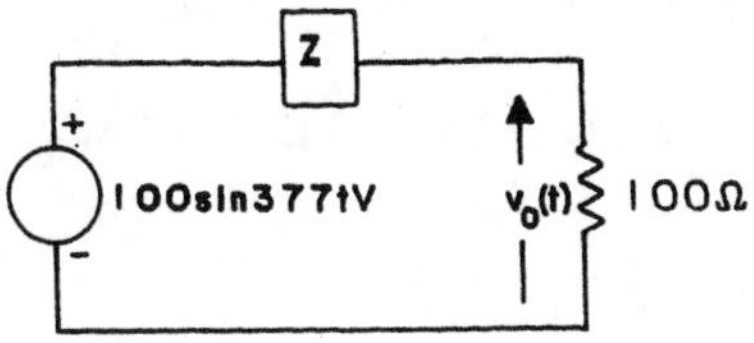

Solution:
(a) $j\omega L = 188.5\angle 90^\circ$

$$\mathbf{V}_o = 100\angle -90^\circ \left(\frac{100}{100+j188.5}\right) = 46.86\angle -152^\circ \text{V}$$

(b) $\dfrac{1}{j\omega C} = 265\angle -90^\circ$

$$\mathbf{V}_o = 100\angle -90^\circ \left(\frac{100}{100-j265}\right)$$

$$= 35.28\angle -20.7^\circ \text{V}$$

16. Determine the value of the capacitor in the following network in order to create a totally real equivalent impedance. The circuit operates at 60 Hz.

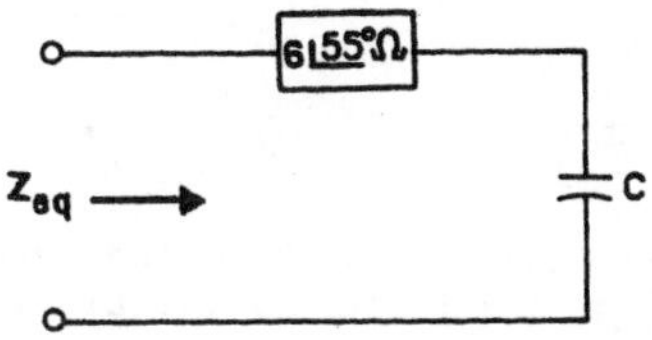

Solution:
$6\angle 55^\circ = 3.44+j4.915$, $j4.915 + \dfrac{1}{j\omega C} = 0$

If $\omega = 377$ then $C = 539.7\mu\text{F}$.

17. Calculate the voltage $\mathbf{V}_o$ and the currents $\mathbf{I}_1$, $\mathbf{I}_2$, and $\mathbf{I}_3$ in the circuit shown below.

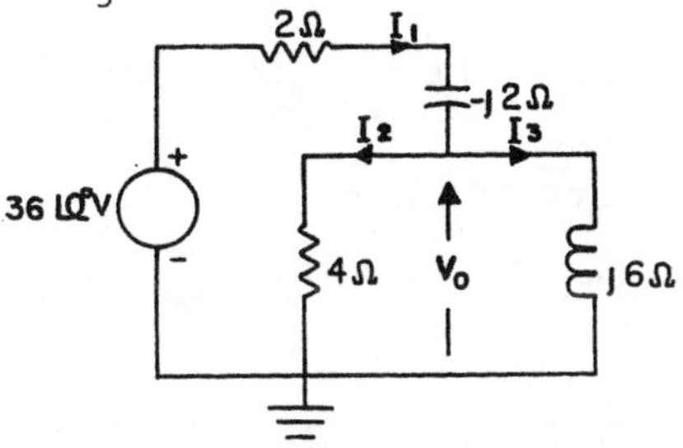

Solution: The total impedance seen by the source is

$$\mathbf{Z}_{eq} = 2-j2+\frac{(4)(j6)}{4+j6} = 4.77\angle -1.8^\circ \Omega$$

Hence

$$\mathbf{I}_1 = \frac{36\angle 0^\circ}{4.77\angle -1.8^\circ} = 7.55\angle 1.8^\circ \text{A}$$

$$\mathbf{I}_2 = \frac{(7.55\angle 1.8^\circ)(6\angle 90^\circ)}{7.21\angle 56.31^\circ} = 6.28\angle 35.49^\circ \text{A}$$

$$\mathbf{I}_3 = \frac{(7.55\angle 1.8^\circ)(4)}{7.21\angle 56.31^\circ} = 4.19\angle -54.51^\circ \text{A}$$

$$\mathbf{V}_o = (4)(6.28\angle 35.49^\circ) = 25.12\angle 35.49^\circ \text{V}$$

18. The current $\mathbf{I}_o$ in the 2-Ω resistor shown below is $\mathbf{I}_o = 4\angle 45^\circ \text{A}$. Calculate $\mathbf{V}_S$.

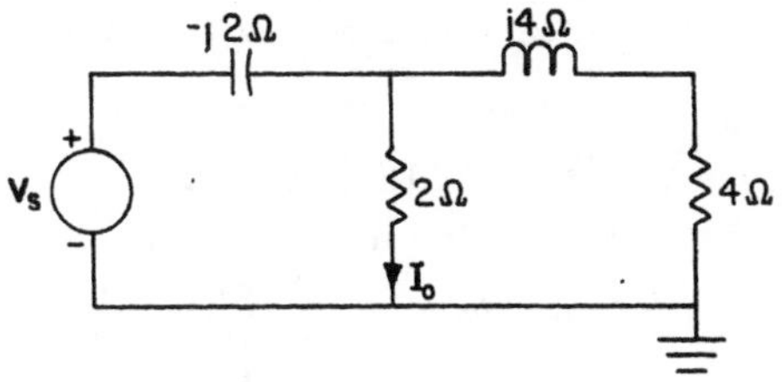

Solution:

$$\mathbf{V}_{2\Omega} = (2)(4\angle 45^\circ) = 8\angle 45^\circ \text{V}$$

$$\mathbf{I}_{RL} = \frac{8\angle 45^\circ}{4+j4} = 1.41\angle 0^\circ \text{A}$$

$$\mathbf{I}_{Total} = \mathbf{I}_o+\mathbf{I}_{RL} = 4\angle 45^\circ+1.41 = 5.10\angle 33.71^\circ \text{A}$$

$$\mathbf{V}_C = (5.10\angle 33.71^\circ)(2\angle -90^\circ) = 10.2\angle -56.29^\circ \text{V}$$

$$\mathbf{V}_S = \mathbf{V}_C+\mathbf{V}_{2\Omega} = (11.31-j2.83)\text{V}$$

19. Find $\mathbf{V}_o$ in the following network.

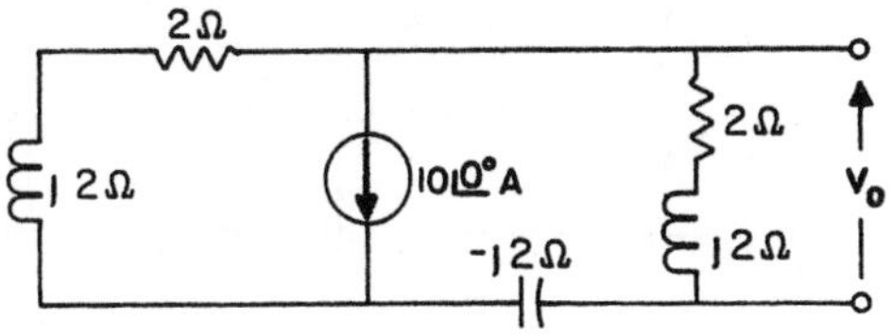

Solution: Employing current division

$$\mathbf{V}_o = -\frac{(10\angle 0^\circ)(2+j2)}{2+j2-j2+2+j2}(2+j2)$$

$$= 17.9\angle -116.56^\circ \text{V}$$

20. Find $\mathbf{V}_S$ in the following network.

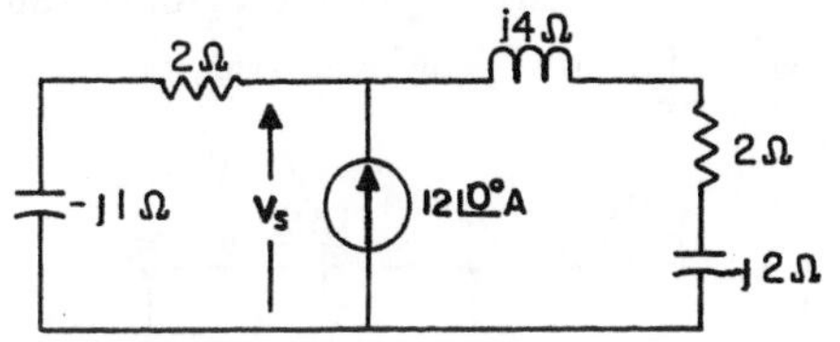

Solution:

$$\mathbf{V}_S = 12\angle 0^\circ \left(\frac{(2-j1)(2+j2)}{4+j1}\right)$$

$$= 18.42\angle 4.4^\circ \text{V}$$

21. Find V_S in the network below if $V_o = 4\underline{/0°}$ V.

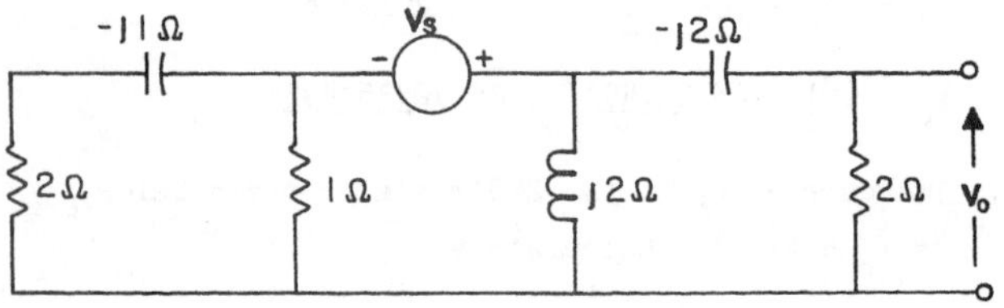

Solution: The network is labelled as follows.

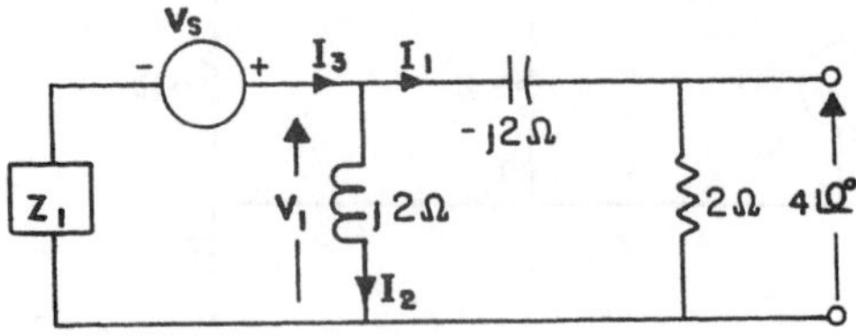

$$I_1 = \frac{4\underline{/0°}}{2} = 2\underline{/0°}\text{A and } V_1 = (2\underline{/0°})(2-j2) = 4-j4\text{V}.$$

$$I_2 = \frac{V_1}{j2} = -j2-2 \text{ and } I_3 = I_1+I_2 = -j2\text{A}$$

$$V_S = V_1+I_3Z_1 = 6.6\underline{/-54.87°}\text{V where } Z_1 = \frac{(1)(2-j1)}{1+2-j1}$$

$= 0.7-j0.1\Omega$.

22. Find V_S in the following network if $V_1 = 4\underline{/0°}$V.

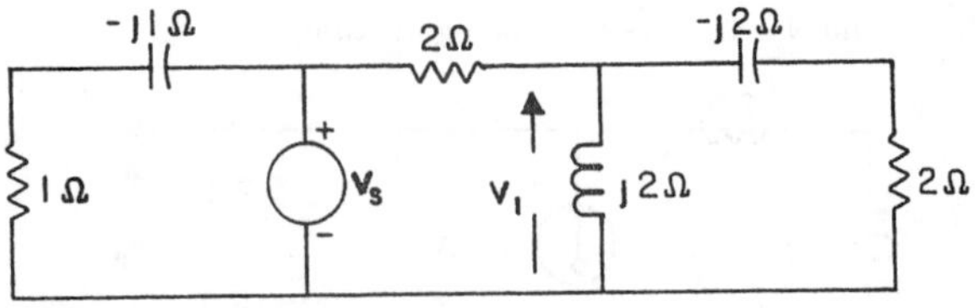

Solution: Using a voltage divider

$$V_1 = 4\underline{/0°} = V_S\left[\frac{\frac{(2-j2)(j2)}{2}}{2+\frac{(2-j2)(j2)}{2}}\right]$$

Therefore $V_S = 6.32\underline{/-18.43°}$V

23. Determine the voltage V_2 in the circuit shown below using linearity, by assuming that it is equal to $1\underline{/0°}$V.

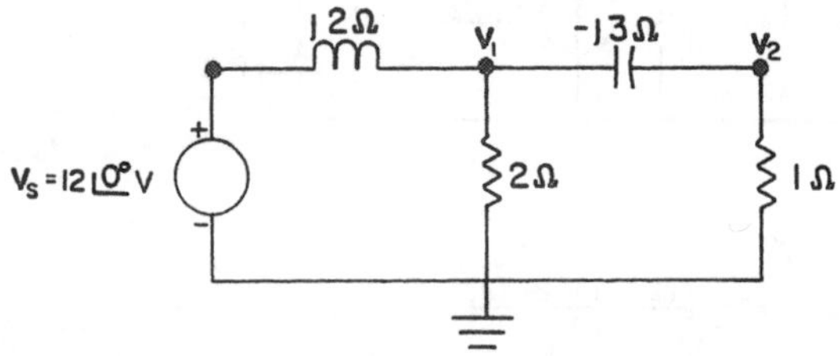

Solution: If $V_2 = 1\underline{/0°}$V then $I_{1\Omega} = 1\underline{/0°}$A.

$V_1 = (1\underline{/0°})(1-j3) = 1-j3$ V and $I_{2\Omega} = \frac{1-j3}{2} =$

$\frac{1}{2} - j\frac{3}{2}$ A.

Using KCL $I_L = I_{1\Omega}+I_{2\Omega} = \frac{3}{2} - j\frac{3}{2}$ A.

Then $V_S = I_L(j2)+V_1 = 4\underline{/0°}$V.

Therefore $\frac{V_2}{V_S} = \frac{1\underline{/0°}}{4\underline{/0°}} = \frac{V_2}{12\underline{/0°}}$ and $V_2 = 3\underline{/0°}$V.

24. Determine V_2 in Problem **23** using nodal analysis.

Solution: Using nodal analysis we can find V_1 from the equation $\frac{V_1-V_S}{j2} + \frac{V_1}{2} + \frac{V_1}{1-j3} = 0$ where $V_S = 12\underline{/0°}$.

Solving for V_1 we obtain $V_1 = \frac{30\underline{/-90°}}{3-j1}$ then we can determine V_2 using voltage division as

$$V_2 = \frac{\frac{30\underline{/-90°}}{3-j1}}{1-j3}(1) = 3\underline{/0°}\text{V}.$$

25. Use nodal equations to find the current in the inductor in the following circuit.

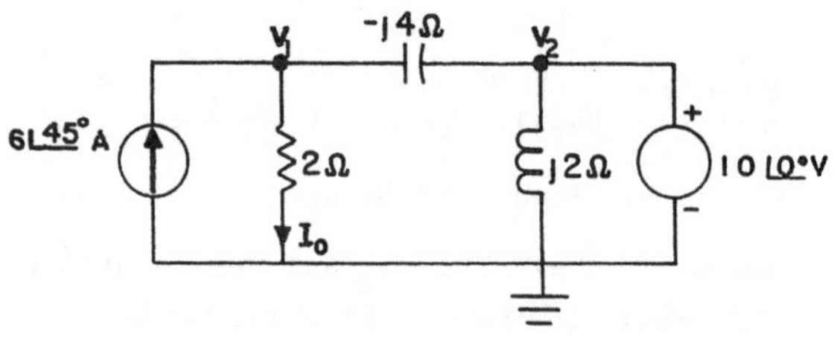

Solution: The current in the inductor is

$$I_L = \frac{10\underline{/0°}}{j2} = 5\underline{/-90°}\text{A}.$$

26. Use nodal analysis to find I_o in the following network.

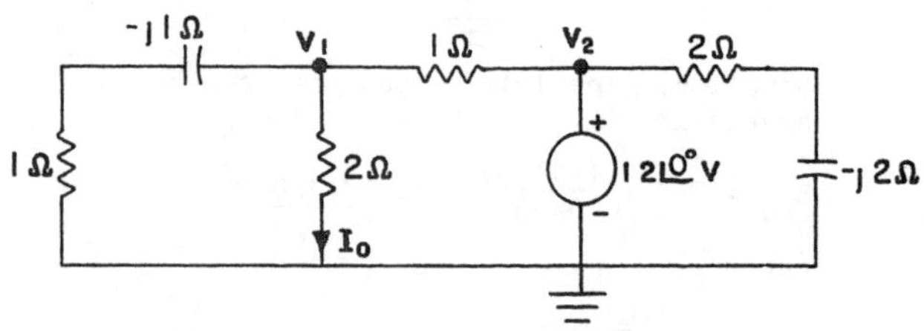

Solution: V_2 is known and the node equation at V_1 is

$$\frac{V_1}{1-j1} + \frac{V_1}{2} + \frac{V_1-12\underline{/0°}}{1} = 0 \text{ or } V_1 = 5.82\underline{/-14°}\text{V}$$

and hence $I_o = \frac{V_1}{2} = 2.91\underline{/-14°}$A.

27. Find I_o in the following network using nodal analysis.

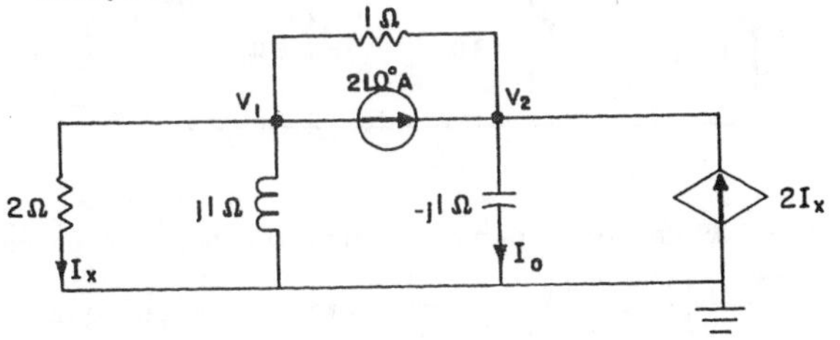

Solution: The nodal equations are

$$\frac{V_1}{2} + \frac{V_1}{j1} + 2\underline{/0^\circ} + \frac{V_1 - V_2}{1} = 0$$

$$\frac{V_2 - V_1}{1} - 2\underline{/0^\circ} + \frac{V_2}{-j1} - 2I_x = 0$$

where $I_x = \frac{V_1}{2}$. Solving for V_2 we obtain

$V_2 = -3.15\underline{/18.43^\circ}$V and hence

$$I_o = \frac{V_2}{-j1} = 3.15\underline{/-71.57^\circ}\text{A}.$$

28. Determine I_x in the circuit below using nodal analysis.

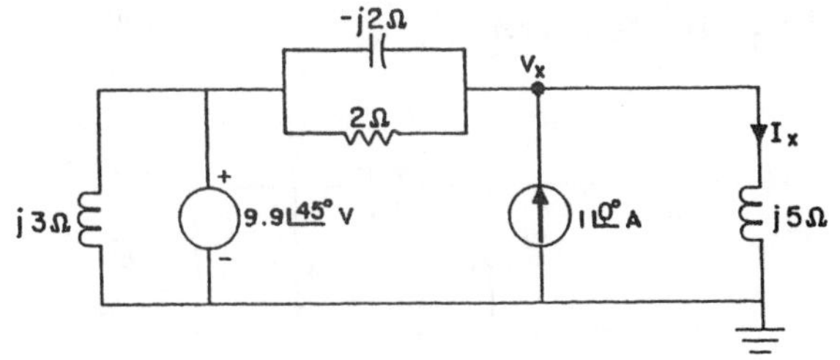

Solution: The node equation for V_x is

$$\frac{V_x - 9.9\underline{/45^\circ}}{-j2} + \frac{V_x - 9.9\underline{/45^\circ}}{2} + \frac{V_x}{j5} = 1\underline{/0^\circ}$$

Therefore $V_x = 12.13\underline{/50.9^\circ}$V and $I_x = \frac{V_x}{j5} =$

$2.43\underline{/-39.1^\circ}$A.

29. Calculate the inductor current in the circuit shown below using nodal analysis.

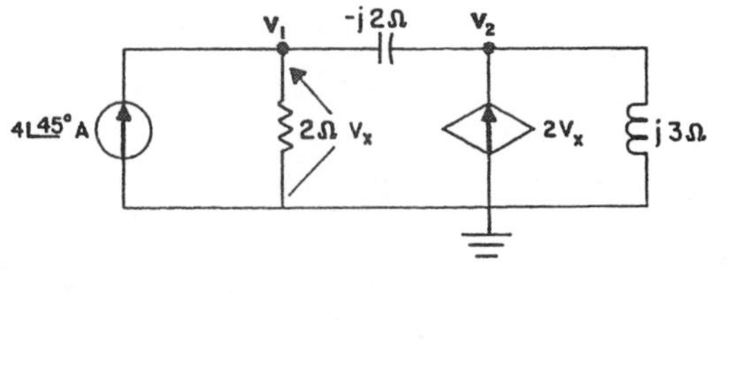

Solution: The nodal equations for the network are

$$\frac{V_1}{2} + \frac{V_1 - V_2}{-j2} = 4\underline{/45^\circ}$$

$$\frac{V_2 - V_1}{-j2} + \frac{V_2}{j3} = 2V_1$$

Solving the equations yields $V_2 = 8.85\underline{/138.7^\circ}$V

and hence $I_L = \frac{V_2}{j3} = 2.95\underline{/48.73^\circ}$A.

30. Find V_2 in Problem 23 using Mesh analysis.

Solution: The Mesh equations are

$$I_1(2+j2) - I_2(2) = 12\underline{/0^\circ}$$

$$-I_1(2) + I_2(3-j3) = 0$$

Solving the equations yields $I_2 = 3\underline{/0^\circ}$A and hence $V_2 = 3\underline{/0^\circ}$V.

31. Solve for I_x in Problem 28 using loop analysis.

Solution: The network can be labelled as follows.

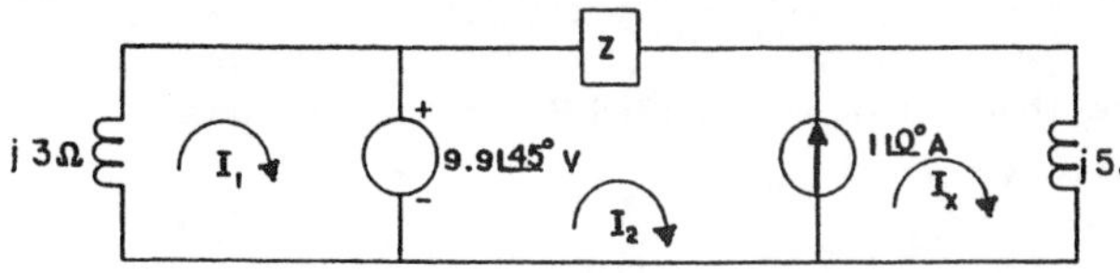

The loop equations are

$$I_1(j3) = -9.9\underline{/45^\circ}$$

$$I_2(1-j1) + I_x(j5) = 9.9\underline{/45^\circ}$$

$$I_x - I_2 = 1\underline{/0^\circ}$$

Solving the equations yields $I_x = 2.43\underline{/-39.1^\circ}$A.

32. Use loop equations to find V_o in the following network.

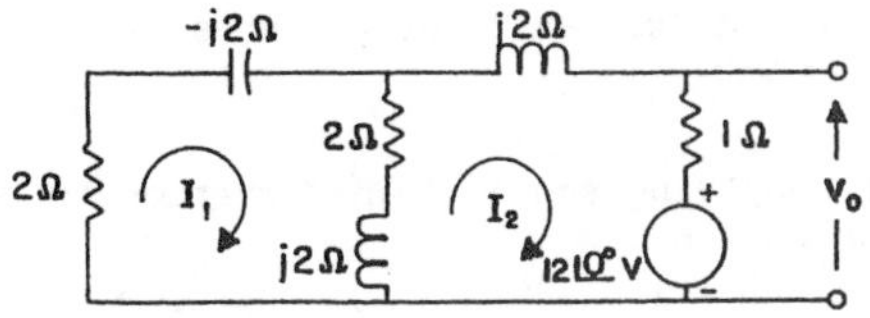

Solution: The loop equations are

$4I_1-(2+j2)I_2 = 0$

$-(2+j2)I_1+(3+j4)I_2 = -12\underline{/0^\circ}$

Solving the equations yields $I_2 = \dfrac{-48\underline{/0^\circ}}{12+j8}$ A and therefore $V_o = 12\underline{/0^\circ}+(1)I_2 = 9.41\underline{/11.31^\circ}$V.

33. Use loop equations to solve for I_o in the following circuit.

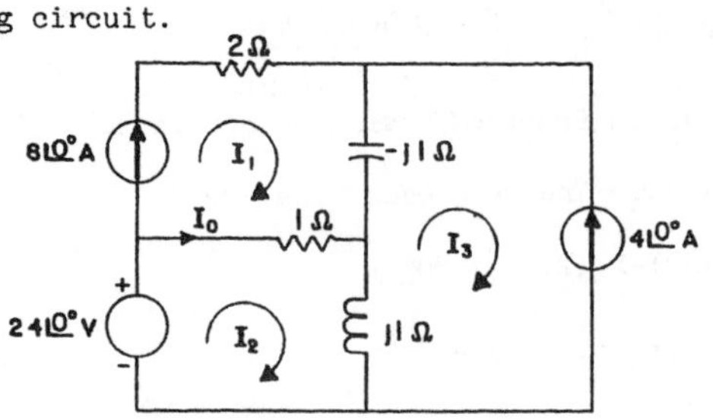

Solution: The loop equations are

$I_1 = 8\underline{/0^\circ}$A, $I_2 = -4\underline{/0^\circ}$A and

$1(I_2-I_1)+j1(I_2-I_3) = 24\underline{/0^\circ}$

These equations yield $I_2 = 22.8\underline{/-52.13^\circ}$A and hence $I_o = I_2-I_1 = 18.97\underline{/-71.57^\circ}$A.

34. Use loop equations to find V_o in the following network.

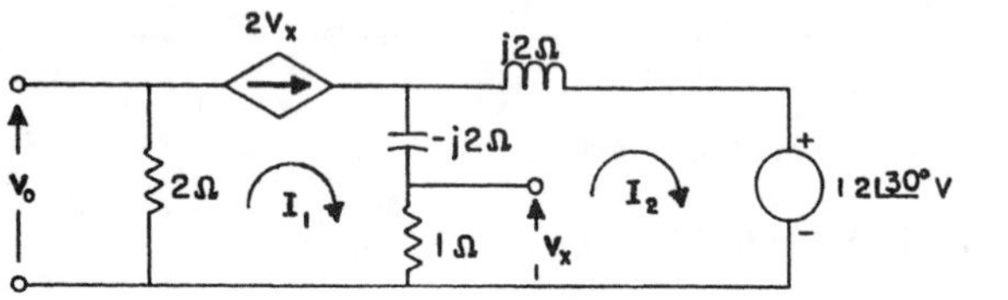

Solution: The loop equations are

$I_1 = 2V_x = -2(I_2-I_1)$

$(I_2-I_1)(1-j2)+I_2(j2) = -12\underline{/30^\circ}$

Solving the equations yields $I_1 = 5.82\underline{/106^\circ}$A and $V_o = -2I_1 = 11.64\underline{/-74^\circ}$V.

35. Determine V_2 in Problem 23 using source transformation.

Solution: Applying source transformation to the network we obtain

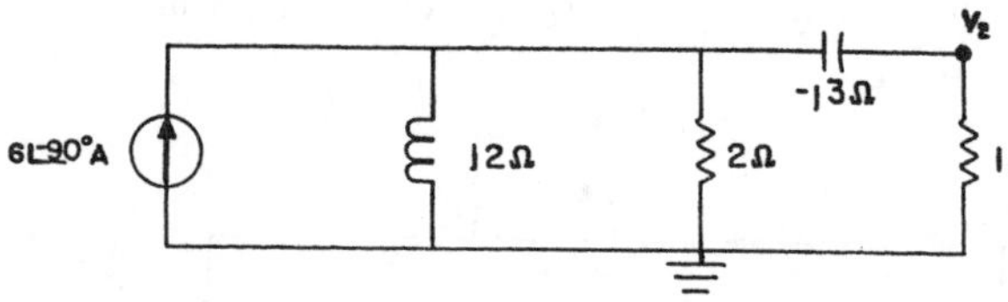

The impedance in parallel with the current source is $\dfrac{(2)(j2)}{2+j2} = \dfrac{4j}{2+j2}$. Transforming the current source and parallel impedance back to a voltage source and series impedance we obtain

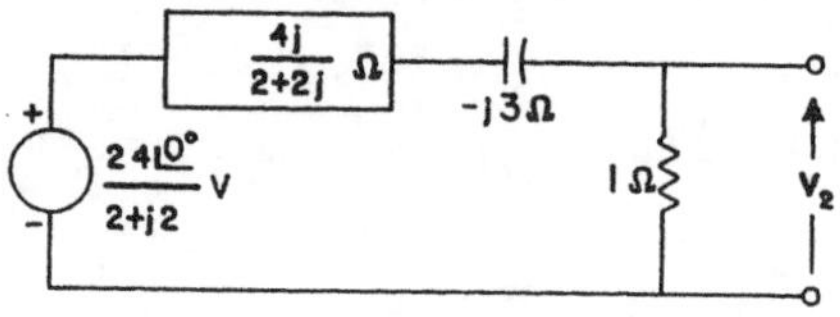

Then

$$V_2 = \frac{\dfrac{24\underline{/0^\circ}}{2+j2}}{\dfrac{j4}{2+j2}-j3+1} = 3\underline{/0^\circ}\text{V}.$$

36. Use source transformation to solve for I_o in the following network.

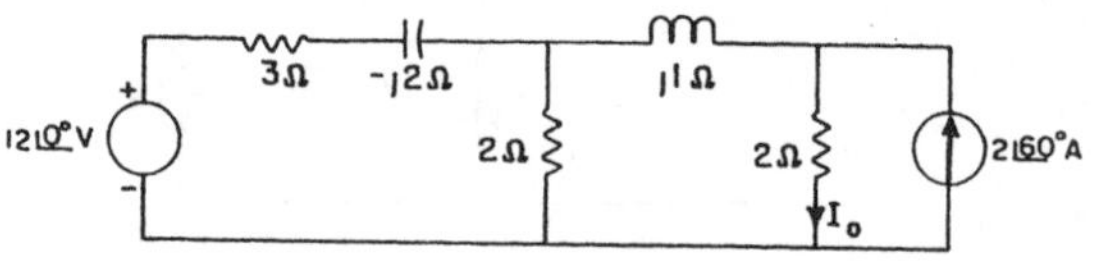

Solution: The following sequence of figures illustrate the transformation techniques

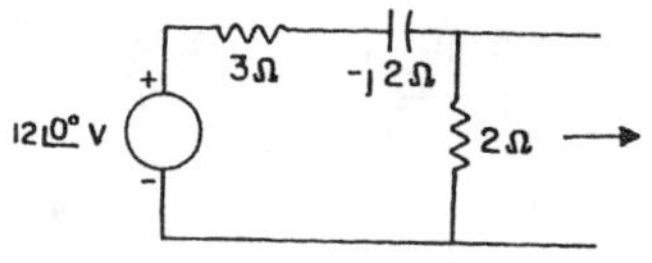

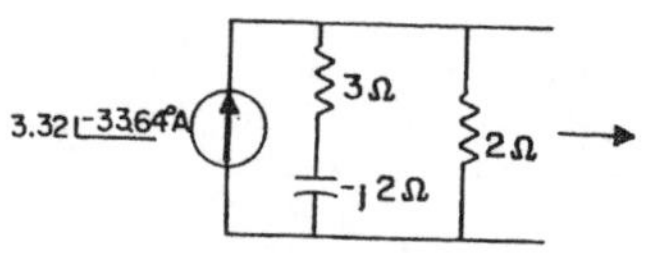

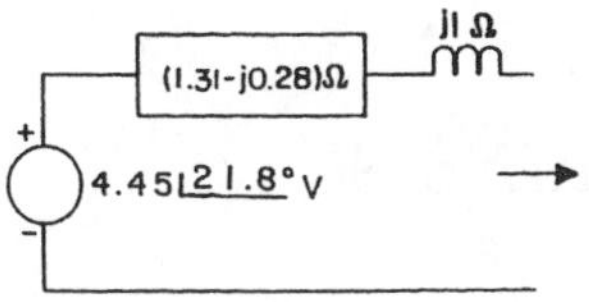

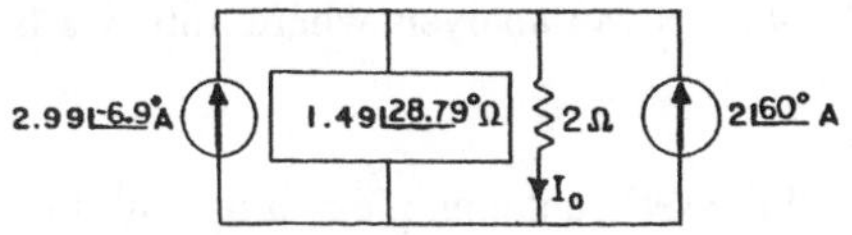

Using current division

$$I_o = \frac{(1.49\underline{/28.79°}+2\underline{/60°})(1.49\underline{/28.79°})}{(1.49\underline{/28.79°})+2}$$

$= 1.89\underline{/35.59°}\text{A}$

37. Find I_o in Problem 25 using superposition.

Solution: The following networks yield the solution.

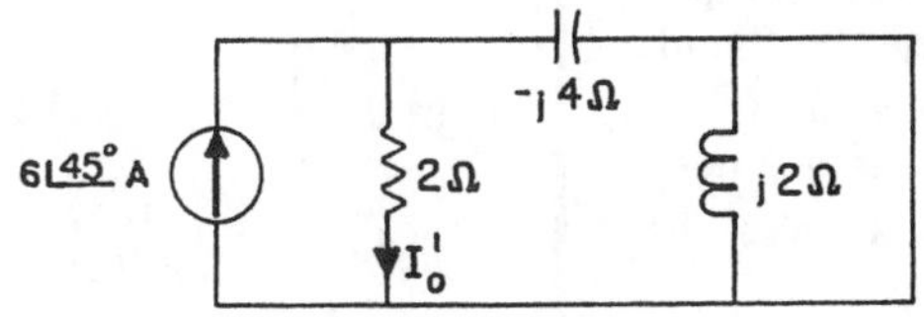

$$I_o^1 = \frac{(6\underline{/45°})(-j4)}{2-j4} = \frac{24\underline{/-45°}}{2-j4}\ \text{A}$$

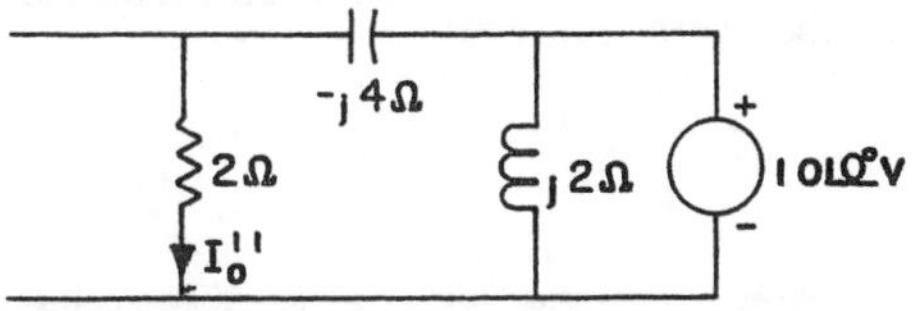

$$I_o^{11} = \frac{10\underline{/0°}}{2-j4}\ \text{A}$$

Hence

$$I_o = I_o^1 + I_o^{11} = \frac{24\underline{/-45°}}{2-j4} + \frac{10\underline{/0°}}{2-j4} = 7.13\underline{/31.25°}\text{A}$$

38. Solve for I_x in Problem 28 using superposition.

Solution: The two networks below are used to solve the problem.

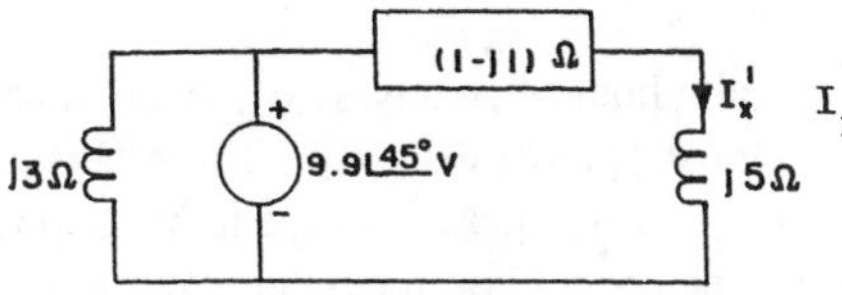

$$I_x^1 = \frac{9.9\underline{/45°}}{1-j1+j5}$$

$= 2.059-j1.235\text{A}$

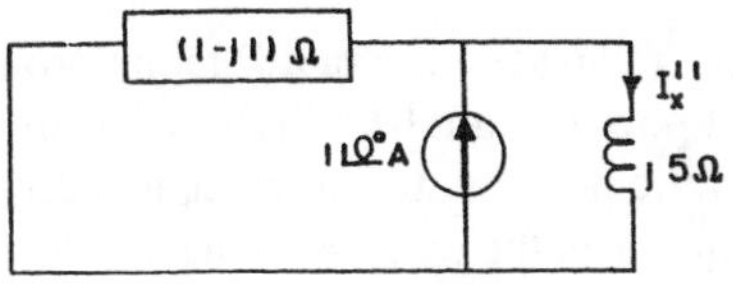

$$I_x^{11} = \frac{(1\underline{/0°})(1-j1)}{1-j1+j5}$$

$= -.177-j.294\text{A}$

Hence $I_x = I_x^1 + I_x^{11} = 2.43\underline{/-39.1°}\text{A}$

39. Use Thevenin's theorem to find V_2 in Problem 23.

Solution: Breaking the network to the right of the 2Ω resistor yields

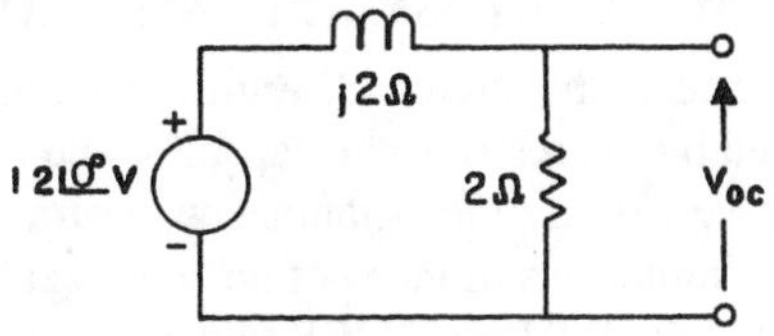

$$V_{oc} = \frac{(12\underline{/0°})}{2+j2}(2) = \frac{24}{2+j2}\ \text{V}$$

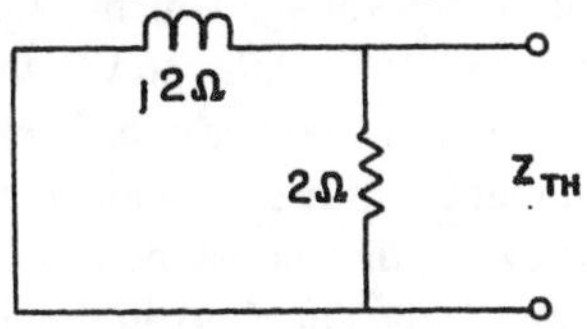

$$Z_{TH} = \frac{(2)(j2)}{2+j2} = \frac{j4}{2+j2}\ \Omega$$

Then

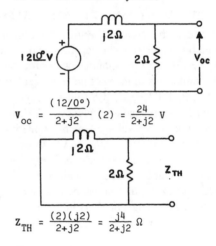

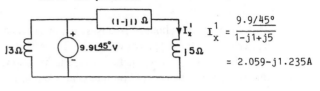

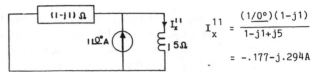

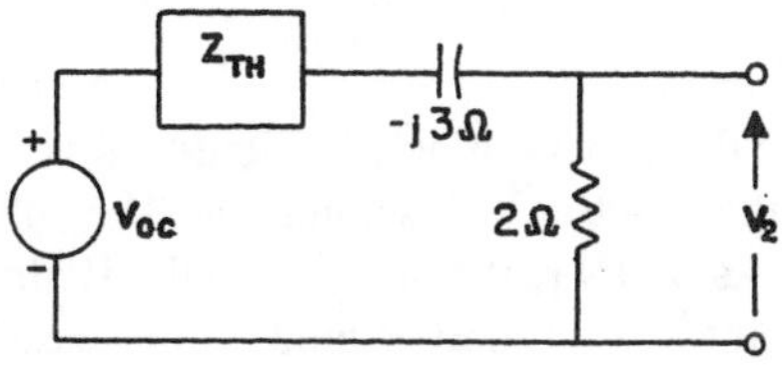

And

$$V_2 = \frac{\frac{24}{2+j2}}{\frac{j4}{2+j2} - j3 + 1}(1) = 3\underline{/0°}\text{V}$$

Chapter 9 STEADY-STATE POWER ANALYSIS

This chapter is all about ac steady-state power. The origins of ac steady-state power analysis are found in the power utility industry; right where the BECA text properly focuses our attention. Power levels in the kilo-watts are quite common. There is however, another application of steady-state power analysis where milli-watts and micro-volts are the norm - communication circuitry.

Consider a simple radio set. The radio station broadcasts energy and the set's antenna picks it up. While the station broadcast power might be millions of watts, the voltage on the antenna is typically in the micro-Volt range. With so little power at the antenna, it is critical that we process the signal efficiently. Thus, engineers design the subcircuits in the radio such that maximum power is transferred between them. Even though the power levels are orders of magnitude lower that in power utility analysis, the method for optimizing power transfer is the same. Namely, the load impedance must be the complex conjugate of the circuit's output impedance.

PSPICE SIMULATIONS

SIMULATION ONE - INSTANTANEOUS VS AVERAGE POWER

Let's simulate the circuit in Figure 9.1 for the instantaneous and average power absorbed by the resistor, the capacitor and the inductor. To find instantaneous power, we will use the instantaneous wattmeter in the BECA library. (Proper use and limitations of the wattmeter are explained in the Wattmeter.txt file in the BECA PART READMES folder.) Figure 9.3 shows the *Schematics* circuit where a sinusoidal frequency of 1 Hz has been chosen. Note that the wattmeter output is a voltage which we interpret as power. Simulation results are shown in Figure 9.4 for the power absorbed by each of the passive elements. Note that the power in each element is sinusoidal at 2 Hz - twice the excitation frequency. Since the inductor and capacitor power waveforms are centered about zero, they consume zero average power, as expected. However, from the data markers in Figure 9.4, we see that the average power absorbed by the resistor is

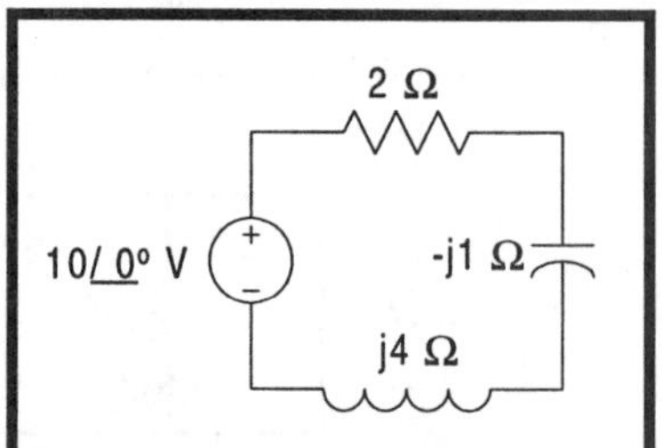

Figure 9.1. Circuit for first PSpice simulation.

$$P_R = \frac{15.37 - 0}{2} = 7.69 \text{ W}$$

Since the x-axis variable in Figure 9.4 is time, we know that this is a transient simulation. The voltage source, V1, is a VSIN part from the SOURCE library. Its attribute box, properly edited for this case is shown in Figure 9.2. The attribute fields bear some explanation. The DC field is the voltage used for dc simulations, the AC field is the phasor magnitude used in AC Sweep simulation and the rest of the fields are used in Transient simulations. Obviously, the VSIN part can be used in any kind of simulation.

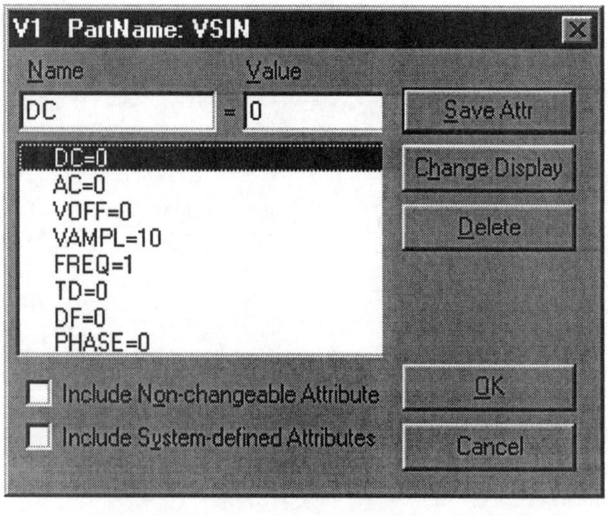

Figure 9.2. The VSIN part attribute box.

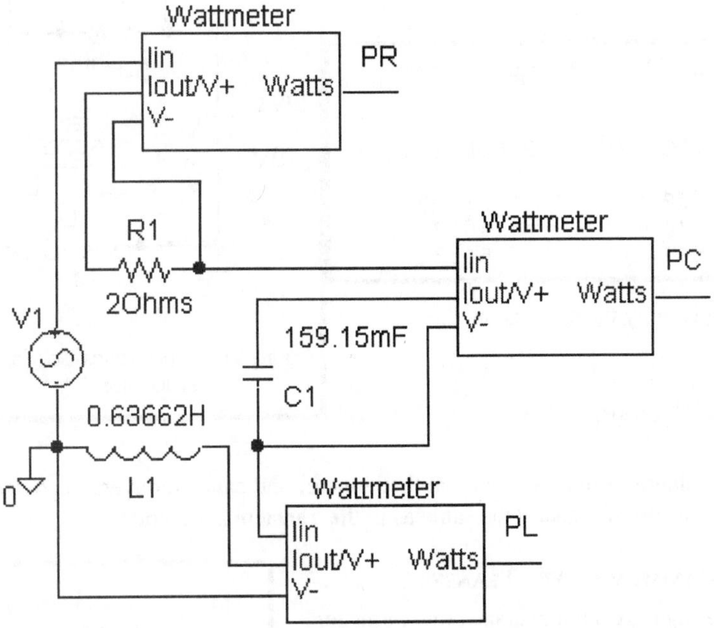

Figure 9.3. The *Schematics* diagram for the instantaneous power simulation.

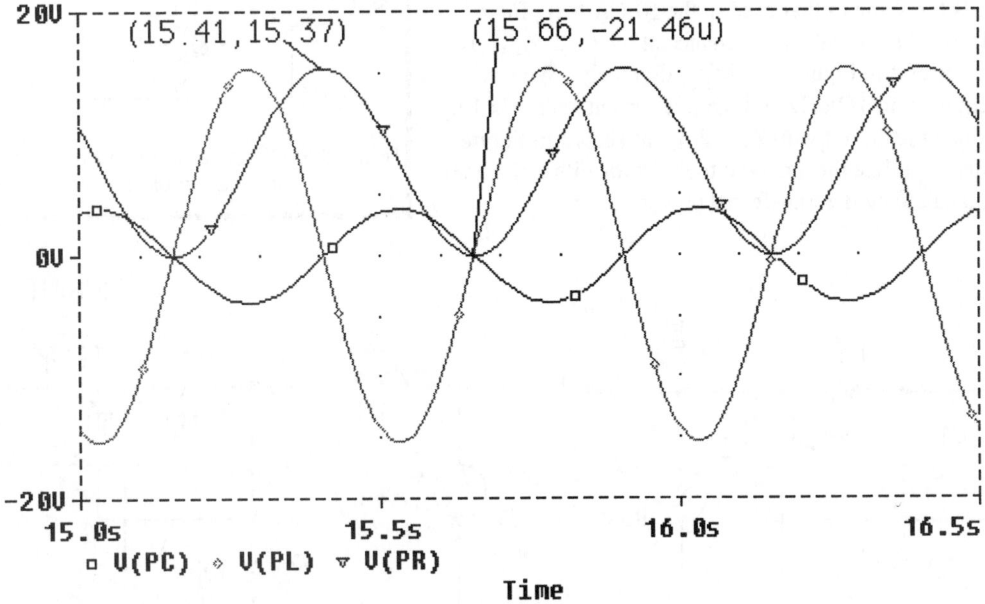

Figure 9.4. The PROBE plot of the absorbed power in the passive elements.

Next, we perform a single frequency ac simulation at $1/2\pi$ Hz using the circuit in Figure 9.5 where the VPRINT2 and IPRINT parts are set to gather magnitude and phase data.

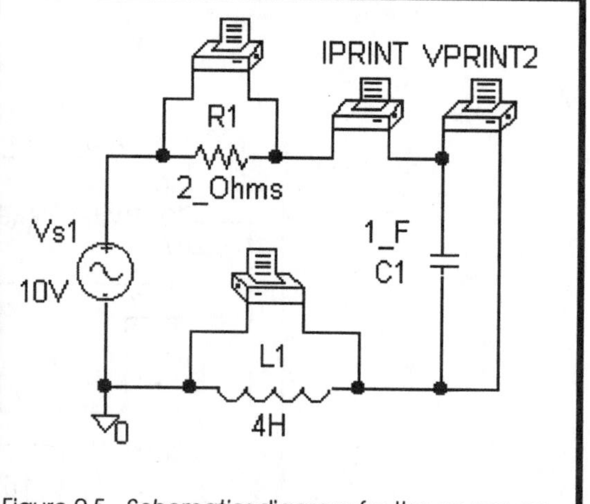

Figure 9.5. *Schematics* diagram for the ac power simulation.

```
LOOP CURRENT = 2.774 A @ -56.31 deg.

COMPONENT     VOLTAGE (V)     PHASE (deg.)

Inductor      11.09             33.69
Capacitor     2.774           -146.3
Resistor      5.547            -56.31
```

The average power is given by the expression

$$P = \frac{V_M I_M}{2}\cos(\theta_V - \theta_I) \qquad (9.1)$$

Using (9.1) and the simulation results we find that the inductor and capacitor average power is zero while the resistor average power is 7.69 W, exactly the same as in the transient simulation.

Simulation Two - Maximum Power Transfer

One method for simulating maximum average power transfer is detailed in this example. We will find the load impedance for the circuit in Figure 9.6 that yields maximum power transfer as well as the transferred power. The simulation procedure is the same as that used in the BECA text. First, find the Thevenin equivalent impedance at the load, $\mathbf{Z_{TH}}$. The *Schematics* circuit in Figure 9.8 should do nicely. We have set the frequency to $1/2\pi$ Hz and the test current source is $1\underline{/0°}$ A. Next, load the circuit with $\mathbf{Z_L} = \mathbf{Z_{TH}}^*$ as shown in Figure 9.7. Finally, simulate the loaded circuit for the either the load current and calculate the transferred power.

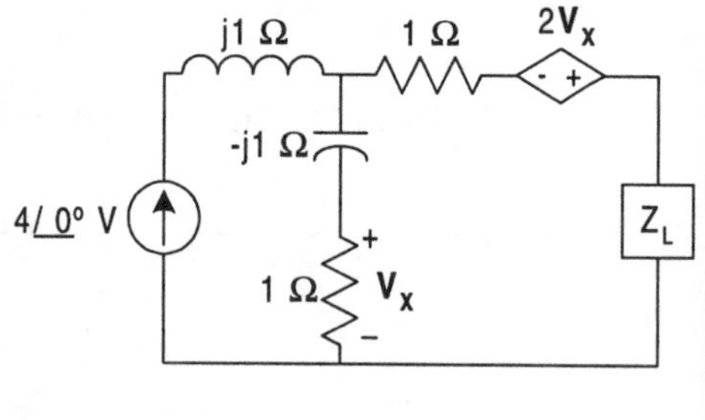

Figure 9.6. Circuit for the maximum power transfer simulation.

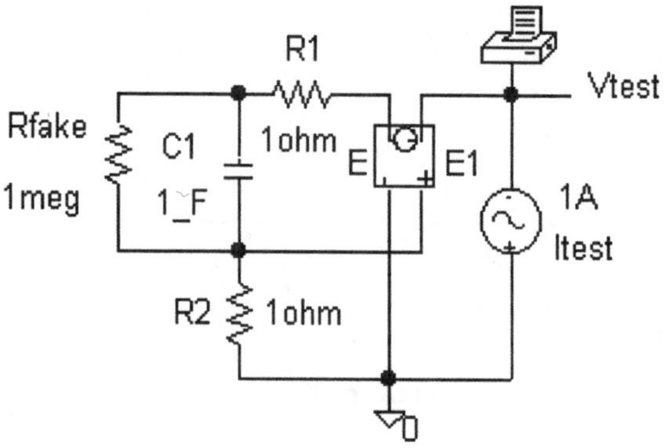

Figure 9.8. *Schematics* circuit for finding Z_{TH}.

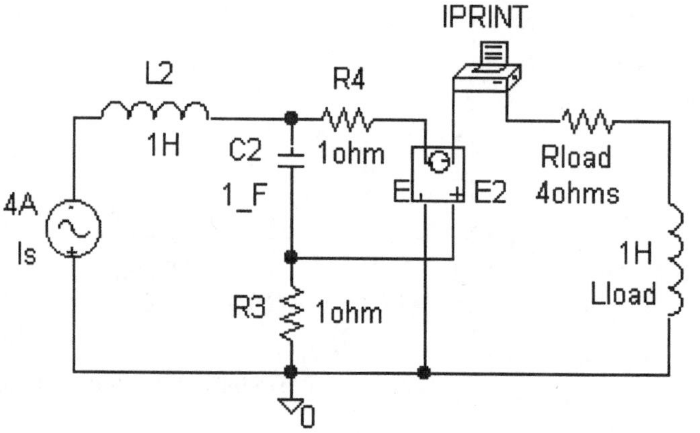

Figure 9.7. *Schematics* diagram for finding the transferred power.

Results for the $\mathbf{Z_{TH}}$ simulation are

```
TEST RESULTS

MAG - Vtest    Phase - Vtest

4.123 V      -14.04 degrees
```

Therefore

$$\mathbf{Z_{TH}} = \frac{\mathbf{V_{TEST}}}{\mathbf{I_{TEST}}} = \frac{4.123\angle{-14.04}}{1\angle 0} = 4 - j1\ \Omega$$

The load impedance for maximum power transfer is

$$\mathbf{Z_{LOAD}} = \mathbf{Z_{TH}}^* = 4 + j1\ \Omega$$

Simulations for the loaded circuit yield

```
MAXIMUM POWER RESULTS

MAG - Iload    Phase - Iload

1.581 A      -18.44 degrees
```

The maximized load power is therefore,

$$P_{LOAD} = \left(\frac{I_M}{2}\right)^2 R_{LOAD} = \left(\frac{1.581}{2}\right)^2 (4) = 2.5\ \text{W}$$

SIMULATION THREE - THE LOAD PART

Power systems are often specified in wattage, voltage and power factor rather than Ohms and Henries. To create appropriate *Schematics* diagrams, we must calculate the equivalent load resistances and inductances. For those occasions, we have created a part in the BECA library called LOAD. Figure 9.9 shows the LOAD part and its Attribute box wherein we can enter the rated voltage, the power factor, the frequency of the simulation and the rated power. PSpice uses this data to calculate the equivalent resistance and inductance of the load. This means of course that the power factor is always lagging for the LOAD part. Also, errors in the simulated load current will occur if the actual load voltage in the simulation does not equal the rated value entered in the Attribute box. However, since any efficient power delivery system should have minimum transmission losses, these errors should be low.

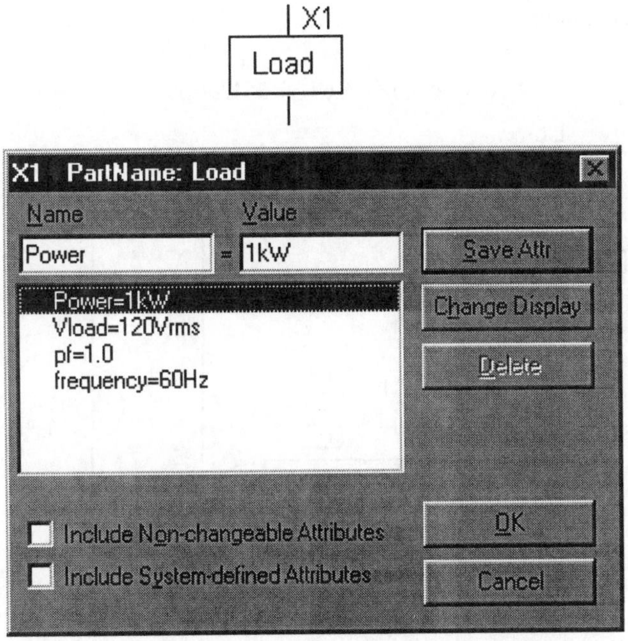

Figure 9.9. The LOAD part and its Attribute box.

We will introduce the LOAD part, by performing a 60 Hz single-frequency simulation on the circuit in Figure 9.10a to find the line current. The *Schematics* diagram is given in Figure 9.10b where the rated voltage of each LOAD part is set to 220 V_{RMS}. From the output file, the simulation results are

```
SIMULATION RESULTS FOR Iline

   Magnitude        Phase
    318.5 A     -31.09 degrees
```

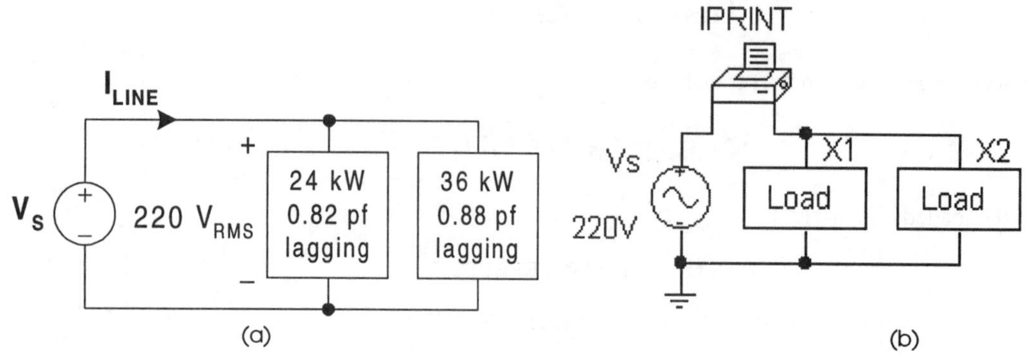

Figure 9.10. Circuits for the third PSpice simulations, (a) the circuit diagram and (b) the *Schematics* diagram.

EWB SIMULATIONS

Let's use EWB to find the maximum power transferred for the circuit in Figure 9.11. The procedure is the same as in PSpice Simulation Two. The EWB circuit for finding the Thevenin equivalent impedance is shown in Figure 9.12. Using a test current source of $1\,\underline{/0^\circ}$ A produces the open circuit voltage plot in Figure 9.13. Thus, the Thevenin impedance is,

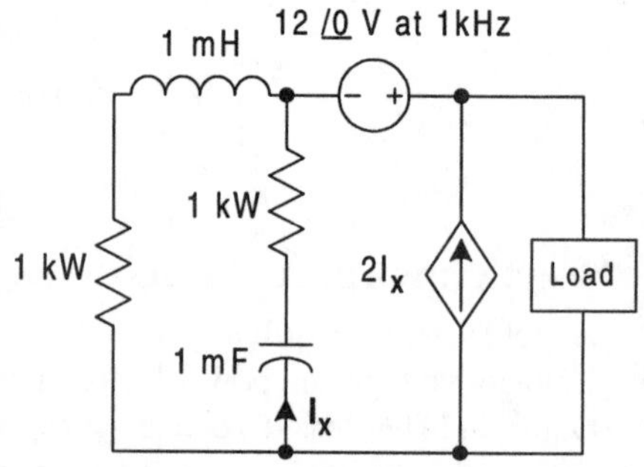

Figure 9.11. Circuit for EWB maximum power transfer simulation.

$$\mathbf{Z_{TH}} \approx 253.0\underline{/-6.67^\circ} = 251.2 - j29.4\ \Omega$$

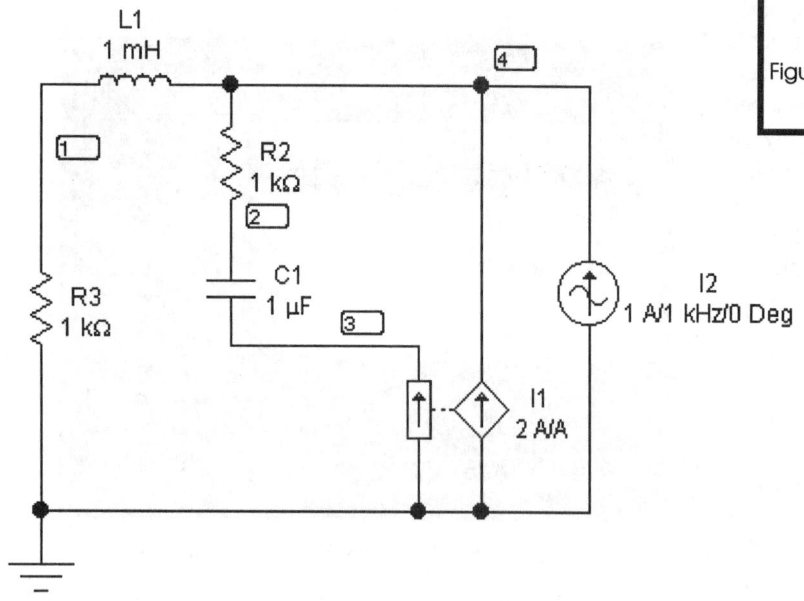

Figure 9.12. EWB diagram for finding Z_{TH}.

Figure 9.14 shows the circuit with a matched load of $\mathbf{Z_L} = \mathbf{Z_{TH}}^* = 251.2 + j29.4\ \Omega$. The inductor value is found using the relationship

$$L = \frac{29.4}{\omega} = \frac{29.4}{2\pi f} = \frac{29.4}{2000\pi} = 4.68 \text{ mH}$$

The dependent current source and resistor in Figure 9.14 make a milli-ammeter. Plotting the voltage at node 8 is in effect plotting the load current in mA! Simulation results are in Figure 9.15, where, at 1 kHz, the load current is 23.88 mA. Now we can calculate the load power to be

$$P_L = \frac{I_L^2 R_L}{2} = \frac{0.02388^2(251.2)}{2} = 71.6 \text{ mW}$$

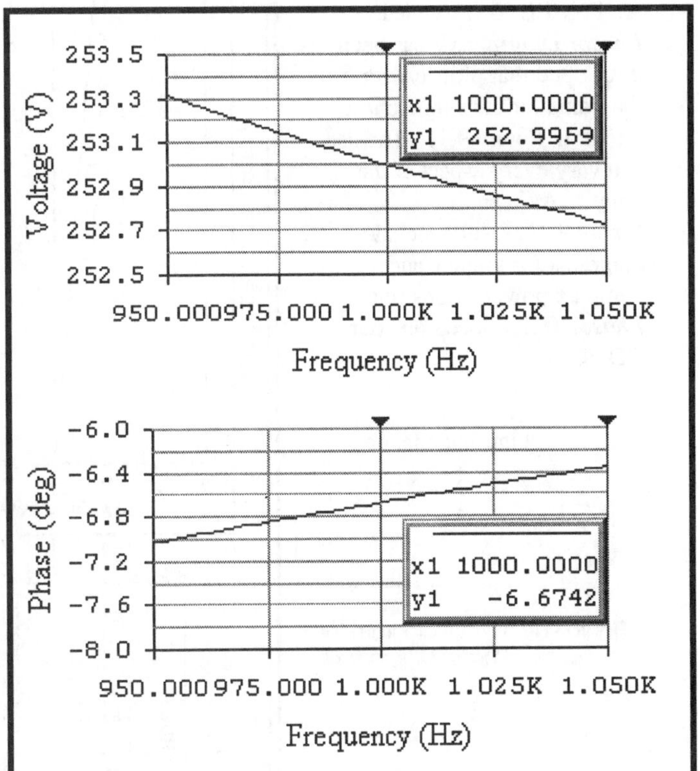

Figure 9.13. EWB simulation results for Z_{TH}.

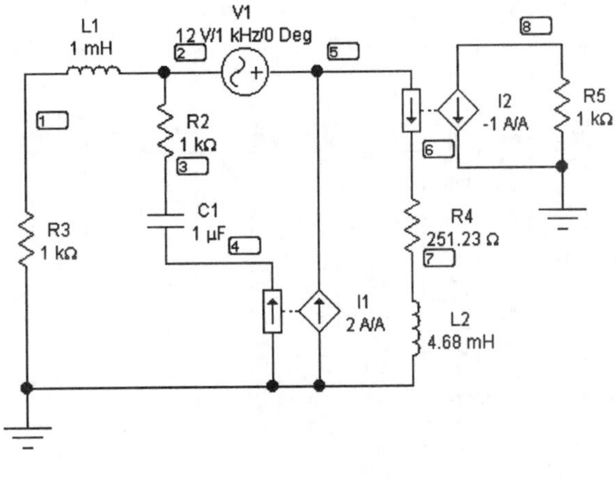

Figure 9.14. EWB circuit for finding the matched load current.

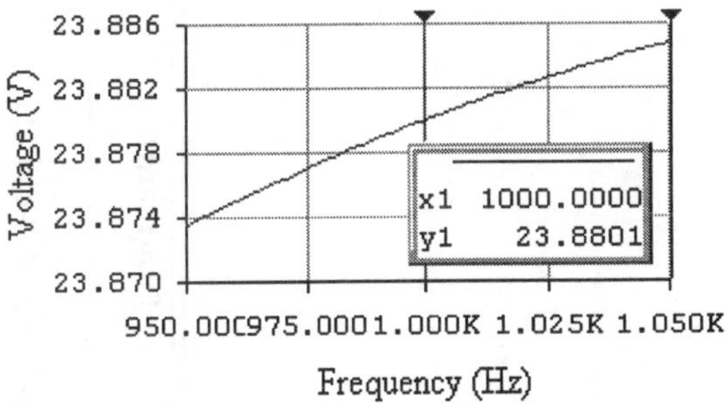

Figure 9.15. EWB results for load current.

EXCEL DEMO

An EXCEL 95 file called *Phasor Diagram.xls* has been developed that plots phasor diagrams for the single-phase circuit in Figure 9.17. The user can vary every aspect of the circuit. As seen in Figure 9.16, results include diagrams for voltage and complex power. Look for *Phasor Diagram.xls* on your CD_ROM.

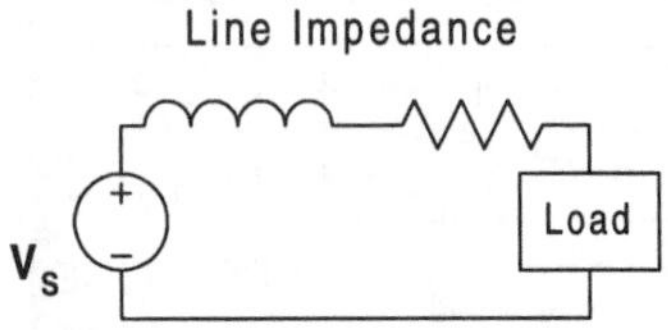

Figure 9.17. Circuit diagram for *Phasor Diagram.xls*

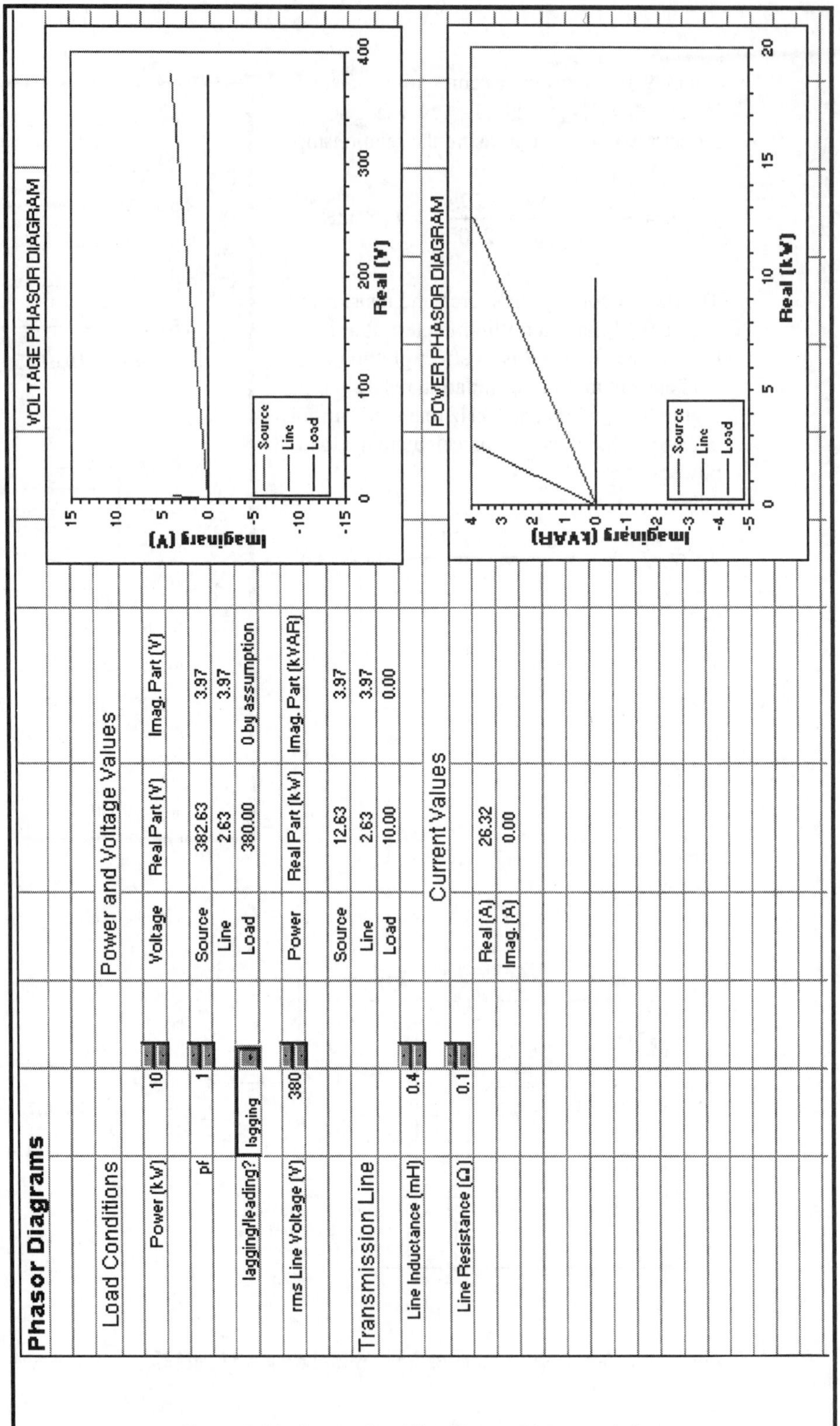

Figure 9.16. A sample of the *Phasor Diagram.xls* file.

PROBLEM SOLVING EXAMPLES

1. At the input of an ac circuit operating at f = 60 Hz, the voltage and current are

$\mathbf{V} = 12\underline{/30^\circ}\text{V}$

$\mathbf{I} = 2\underline{/10^\circ}\text{A}$

Determine the expression for the instantaneous power.

Solution: $v(t) = 12\cos(377t+30^\circ)\text{V}$, $i(t) = 2\cos(377t+10^\circ)\text{A}$, therefore $p(t) = 12\cos(20^\circ) + 12\cos(754t+40^\circ)\text{W}$.

2. Given the information in Problem 1 compute the average power supplied to the circuit.

Solution: $P_{ave} = 12\cos 20^\circ = 11.28\text{W}$.

3. Compute the instantaneous and average power that is supplied to the load $\mathbf{Z}_L$ for each of the circuits below if $v(t) = 10\cos(377t+20^\circ)\text{V}$ and $\mathbf{Z}_L = 40\underline{/45^\circ}\Omega$.

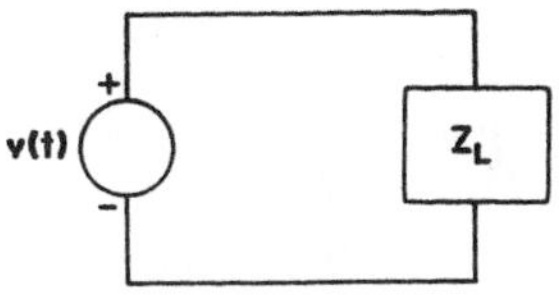

(a)

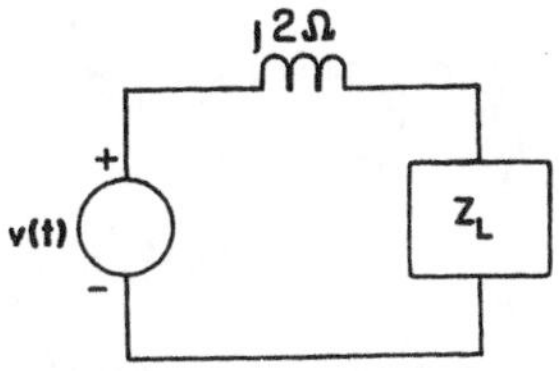

(b)

Solution: (a) $\mathbf{I} = \dfrac{10\underline{/20^\circ}}{40\underline{/45^\circ}} = 0.25\underline{/-25^\circ}\text{A}$, hence

$i(t) = 0.25\cos(\omega t-25^\circ)\text{A}$.

$$p(t) = \frac{V_M I_M}{2}\left[\cos(\theta_v-\theta_i)+\cos(2\omega t+\theta_v+\theta_i)\right]$$

$$= .8839+1.25\cos(754t-5^\circ)\text{W} \qquad P_{ave} = 0.8839\text{W}$$

(b) $\mathbf{I} = \dfrac{10\underline{/20^\circ}}{j2+40\underline{/45^\circ}} = 0.241\underline{/-27^\circ}\text{A}$

$$\mathbf{V}_L = 10\underline{/20^\circ}\left(\frac{40\underline{/45^\circ}}{j2+40\underline{/45^\circ}}\right) = 9.65\underline{/18^\circ}\text{V}$$

Hence

$$p(t) = \frac{(9.65)(0.241)}{2}\left[\cos(18^\circ+27^\circ)\right.$$

$$\left.+\cos(754t+18^\circ-27^\circ)\right]$$

$$= .8224+1.16\cos(754t-9^\circ)\text{W}$$

$P_{ave} = 0.8224\text{W}$

4. Compute the instantaneous and average power that is supplied to the load $\mathbf{Z}_L$ for each of the following circuits if $i(t) = 2\cos(377t+30^\circ)\text{A}$ and $\mathbf{Z}_L = 50\underline{/-20^\circ}\Omega$.

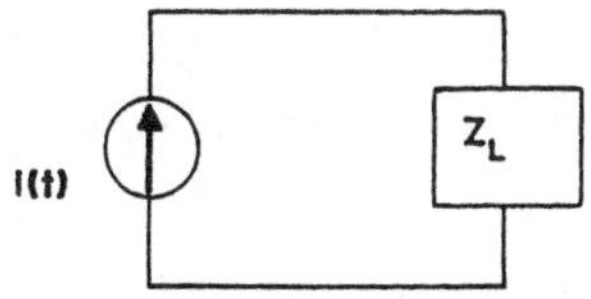

(a)

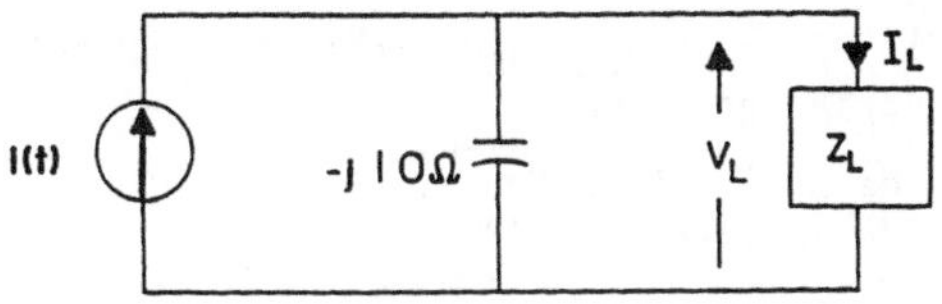

(b)

Solution: (a) $\mathbf{V}_L = (2\underline{/30^\circ})(5\underline{/-20^\circ}) = 100\underline{/10^\circ}\text{V}$

$$p(t) = \frac{V_M I_M}{2}\left[\cos(\theta_v-\theta_i)+\cos(2\omega t+\theta_v+\theta_i)\right]$$

$$= 93.97+100\cos(754t+40^\circ)\text{W}, \quad P_{ave} = 93.97\text{W}$$

$$\text{(b) } \mathbf{V}_L = 2\underline{/30^\circ}\left[\frac{(50\underline{/-20^\circ})(10\underline{/-90^\circ})}{50\underline{/-20^\circ}+10\underline{/-90^\circ}}\right] = 18.44\underline{/-50^\circ}\text{V}$$

$$\mathbf{I}_L = 2\underline{/30^\circ}\left[\frac{-j10}{-j10+50\underline{/-20^\circ}}\right] = .3687\underline{/-30^\circ}\text{A}$$

$$p(t) = \frac{V_M I_M}{2}\left[\cos(\theta_v-\theta_i)+\cos(2\omega t+\theta_v+\theta_i)\right]$$

$$= 3.19+3.4\cos(754t-80^\circ)\text{W} \qquad P_{ave} = 3.19\text{W}$$

5. Determine the average power absorbed by the resistor in the following circuit.

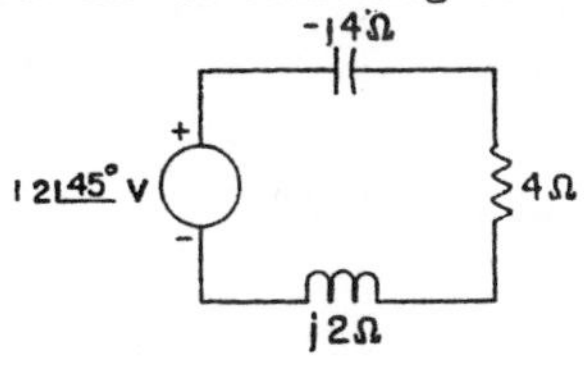

Solution: $I = \dfrac{12\angle 45^\circ}{4-j2} = 2.68\angle 71.57^\circ A$

$P = \frac{1}{2} I_M^2 R = \frac{1}{2}(2.68)^2(4) = 14.36W$

6. Determine the average power supplied to the following network.

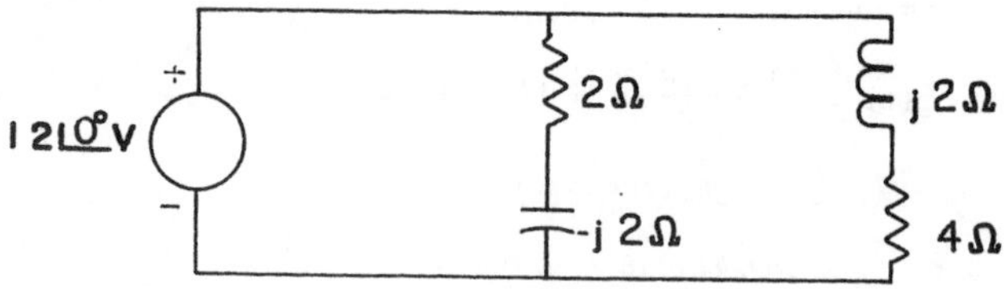

Solution: $I_{RC} = \dfrac{12\angle 0^\circ}{2-j2} = 4.24\angle 45^\circ A$

$I_{RL} = \dfrac{12\angle 0^\circ}{4+j2} = 2.68\angle -26.57^\circ A$, $P_{2\Omega} = \frac{1}{2}(4.24)^2(2) =$ 18W. $P_{4\Omega} = \frac{1}{2}(2.68)^2(4) = 14.4W$. Since power supplied is equal to the power absorbed $P_s =$ $18+14.4 = 32.4W$.

7. Determine the expression for the instantaneous power supplied to the network in Problem 6.

Solution: $I_T = I_{RL}+I_{RC} = 4.24\angle 45^\circ + 2.68\angle -26.57^\circ$ $= 5.69\angle 18.43^\circ$. Then $P_{sup} =$ $(12 \cos \omega t)[5.69 \cos(\omega t+18.43^\circ)] =$ $32.4+32.4 \cos(2\omega t+18.43^\circ)W$

8. Determine the average power absorbed by the 2Ω resistor in the circuit below.

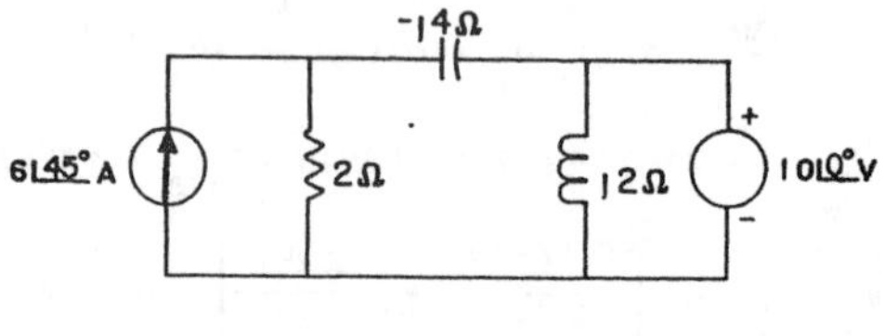

Solution: The voltage across the 2Ω resistor can be found using a node equation.

$$\frac{V}{2} + \frac{V-10\angle 0^\circ}{-j4} = 6\angle 45^\circ \text{ and hence } V = 14.21\angle 31.27^\circ V.$$

Therefore $I_{2\Omega} = 7.11\angle 31.27^\circ A$ and $P_R =$ $\frac{1}{2}(7.11)^2(2) = 50.55W$.

9. Find the power absorbed by the following network.

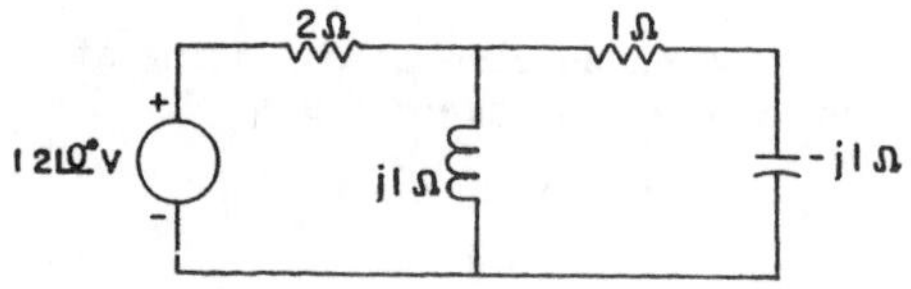

Solution: The clockwise current in the 2Ω resistor is $I_1 = \dfrac{12\angle 0^\circ}{2+j1||(1-j1)} = 3.80\angle -18.43^\circ A$.

The clockwise current in the 1Ω resistor is

$$I_2 = \frac{\frac{12\angle 0^\circ}{3+j1}(j1)}{1+j1-j1} = 3.80\angle 71.57^\circ A .$$

Therefore

$$P_{ABS} = \frac{1}{2}(3.80)^2(2) + \frac{1}{2}(3.80)^2(1) = 21.63W$$

$$P_{sup} = \frac{1}{2}(12)(3.80)\cos[0-(-18.43^\circ)] = 21.63W$$

10. Find the total power absorbed in the circuit below.

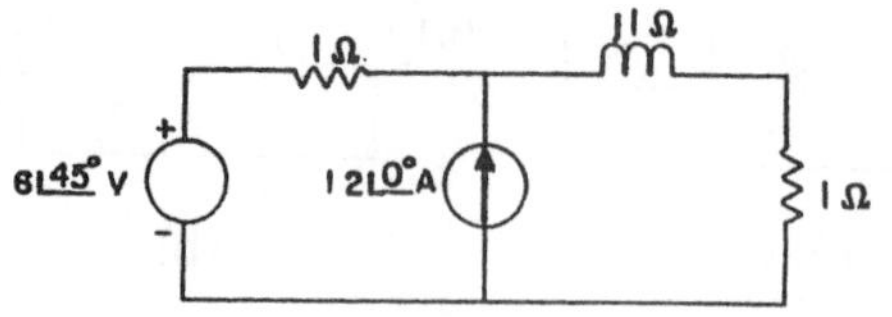

Solution: Consider the following diagram

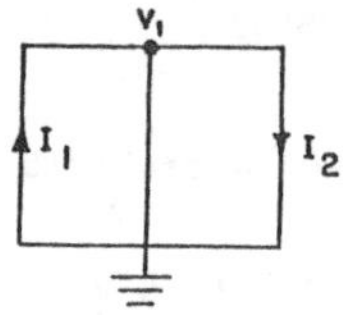

The node equation for V_1 is

$$\frac{V_1-6\angle 45^\circ}{1} + \frac{V_1}{1+j1} = 12\angle 0^\circ. \text{Hence } V_1 = 10.62\angle 33.06^\circ V.$$

Hence

$$I_1 = \frac{6\angle 45^\circ - 10.62\angle 33.06^\circ}{1} = 4.91\angle -161.6^\circ A$$

$$I_2 = \frac{10.62\angle 33.06^\circ}{1.41\angle 45^\circ} = 7.53\angle -11.94^\circ A$$

$$P_R = \frac{1}{2}(4.91)^2(1) + \frac{1}{2}(7.53)^2(1) = 40.41W \text{ ABS}$$

$P_{\underline{6/45°}} = (-\frac{1}{2})(6)(4.91)\cos[45° - (-161.6°)]$

$= 13.11W$ ABS

$P_{\underline{12/0°}} = (-\frac{1}{2})(10.62)(12)\cos(33.06° - 0°)$

$= -53.52W$ ABS

Therefore

Total $P_{ABS} = 40.41 + 13.11 = 53.52W$

Total $P_{sup} = -53.52W$

11. Calculate the value of Z_L for maximum average power transfer in the following circuit.

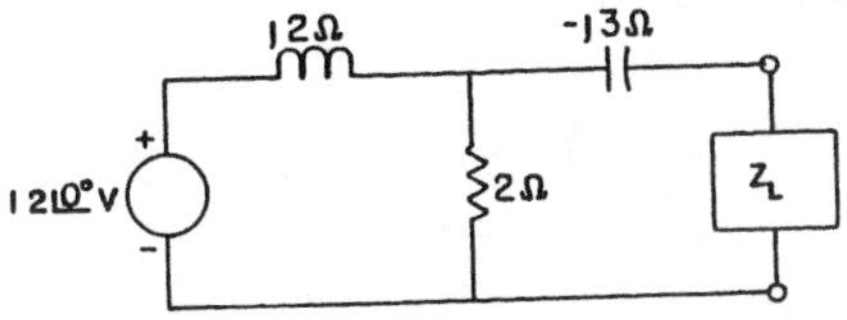

Solution: Determining the Thevenin equivalent impedance at the load we obtain $Z_{TH} = \frac{2(j2)}{2+j2} - j3 = 1.00 - j2\Omega$. Therefore $Z_L = 1.00 + j2\Omega$.

12. What is the value of the maximum average power that can be transferred to the load in the circuit below.

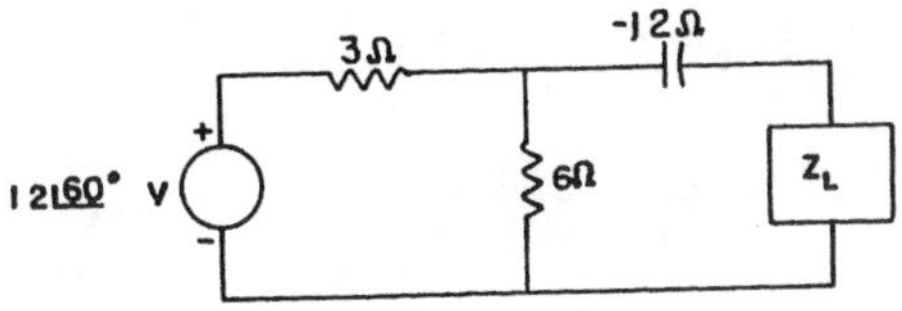

Solution: Forming a Thevenin equivalent we obtain

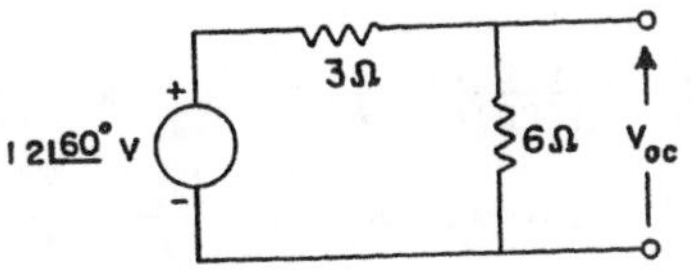

$V_{oc} = 8\underline{/60°}V$

$Z_{TH} = 2\Omega$

Therefore the network reduces to

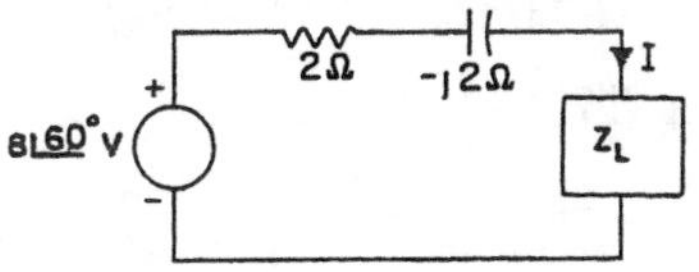

$Z_L = 2 + j2\Omega$

$I = \frac{8/60°}{2+2} = 2\underline{/60°}A$

and $P_L = \frac{1}{2}(I_M)^2 R_L = \frac{1}{2}(2)^2(2) = 4W$.

13. Determine the value of Z_L for maximum power transfer in the following circuits.

(a)

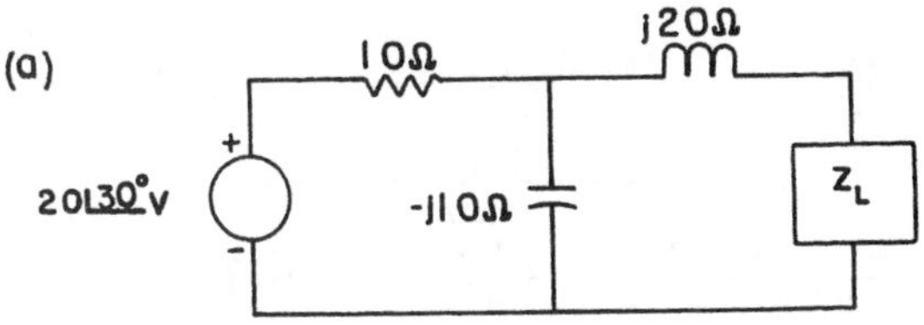

(b)

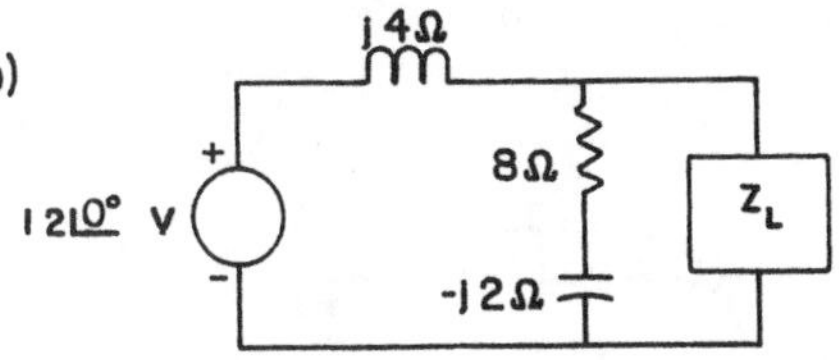

Solution:

(a) $Z_{TH} = (10||-j10) + j20 = 15.81\underline{/71.6°}\Omega$

Therefore $Z_L = Z^*_{TH} = 15.81\underline{/-71.6°}\Omega$.

(b) $Z_{TH} = (8 - j2)||j4 = 4\underline{/61.9°}\Omega$

Hence $Z_L = Z^*_{TH} = 4\underline{/-61.9°}\Omega$.

14. Determine the impedance Z_L for maximum average power transfer and the value of the maximum power transferred for the circuit below.

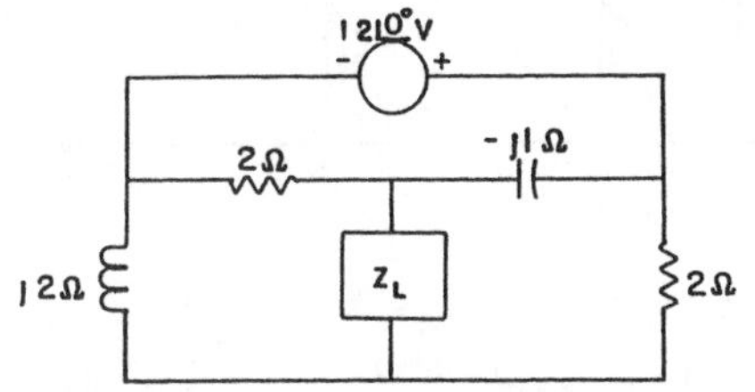

Solution: The following network can be used to determine the open circuit voltage.

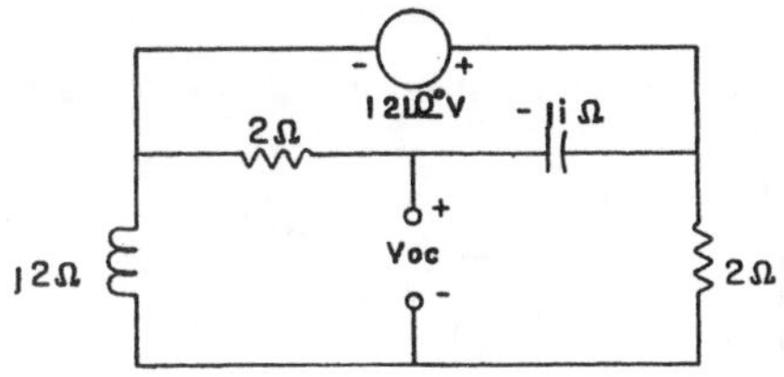

$V_{oc} = \frac{12/0°}{2-j1}(2) - \frac{12/0°}{2+j2}(j2) = 3.78\underline{/-18.67°}V$

Z_{TH} is computed as follows

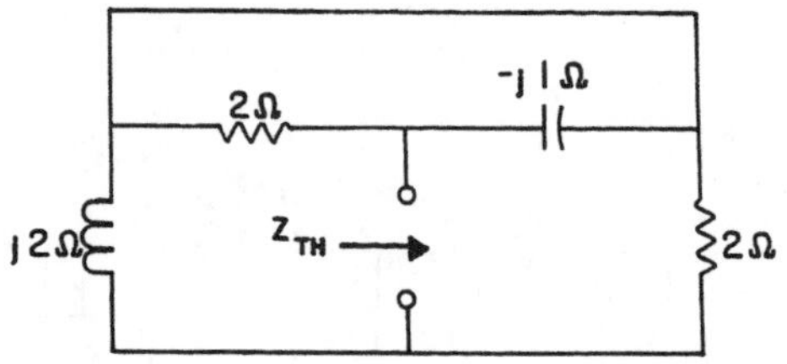

$$Z_{TH} = \frac{(2)(-j1)}{2-j1} + \frac{(2)(j2)}{2+j2} = 1.40+j0.2\Omega$$

Hence $Z_L = Z_{TH}^* = 1.40-j0.2\Omega$.

Therefore

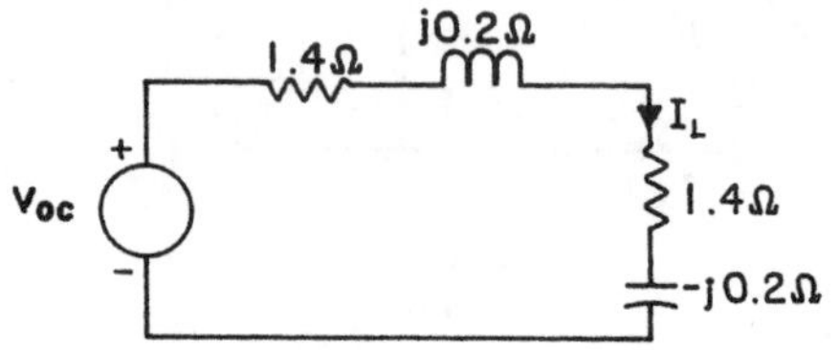

$$I_L = \frac{3.78\underline{/-18.67^\circ}}{2.8} = 1.35\underline{/-18.67^\circ}\text{A}$$

Hence

$$P_L = \frac{1}{2}(1.35)^2(1.4) = 1.28\text{W} .$$

15. Compute the rms value of the current i(t) given by the waveform shown below.

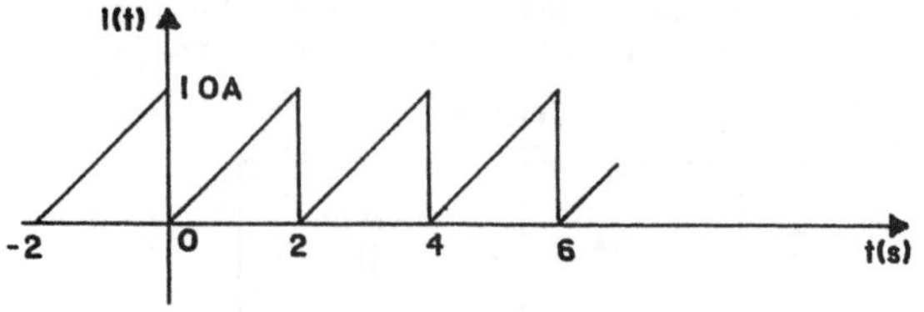

Solution:

$$I_{rms} = \left[\frac{1}{2}\int_0^2 (5t)^2 dt\right]^{1/2} = \left[\frac{1}{2}\left(\frac{25t^3}{3}\right)\Big|_0^2\right]^{1/2}$$

$$= \frac{10\sqrt{3}}{3}\ \text{A}$$

16. Calculate the rms value of the waveform shown below.

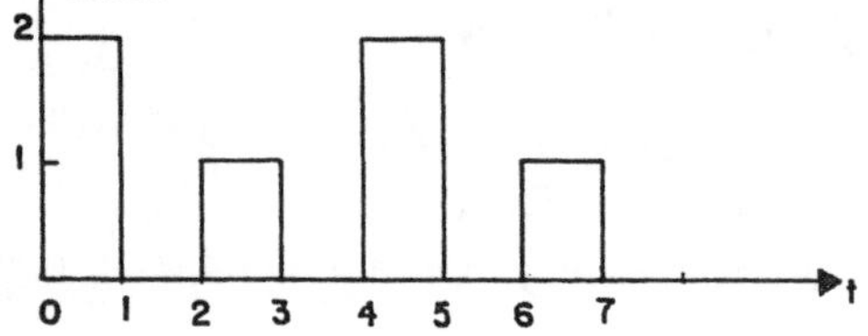

Solution:

$$V_{rms} = \left[\frac{1}{4}\left(\int_0^1 2^2 dt + \int_1^2 0^2 dt + \int_2^3 1^2 dt\right)\right]^{1/2}$$

$$= \frac{5}{4}\ \text{V}$$

17. Given the following information, determine the power factor of the load shown below.

(a) $I_{rms} = 6\text{A}$, $V_{rms} = 120\text{V}$

$P_L = 600\text{W}$, $\theta_Z<0$

(b) $I_{rms} = 5\underline{/60^\circ}\text{A}$, $V_{rms} = 100\underline{/45^\circ}\text{V}$

(c) $i(t) = 1.414\cos(\omega t+10^\circ)\text{A}$

$v(t) = 99\cos(\omega t+20^\circ)\text{V}$

(d) $I = 4.5\underline{/30^\circ}\text{A}_{rms}$

$Z_L = 330\underline{/50^\circ}\Omega$

(e) $P_L = 2000\text{W}$, $V_{rms} = 500\text{V}$

$Re(Z_L) = 80\Omega$ $\theta_Z>0$

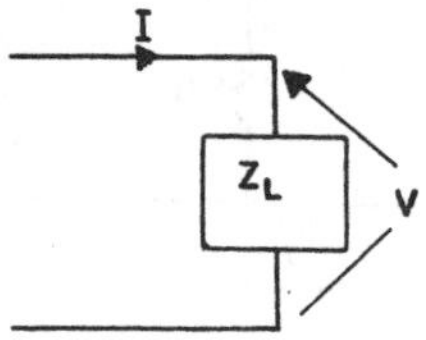

Solution:

(a) $PF = \frac{600}{(6)(120)} = 0.833$ leading

(b) $PF = \cos(\theta_v-\theta_i) = \cos(45^\circ-60^\circ) = 0.966$ leading

(c) $PF = \cos(\theta_v-\theta_i) = \cos(20^\circ-10^\circ) = 0.985$ lagging

(d) $PF = \cos(50^\circ) = 0.643$ lagging

(e) $P_L = I_M^2 R$. Hence $I_M = \sqrt{\frac{P_L}{R}} = 5\text{A rms}$

$$PF = \frac{P_L}{V_M I_M} = \frac{2000}{(5)(500)} = 0.8 \text{ lagging}$$

18. An industrial load operates at 40KW, 0.77PF lagging. The load voltage is $200\underline{/0^\circ}$ V rms at 60Hz. The line impedance is $0.1+j0.2\Omega$. Determine the magnitude of the input voltage to the line and the PF of the input source.

Solution: $I_L = \frac{(40)(10^3)}{(200)(0.77)} = 259.74\text{A}$

$\theta = \cos^{-1}0.77 = 39.65^\circ$. Hence $I_L = 259.74\underline{/-39.65^\circ}\text{A}$.

Hence

$$V_{Input} = 200\underline{/0^\circ}+(259.74\underline{/-39.65^\circ})(0.1+j0.2)$$

$$= 254.23\underline{/5.287^\circ}\text{V}$$

V_{rms} at Input = 254.23V

$\theta_v-\theta_i = 5.287^\circ-(-39.65^\circ) = 44.94^\circ$

$PF_{Input} = \cos 44.94^\circ = 0.708$ lagging

Chapter 10 POLYPHASE CIRCUITS

For some reason, three-phase circuit analysis gives some students a lot of trouble. Personally, I think the jargon is largely to blame. To avoid confusion, it is good to have a few simple definitions in hand. It is even better if those definitions do not depend the topology - delta vs. wye. The power company has this perspective. They are concerned with power delivery to customers. Whether a motor is wye or delta connected is of little interest. It is not surprising then that the terms used to define power delivery, listed in Table 10.1 are topology independent. Notice that these terms actually define the power line - its voltage, current and impedance. The terms in Table 10.1 are well worth memorizing as they serve as a point of reference for other three phase terminology.

Table 10.1

Topology-independent Terms in 3-θ Jargon

Term	Description	BECA Equivalent
Line impedance	Resistance and inductance of the actual power lines	$\mathbf{Z}_{line}$
Line current	Current in the actual power lines	**IaA**, etc.
Line voltage	The line-to-line voltage	$\mathbf{V}_{AB}$, etc.
Line-to-Neutral Voltage	The voltage from a power line to the neutral	$\mathbf{V}_{AN}$, etc.

When discussing the voltages and currents in a particular load or source, the topology affects the meaning of the terms we use. In particular, the terms *phase current* and *phase voltage* are completely dependent on connection scheme. Do not confuse them with the terms in Table 10.1 - they are not synonymous. However, if we look at these two terms from the perspective of the load rather than the line, we find they are similar. The phase voltage is the voltage across any one of the three load components, while the phase current is the current through a load component. This is shown in Figure 10.1 for both delta and wye connections.

Notice that the line current (the actual transmission line current), is the same as the wye-connected phase current. As a result, the line impedances are in series with the load components. This is the motivation for converting balanced loads and sources to a wye-wye format. It allows us to combine source, line and load particulars into a single-phase analysis whose solution yields the currents in the physical transmission lines. So, think wye-wye.

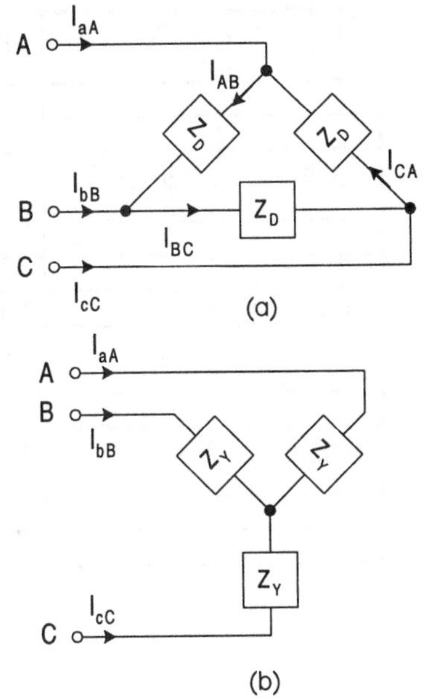

Figure 10.1. Generic (a) delta and (b) wye load connections.

SIMULATING THREE-PHASE CIRCUITS IN PSPICE

Many power system loads are specified in rated power, voltage and power factor rather than resistance and inductance. This makes circuit simulators like PSpice and EWB less than optimum for power system simulations. In fact, we will not include any EWB simulations in this chapter. Another drawback is that drawing balanced three-phase loads in *Schematics* is a bit redundant since the only difference between phases is a 120° phase shift. To speed things along, balanced loads and sources for both wye and delta configurations have been created and included in the BECA library. Names and attributes of the parts are discussed in the next section.

In this chapter, we will perform four simulations. In simulation #1, the loads are given in resistance and inductance. In simulation #2, we will make use of the custom parts mentioned above. In simulation #3, we will investigate shortcomings of our custom loads. Finally, in simulation #4, we will demonstrate how to find the required capacitance values for power factor correction applications.

THREE-PHASE PARTS IN THE BECA LIBRARY

As mentioned earlier, six balanced 3-θ parts have been created for this study manual. They are shown in Table 10.2 and included in the BECA library for your use. While the editable attributes of each part are listed in Table 10.2, detailed descriptions of each part are in the READMES folder on the CE-ROM. One point of interest regarding the loads, the part attributes are used to determine effective resistance and inductance/capacitance of the load. If the actual load voltage and the rated voltage attribute are not equal, errors will result. This will be discussed in simulation #3.

Table 10.2

Parts List And Attributes For The BECA Library 3-Q Parts

Part Name	Function	Attributes
V_delta	Delta connect. Source	Vab = Line-to-line voltage
		Phaseab = phase of V_{ab} in degrees
		Sequence = +1 for abc; -1 for cba
		Freq = Source frequency in Hz
V_wye	Wye connect. Source	Van = Line-to-neutral voltage
		Phasean = phase of V_{an} in degrees
		Sequence = +1 for abc; -1 for cba
		Freq = Source frequency in Hz
Delta_Load _Inductive	Lagging Δ Load	Vab = Line to line voltage
		VA = Total load volt-ampere rating
		pf = Power factor
		frequency = Source frequency
Delta_Load _Capacitive	Leading Δ Load	Vab = Line to line voltage
		VA = Total load volt-ampere rating
		pf = Power factor
		frequency = Source frequency
Wye_Load _Inductive	Lagging Y Load	Van = Line to neutral voltage
		VA = Total load volt-ampere rating
		pf = Power factor
		frequency = Source frequency
Wye_Load _Capacitive	Leading Y Load	Van = Line to neutral voltage
		VA = Total load volt-ampere rating
		pf = Power factor
		frequency = Source frequency

PSPICE SIMULATIONS

Simulation One

A particular 60 Hz balanced delta-wye system has a 220-V_{RMS} abc sequence source with $\underline{/\ V_{ab}} = 20°$. The load impedance per phase is 8 + j6 Ω and the line impedance is 0.4 + j0.8 Ω. Let's use PSpice to find the line currents and phase voltages at the load in phasor form. The circuit in Figure 10.2 will do nicely. We have used the V_delta part in the BECA library for the balanced source, and, while the plethora of VPRINT1 and IPRINT parts is visually stunning, there are only three of each - one for each phase. Also, each VPRINT1 and IPRINT part have been set to acquire magnitude and phase data.

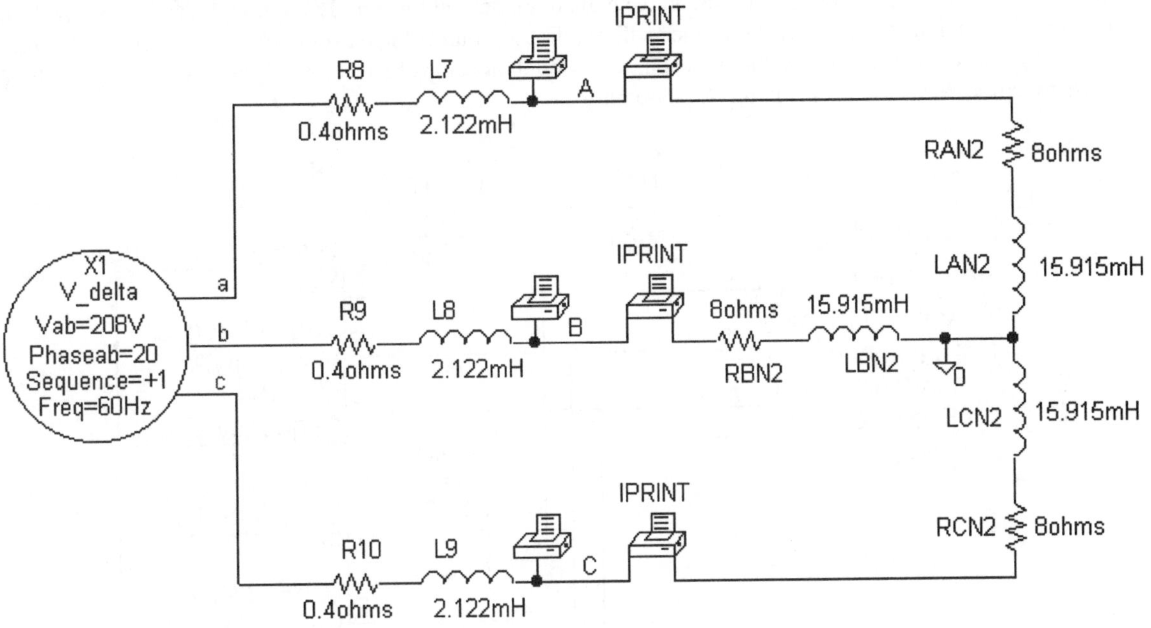

Figure 10.2. The *Schematics* diagram for Simulation One employing the V_delta source and VPRINT1 / IPRINT data gathering parts.

The simulation results are listed below.

```
                    ****      AC ANALYSIS    ****

  PHASE                                       LINE
VOLTAGES  MAGNITUDE(RMS)    PHASE          CURRENTS   MAGNITUDE(RMS)      PHASE

   Van       1.111E+02   -1.212E+01          IaA       1.111E+01      -4.899E+01

   Vbn       1.111E+02   -1.321E+02          IbB       1.111E+01      -1.690E+02

   Vcn       1.111E+02    1.079E+02          IcC       1.111E+01       7.101E+01
```

Note that the phase voltages (line-to-neutral for the wye-load) are equal in magnitude and 120° out of phase - just they ought. The same can be said of the line currents.

Simulation Two

Two industrial plants receive their power from the same substation where the line voltage is 4.6 kV_{RMS} and the phase sequence is *abc*. Plant 1 is rated at 300 kVA and pf = 0.8 lagging, while plant 2 is rated at 350 kVA at 0.64 lagging. The source and loads are balanced. Let's find the line currents at the source and into each plant.

Since no information is given regarding delta or wye connections, we will use a delta-delta connection. In this way, we do not have to convert line voltage to line-to-neutral voltage. Furthermore, as seen in Figure 10.3, we will use the V_delta and Delta_Load_Inductive parts to model the system. Even though the delta parts in the diagram have no neutral connection, PSpice still requires a grounded node somewhere. We arbitrarily chose to put it at phase C. Also, since we know the phase angle relationships between the phases, we will use IPRINT parts to extract the line currents in the A phase only.

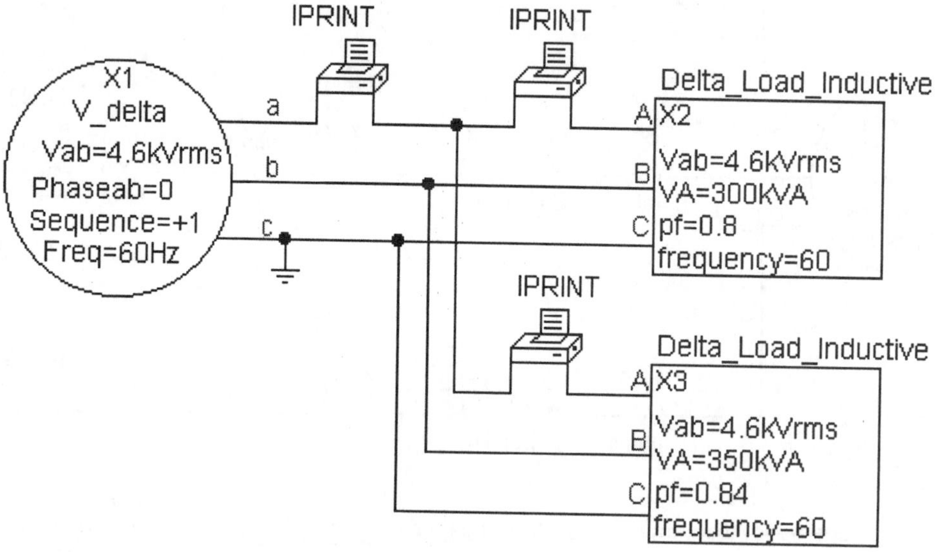

Figure 10.3. *Schematics* circuit for Simulation Two using the delta load and source parts in the BECA library.

The simulation results are listed below. We can check the simulation results by add the currents in the loads: it should equal the source current.

$$I_{LOAD1} + I_{LOAD2} = 37.65\underline{/-66.87^\circ} + 43.93\underline{/-62.83^\circ} = 14.79 + 20.03 - j(34.62 + 39.09) = 81.53\underline{/-64.71^\circ} \quad Q.E.D.$$

**** AC ANALYSIS ****

PHASE A LINE CURRENTS	MAGNITUDE (ARMS)	PHASE (DEGREES)
Source	81.53	-64.71
Load 1	37.65	-66.87
Load 2	43.93	-62.86

Simulation Three

As mentioned earlier, the attributes of a BECA 3-θ load are used to calculate equivalent load resistance and inductance / capacitance values which are sent to PSpice as circuit elements. Thus, the circuit PSpice eventually solves looks very similar to Figure 10.2. There is a problem however when the actual load voltage is not the same as the attribute value.

As an example, consider a 60-Hz, balanced wye-wye connected system where the source voltage is known to be 208V_{RMS} and the line resistance and inductance of 0.1 Ω and 1 mH respectively. Also, the load is *supposed* to operate at 208 V and 18 kVA at 0.75 power factor lagging. We know that with line impedance included in the simulation, it is impossible for the both the source and load voltages to be 208 V_{RMS}. We will assume that the rated power and power factor are maintained even though the load voltage will be less than 208 V_{RMS}. To obtain accurate simulations, we must find the magic value of the load part attribute, Van, such that the source voltage and load power/power factor are satisfied. We can do this in PSpice very easily.

We will sweep the Van attribute and plot the load complex power. With the cursors, we can find the required Van value. Using that value in a new simulation, we can obtain accurate simulations. The appropriate *Schematics* diagram is shown in Figure 10.4 where the PARAM part creates the variable, Vsweep, and the wye load attribute, Van, has been set to {Vsweep}. The braces inform PSpice that the value of Van is a function rather than a number. The required Parametric setup is shown in Figure 10.5. After simulating and PROBE opens, we want to plot the load power. Select Add from the Traces menu to open the window in Figure 10.6. Notice the Trace expression at the bottom of the figure. It is the total complex power at the load. The resulting PROBE plot is in Figure 10.7 with the Vsweep value required for operation at 18 kVA - namely, 198.65 V_{RMS}. Returning to the *Schematics* diagram, we change Van from {Vsweep} to 198.65V_{RMS} and disable the Parametric sweep. From the resulting output file, the line current and line-to-neutral voltage are

**** AC ANALYSIS ****

Simulated Quantity	Magnitude (rms)	Phase (degrees)
Line Current	30.21 A	-38.88
Line to Neutral Voltage	198.6 V	-2.010

And, of course, these values yield a total complex power of

$$S_{3\phi} = 3V_{AN}I_{aA} = 3*30.30*198.6 = 18.00 \text{ kVA}$$

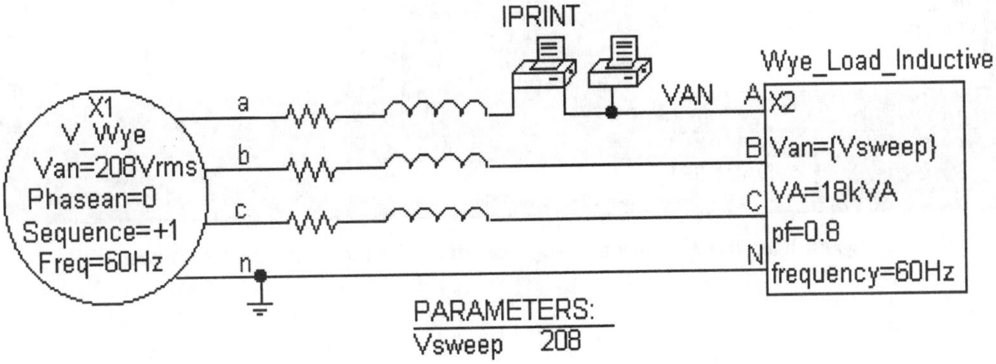

Figure 10.4. The Schematics diagram for Simulation Three.

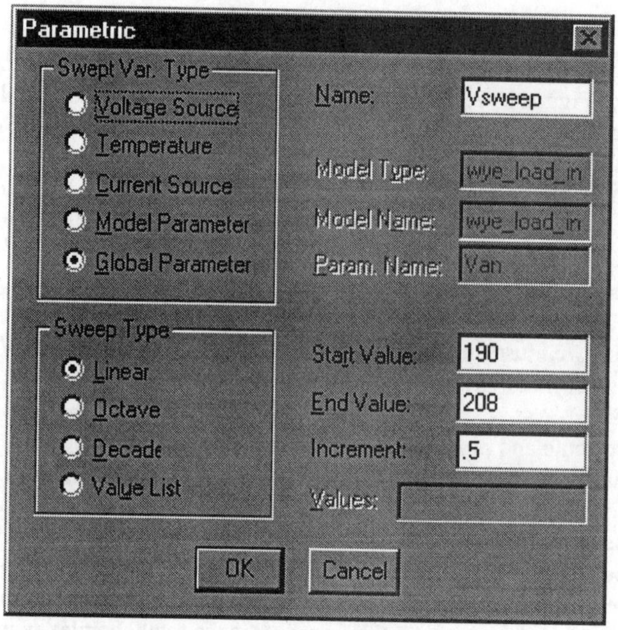

Figure 10.5. The Parametric Setup box edited to vary Vsweep between 190 and 208 V.

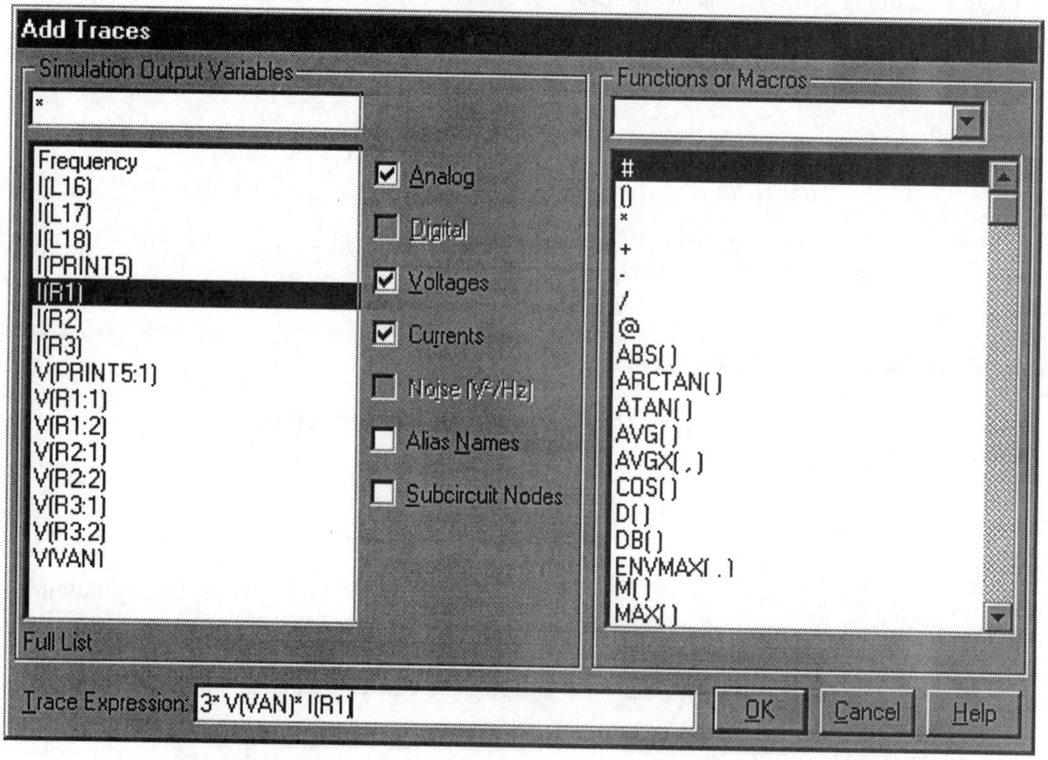

Figure 10.6. The Add Trace dialog box set to plot the total complex power.

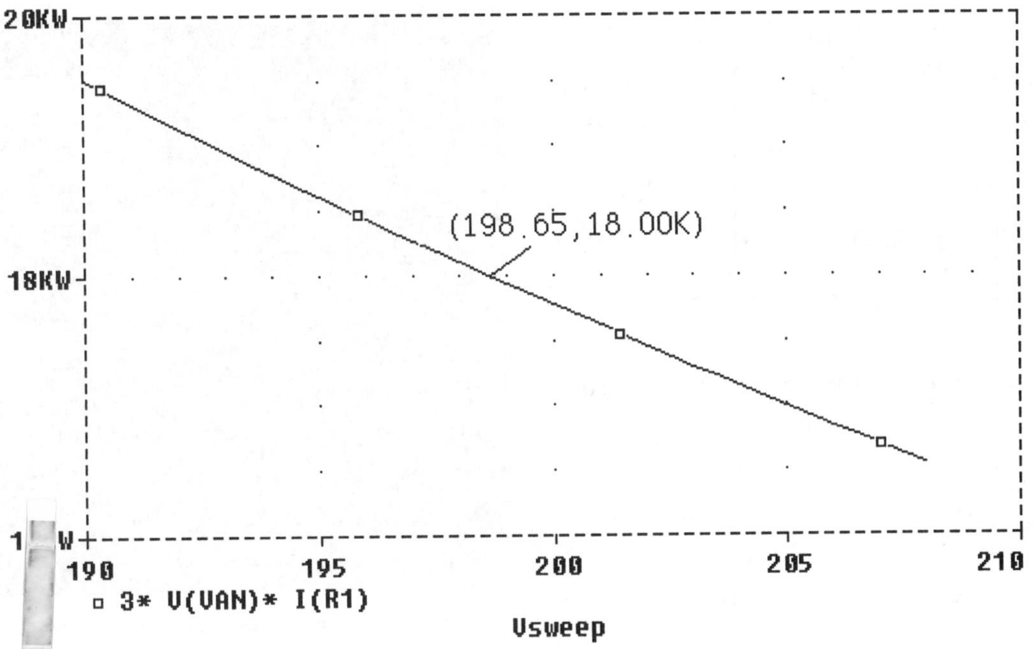

ure 10.7. PROBE plot showing that when Vsweep = 198.65 V_{RMS}, the load complex power is 18 kVA.

Simul ion Four

As me ioned in the BECA text, power factor correction is a real concern of the power company. The inductive loads esented by industries (lots of motors) require the utility to generate more energy to supply the same powe) those industries. To improve the situation, customers with poor power factors pay a higher \$/kW-hour rate. the rate increase is large enough, the customer will purchase power factor correction capacitors.

In thi imulation, we will show how PSPICE can be used to solve power factor correction problems. Consider a wy -vye, 60 Hz, balanced 3-θ system with a line-to-neutral voltage of 440 V_{RMS}, a line impedance consisting of 5 Ω resistance and 10 mH inductance and an equivalent load of 12 Ω and 100 mH. A power factor between 0.95 and 1.0 lagging is desired. What is the range of the power factor correction capacitor value? The single-phase equivalent *Schematics* diagram is shown in Figure 10.9 where a power factor correction cap have been added. Since we wish to plot the power factor versus the capacitor value, the power factor correction capacitor value is set equal to the variable defined in the PARAM part as Cpf. The Parametric setup is shown in Figure 10.8 and, of course, a 60- Hz AC Sweep analysis is requested. After simulation, to plot the power factor in PROBE, we enter the following Trace Expression in the Add Traces window.

We should point out that in PROBE signal phases are in degrees but trigonometric function arguments are in radians, which explains the π/180 factor.

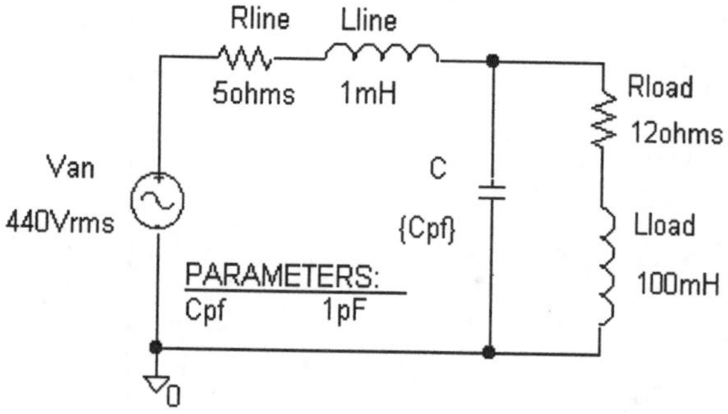

Figure 10.9. *Schematics* diagram for Simulation Four.

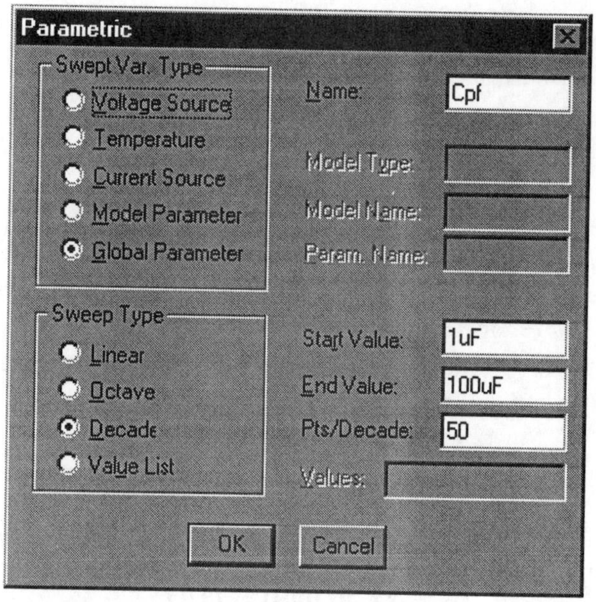

Figure 10.8. Parametric setup to sweep Cpf.

From the simulation results in Figure 10.10, we see that the power factor condition is satisfied for a capacitor value between 57 μF and 63.1 μF. Remember that the simulation is for a single phase. Three capacitors are required for the actual system.

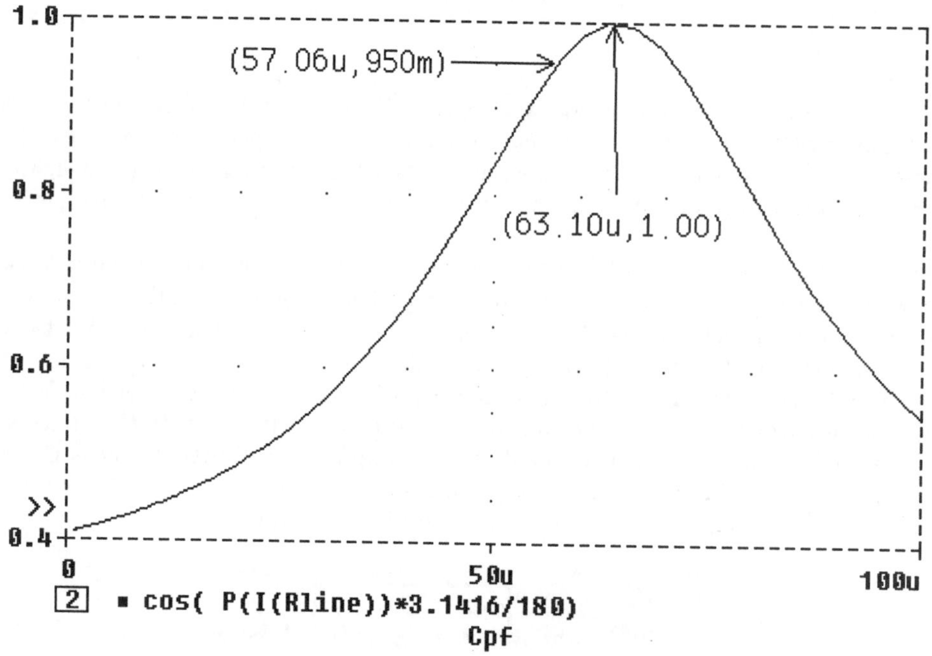

Figure 10.10. Simulation results for Simulation Four.

PROBLEM SOLVING EXAMPLES

1. In the circuit shown below, will a current be present in the neutral line? If so, why? If not, why not?

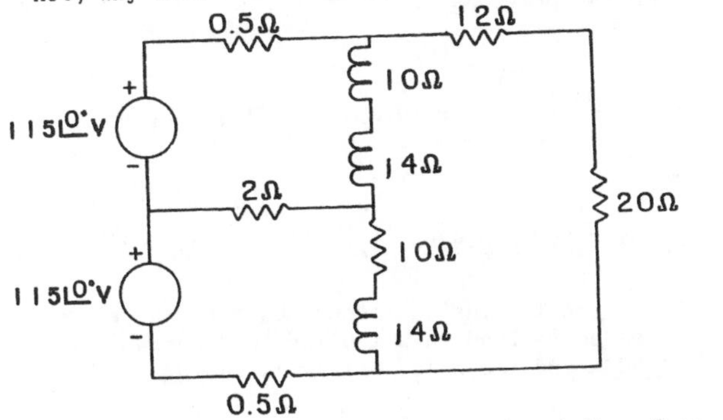

Solution: The circuit is balanced and therefore $I_n = 0$.

2. For a balanced three-phase wye-wye connection with $Z_Y = 50\underline{/20°}\Omega$ determine the phase voltages, line voltages, line currents, and phase currents if $V_{an} = 120\underline{/0°}$ V. Assume positive phase sequence.

Solution:

Phase Voltage	Line Voltage
$V_{an} = 120\underline{/0°}$V	$V_{ab} = 120\sqrt{3}\underline{/30°}$ V
$V_{bn} = 120\underline{/-120°}$ V	$V_{bc} = 208\underline{/-90°}$ V
$V_{cn} = 120\underline{/-240°}$V	$V_{ca} = 208\underline{/-210°}$V

Line Current	Phase Current
$I_a = \frac{V_{an}}{Z_Y} = 2.4\underline{/-20°}$ A	$2.4\underline{/-20°}$ A
	$2.4\underline{/-140°}$ A
$I_b = 2.4\underline{/-140°}$A	$2.4\underline{/-260°}$ A
$I_c = 2.4\underline{/-260°}$A	

3. A positive-sequence three-phase balanced wye voltage source has a line voltage of $V_{bc} = 200\underline{/-30°}$ V. Determine the phase voltages of the source.

Solution: The magnitude of the phase voltage is $\frac{200}{\sqrt{3}}$V. The phase relationships are specified in fig. 12.10. Hence $V_{bn} = \frac{200}{\sqrt{3}}\underline{/-60°}$V, $V_{an} = \frac{200}{\sqrt{3}}\underline{/60°}$V and $V_{cn} = \frac{200}{\sqrt{3}}\underline{/-180°}$V.

4. A positive-sequence balanced three-phase wye-connected source supplies power to a balanced wye-connected load. The line voltage is 100 V. The line impedance is 2Ω per phase, and the load impedance is 50+j30Ω per phase. Determine the load voltages if $\underline{/V_{an}} = 0°$.

Solution: The circuit diagram for the a-phase is

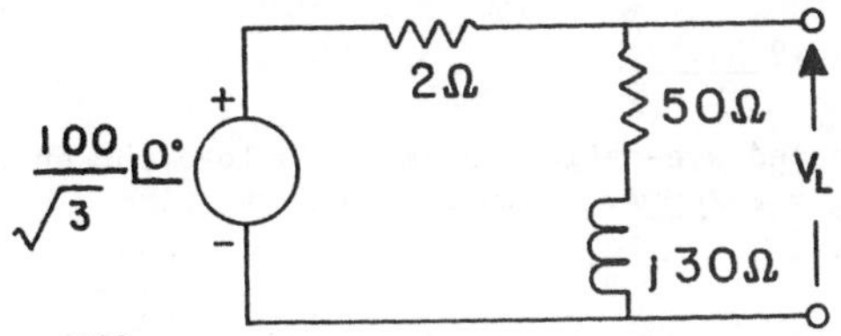

$$I_a = \frac{\frac{100}{\sqrt{3}}\underline{/0°}}{52+j30} = 0.96\underline{/-30°}\text{A}$$

$$V_{La} = (0.96\underline{/-30°})(50+j30) = 55.98\underline{/0.98°}\text{V}$$

$V_{Lb} = 55.98\underline{/-119.02°}$V and $V_{Lc} = 55.98\underline{/-239.02°}$V

5. A balanced three-phase system has a balanced wye load in parallel with a balanced delta load. If the source is a balanced wye connection with a positive phase sequence and the phase voltage $V_{an} = 200\underline{/30°}$V, determine the line currents when the load impedances, $Z_\Delta = 50+j70\Omega$ and $Z_Y = 15+j30\Omega$, are converted to an equivalent delta.

Solution: $Z_Y \rightarrow Z_\Delta = 3(15+j30) = 45+j90\Omega$

The total load impedance is

$$Z_\Delta^1 = (45+j90)||(50+j70) = 46.5\underline{/58.6°}\Omega$$

$$V_{ab} = 200\sqrt{3}\underline{/30°+30°} = 346.4\underline{/60°}\text{V}$$

$$I_{ab} = \frac{V_{ab}}{Z_\Delta^1} = 7.45\underline{/1.4°}\text{A and } I_{ca} = 7.45\underline{/121.4°}\text{A}$$

Therefore $I_a = I_{ab} - I_{ca} = 12.9\underline{/-28.6°}$A

$I_b = 12.9\underline{/211.4°}$A and $I_c = 12.9\underline{/91.4°}$A

6. Repeat problem 5 if the loads are converted to an equivalent wye. Compare the results of both problems.

Solution: $\mathbf{Z}_\Delta \rightarrow \mathbf{Z}_Y = \frac{50}{3} + j\,\frac{70}{3}$. The total impedance is

$$\mathbf{Z}_Y^1 = \left(\frac{50}{3} + j\,\frac{70}{3}\right) || (15+j30) = 15.5\underline{/38.6°}\Omega$$

$$\mathbf{I}_a = \frac{200\underline{/30°}}{\mathbf{Z}_Y^1} = 12.9\underline{/-28.6°}\text{A},\ \mathbf{I}_b = 12.9\underline{/211.4°}\text{A},$$

$$\mathbf{I}_c = 12.9\underline{/91.4°}\text{A}$$

7. A balanced wye-delta system has a load current of $\mathbf{I}_{AB} = 25\underline{/60°}\text{A}$. Determine all the line currents.

Solution: Using the relationships between line and phase currents if $\mathbf{I}_{AB} = 25\underline{/60°}\text{A}$ then $\mathbf{I}_a = 25\sqrt{3}\underline{/30°}\text{A}$, $\mathbf{I}_b = 25\sqrt{3}\underline{/-90°}\text{A}$ and $\mathbf{I}_c = 25\sqrt{3}\underline{/-210°}\text{A}$.

8. In a balanced wye-delta system the load per phase is 20+j10Ω. If the voltage $\mathbf{V}_{ab} = 208\underline{/30°}\text{V}$, determine the line currents in the system.

Solution: If $\mathbf{V}_{ab} = 208\underline{/30°}\text{V}$, then $\mathbf{V}_{AB} = 208\underline{/30°}\text{V}$.

$$\mathbf{I}_{AB} = \frac{208\underline{/30°}}{20+j10} = 9.30\underline{/3.43°}\text{A}.$$ Using the relationships between line and phase currents $\mathbf{I}_a = 9.3\sqrt{3}\underline{/-26.57°}\text{A}$, $\mathbf{I}_b = 9.3\sqrt{3}\underline{/-146.57°}\text{A}$ and $\mathbf{I}_c = 9.3\sqrt{3}\underline{/-266.57°}\text{A}$.

9. A three-phase balanced system has a load consisting of a delta in parallel with a wye. The impedance per phase for the delta is 12+j9Ω and for the wye is 6+j3Ω. The source is a balanced wye with a positive phase sequence. If $\mathbf{V}_{an} = 120\underline{/30°}\text{V}$, determine the line currents.

Solution: $\mathbf{Z}_\Delta \rightarrow \mathbf{Z}_Y = 4+j3\Omega$

The equivalent impedance per phase is

$$\mathbf{Z}_{eq} = (4+j3)||(6+j3) = 2.88\underline{/32.47°}\Omega$$

$$\mathbf{I}_a = \frac{120\underline{/30°}}{2.88\underline{/32.47°}} = 41.72\underline{/-2.47°}\text{A}$$

$\mathbf{I}_b = 41.72\underline{/-122.47°}\text{A}$ and $\mathbf{I}_c = 41.72\underline{/-242.47°}\text{A}$

10. Determine the line currents in Problem 9 if $\mathbf{V}_{ab} = 100\underline{/-60°}\text{V}$.

Solution: If $\mathbf{V}_{ab} = 100\underline{/-60°}\text{V}$ then $\mathbf{V}_{an} = \frac{100}{\sqrt{3}}\underline{/-90°}\text{V}$.

From the previous solution

$$\mathbf{I}_a = \frac{\frac{100}{\sqrt{3}}\underline{/-90°}}{2.88\underline{/32.47°}} = 20.05\underline{/-122.47°}\text{A}$$

Then

$$\mathbf{I}_b = 20.05\underline{/-242.47°}\text{A},\ \mathbf{I}_c = 20.05\underline{/-362.47°}\text{A}$$

11. Calculate the instantaneous power for a balanced three-phase load in which the load current is $2\underline{/-40°}\text{A}$ and the load voltage is $240\underline{/0°}\text{V}$.

Solution: $p(t) = 3\,\frac{V_M I_M}{2}\cos\theta$ where

$V_M = 240\sqrt{2}$, $I_M = 2\sqrt{2}$, and $\theta = 0-(-40°) = 40°$

Hence $p(t) = 3\,\frac{(240\sqrt{2})(2\sqrt{2})}{2}\cos 40° = 1103\text{W}$

12. A three-phase positive sequence wye-connected source supplying 1.20KVA with a power factor of .60 lagging has a line voltage $\mathbf{V}_{ab} = 120\underline{/45°}\text{V}$ and is connected to two balanced wye loads. If the first wye-connected load is purely inductive and uses 800 VAR, find the phase impedance of the second load.

Solution: $\theta = \cos^{-1}(0.6) = 53.1°$

S for the 3ϕ system source is $\mathbf{S}_{3\phi} = 1200\underline{/53.1°} = 720+j960$ VA

$\mathbf{S}_{3\phi}$ for load 2 is $\mathbf{S}_{3\phi\text{ load }2} = 720+j960-j800 = 720+j160$ VA

$\mathbf{S}_{1\phi\text{ load }2} = 240+j53.3$ VA

$$\mathbf{V}_{an} = \frac{120}{\sqrt{3}}\underline{/45°-30°} = 69.3\underline{/15°}\text{V}$$

$$\mathbf{I}^*_{\text{load }2} = \frac{\mathbf{S}}{\mathbf{V}} = 3.55\underline{/-2.47°}\text{A}$$

Then $\mathbf{Z}_{\text{load }2} = \frac{\mathbf{V}}{\mathbf{I}} = 19.5\underline{/12.5°}\Omega$

13. Determine the magnitude of the current flowing through each of the loads in the network described in Problem 12.

Solution: $$\mathbf{I}_L^* = \frac{\mathbf{S}_{1\phi}}{\mathbf{V}_{ab}} = \frac{(720+j960)/3}{69.3\underline{/15°}} = 5.77\underline{/38.1°}\text{A}$$

Therefore $\mathbf{I}_L = 5.77\underline{/-38.1°}$ and from the previous solution $\mathbf{I}_{\text{load }2} = 3.55\underline{/2.47°}\text{A}$. $\mathbf{I}_{\text{load }1} = \mathbf{I}_L - \mathbf{I}_{\text{load }2} = 3.85\underline{/-75°}\text{A}$. Hence $|\mathbf{I}|_{\text{load }1} = 3.85\text{A}$ and $|\mathbf{I}|_{\text{load }2} = 3.55\text{A}$.

14. A delta-connected three-phase source with a positive phase sequence is connected through a transmission line with an impedance of 0.05+j0.1Ω per phase to a Y-connected load with a per phase impedance of 20+j8Ω. The source voltages are $\mathbf{V}_{ab} = 208\angle 40^\circ$V, $\mathbf{V}_{bc} = 208\angle -80^\circ$V and $\mathbf{V}_{ca} = 208\angle -200^\circ$V. Determine the line currents, the magnitude of the line voltage at the load, and the real power loss in the lines.

Solution: Converting the delta source to a wye source yields $\mathbf{V}_{an} = 120\angle 10^\circ$V. The circuit diagram for the a-phase is

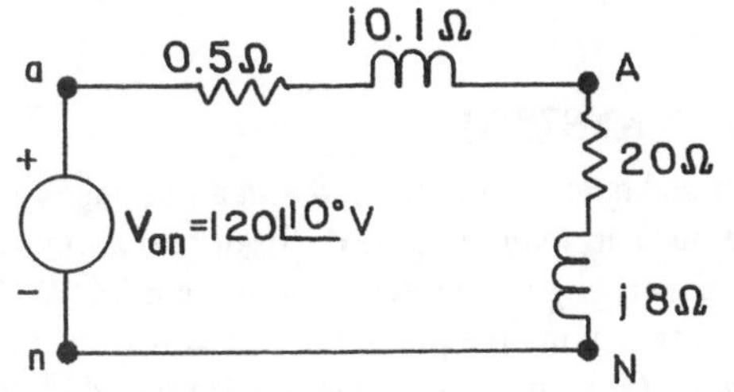

$$\mathbf{I}_{aA} = \frac{120\angle 10^\circ}{20.05+j8.1} = 5.55\angle -12^\circ\text{A},\ \mathbf{I}_{bB} = 5.55\angle -132^\circ\text{A}$$

and $\mathbf{I}_{cC} = 5.55\angle -252^\circ$A

$$\mathbf{V}_{AN} = (5.55\angle -12^\circ)(20+j8) = 119.55\angle 9.8^\circ\text{V}$$

$$|\mathbf{V}_L| = (\sqrt{3})\ V_{AN} = 207.06\text{V}$$

$$P_{Lines} = 3(5.55)^2(.05) = 4.62\text{W}$$

15. Two wattmeters are used to measure the total power in a balanced Y-Y system. The line voltage at the source is 208V. The load is known to be capacitive in nature. Without reversing terminals the wattmeter readings are $P_A = 1200$W and $P_B = 800$W. Determine the load impedance per phase.

Solution:

$$\theta = \tan^{-1}\left(\frac{1200-800}{2000}\right)\sqrt{3} = 19.11^\circ \text{ and } \cos\theta = 0.945$$

$$P_T = 2000\text{W and } P_P = 666.67\text{W} = V_P I_P \cos\theta$$

$$\text{Hence } I_P = \frac{666.67}{(120)(0.945)} = 5.88\text{A and}$$

$$\mathbf{Z}_P = \frac{120\angle -19.11^\circ}{5.88} = 20.41\angle -19.11^\circ\Omega$$

16. A balanced three-phase wye-wye system has two parallel loads. Load 1 is rated at 4000 VA, 0.9 PF lagging, and load 2 is rated at 3600 VA, 0.85 PF lagging. If the line voltage is 208V, determine the magnitude of the line current.

Solution:

$$\mathbf{S}_1 = 4000\angle \cos^{-1}0.9 = 3600+j1743.56\text{VA}$$

$$\mathbf{S}_2 = 3600\angle \cos^{-1}0.85 = 3060+j1896.42\text{VA}$$

$$\mathbf{S}_{Total} = 6660+j3639.98\text{VA}$$

$$|\mathbf{S}_{Total}| = 7589.8 \text{ VA}$$

$$I_L = \frac{7589.8}{208\sqrt{3}} = 21.07\text{A}$$

17. Determine the value of the wattmeter reading in the following circuit.

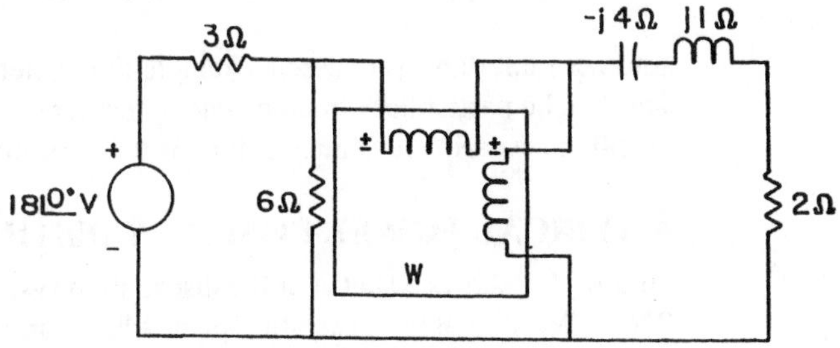

Solution: Using Thevenin's theorem the network can be reduced to

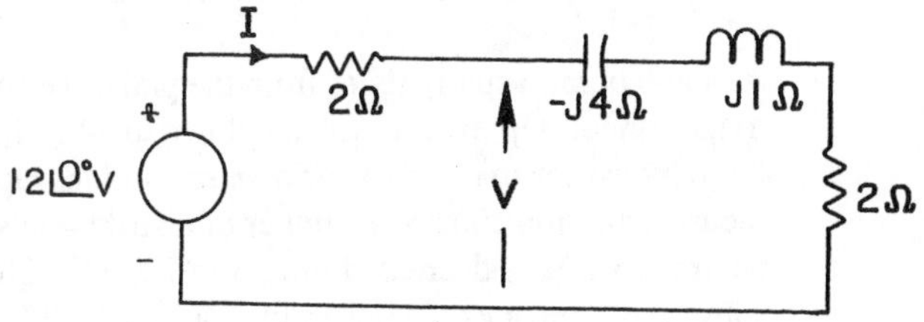

$$\mathbf{I} = \frac{12\angle 0^\circ}{4-j3} = 2.4\angle 36.87^\circ\text{A}$$

$$\mathbf{V} = 12\angle 0^\circ - 4.8\angle 36.87^\circ = 8.65\angle -19.44^\circ\text{V}$$

$$P = |\mathbf{V}||\mathbf{I}|\cos(\theta_V - \theta_I)$$

$$= (8.65)(2.4)\cos(-19.44^\circ - 36.87^\circ)$$

$$= 11.52\text{W}$$

Chapter 11 MAGNETICALLY COUPLED NETWORKS

A power utility distribution system is an excellent example of a magnetically coupled network. The labyrinth of transformers, substations and cabling that we see everyday distributes billions of watts of generated power to millions of customers with high efficiency. In this chapter, we will look at a typical distribution system in detail, discussing the purpose of various substations and pole-mounted equipment.

Another important, yet unseen, example of magnetic coupling between circuits is noise due to magnetic fields. This can be particularly troublesome in instrumentation circuitry. We will consider a prototypical magnetic coupling scenario and simple solutions to the problem.

A TYPICAL POWER COMPANY DISTRIBUTION SYSTEM

Figure 11.1 shows a sketch of the distribution system discussed here. Keep in mind that all voltages are in RMS. We start at the generating plant which can be of the nuclear, hydroelectric or fossil fuel variety. As an example, Figure 11.2 shows an Alabama Power fossil fuel generating plant rated at more than 2 GW. Typical voltages at the generator outputs are around 10 kV. But before leaving the plant, the line voltage is increased to 230 - 500 kV using the transformers in the step-up substation. Of course, the current is decreased by the same amount, which has two benefits. First, lower current means smaller conductors can be used which lowers cost. Second, lower current leads to lower power loss in the transmission lines. Also, transmission is usually done in a delta connection since three conductors are less expensive than four.

At some point, usually miles from the plant, the first of many transmission substations is encountered. Here, transformers step the line voltage down to 44 to 115 kV. Moving down from the transmission sub, we encounter large industries that have their own substations tapping into the power grid at this point. Also downstream from the transmission sub are a number of distribution substations, like the one in Figure 11.3a, where the line-to-neutral voltage is decreased further to 7200 V. The substations you see around your community are distribution subs. Also, most of the power lines are at 7200 V! Businesses such as malls, apartment complexes and restaurants will tie into all three phases, using three transformers as seen in Figure 11.3b to reduce the voltage to usable levels.

For residential customers, there is one final transformer which we will call the step-down center tap transformer that reduces 7200 V to the 120/240 V service at your home. The inner workings of this transformer are depicted in the inset on Figure 11.1. The total secondary voltage is 240 V, but a center tap connection effectively splits the secondary in half. Now we have two sources of 120 V and one source of 240 V. All three secondary conductors are wrapped into one cable, called triplex, shown in Figure 11.3d, that runs to the service entrance at your home. Also, the center tap connection is grounded to the earth both at the power pole and at the service entrance, making it the neutral connection in your home wiring. Once inside your home the conductors are color coded as follows: center tap (neutral) is white, one 120 V line is black and the other is red. A photograph of a 75 kW center-tap transformer is shown in Figure 11.3c.

As you drive about, you will see plenty of center tap transformers and 7200 V, three-phase wiring lining the streets of your neighborhood. You will also find other pole mounted equipment such as the 3-θ capacitor banks shown in Figure 11.4b. These banks provide power factor correction and a small voltage boost.

If you're travelling along a transmission line (44 - 115 kV) you may come across a bank of three regulators, shown in Figure 11.4a. Regulators are variable transformers that automatically adjust their turns ratios to maintain constant line voltage.

Have you ever had the power go out for a few seconds, come back, go out for a few seconds, come back, go out for a minute or so, then come back on and stay on? That's the work of the oil-enclosed, circulating reclosure (OCR) in Figure 11.4c. OCR's are circuit breakers with internal timing circuitry for resetting the breaker. When a short circuit is detected, the OCR opens for a few seconds then recloses. If the short is gone then everything's fine. If not, the OCR opens again. This cycle is usually repeated three times with the last cycle having a longer off time. If the short is still there, the OCR will stay open.

The last piece of pole mounted equipment we will discuss is the arrestor/switch shown in Figure 11.4d. As the name implies, these serve both as manual switches for safety during repairs and as lightning arrestors. When the switch is open, the bar falls down as seen on the right. In this way, a lineman can spot an open switch from his truck.

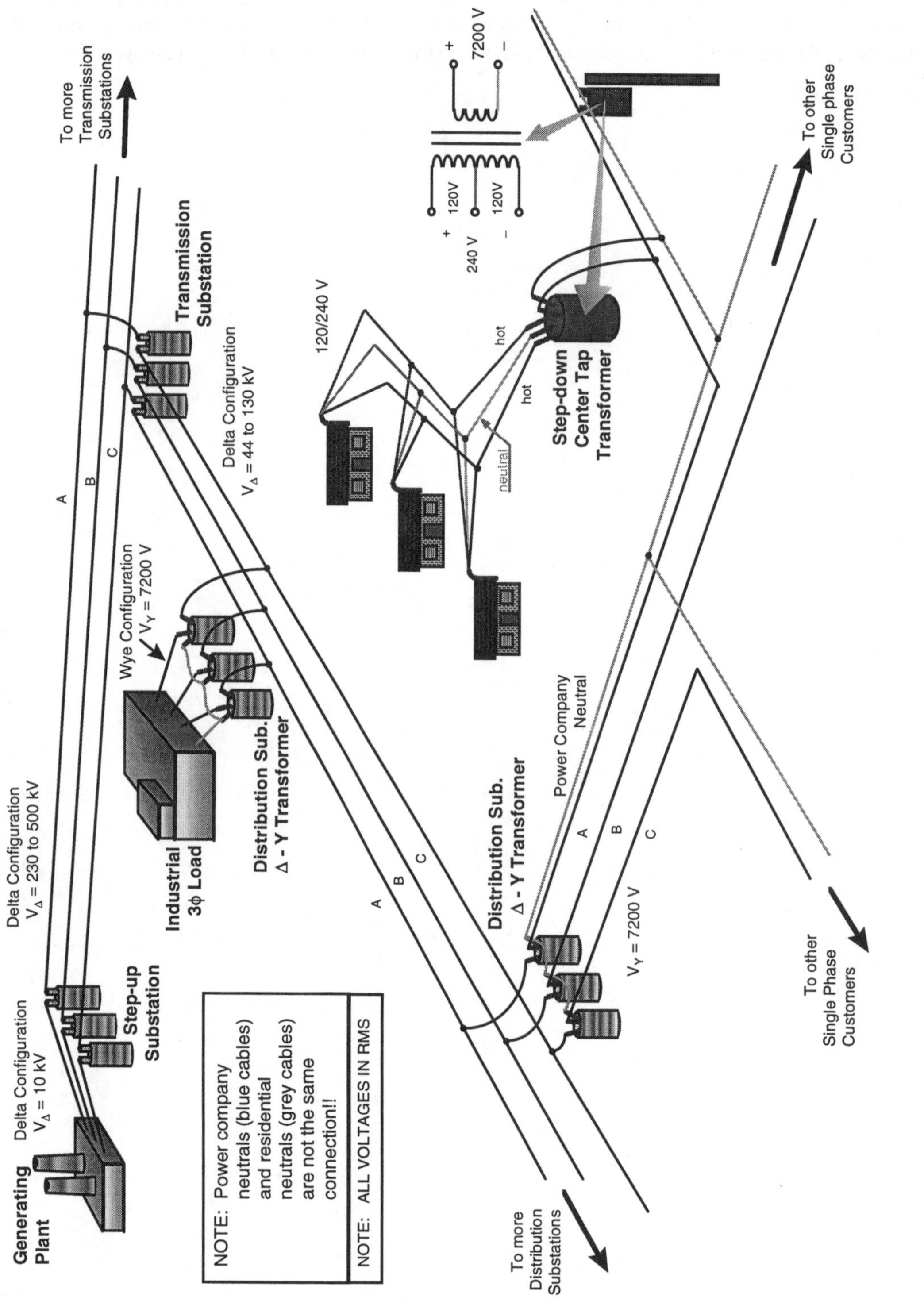

Figure 11.1. A conceptual diagram of a power distribution system from the generating plant to the customer.

Figure 11.2. E.G. Gaston Steam Plant near Wilsonville, Alabama.

Five generating units:

Units 1 - 4: 265 MW nominal per unit, 280 MW maximum. 100 coal cars per unit per day.
Unit 5: 860 MW nominal, 960 MW maximum (high pressure unit) 100 coal cars per day.

Total power output: 1.92 GW nominal, 2.08 GW maximum.

(a)

(b)

(c)

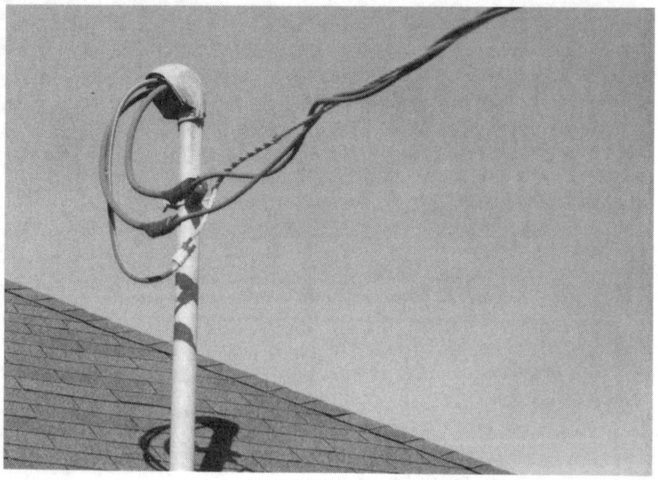

(d)

Figure 11.3. A variety of power distribution equipment: (a) Auburn University's substation, (b) three 37.5 kVA center-tap transformers servicing a 16-plex movie theatre, (c) a 75 kVA center-tap transformer and (d) a residential service triplex drop showing triplex wiring.

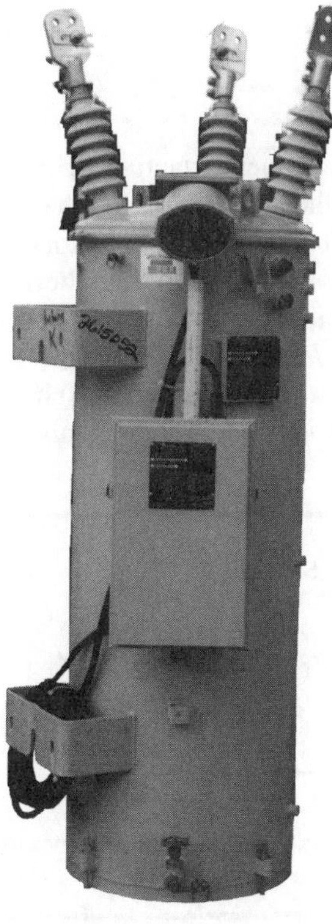

(a)

(b)

(c)

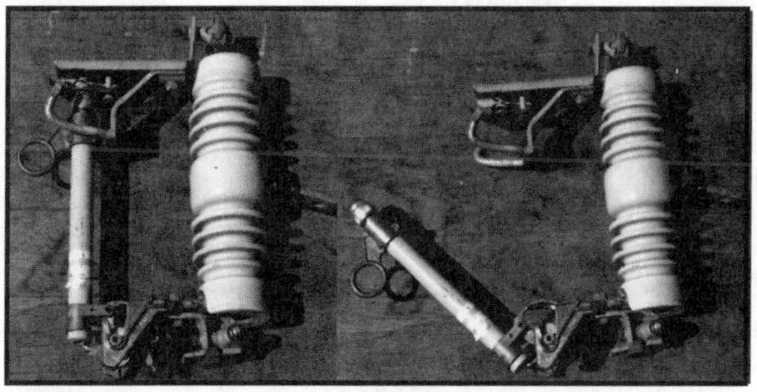

(d)

Figure 11.4. Some additional pole-mounted equipment: (a) automatic regulator (usually found in banks of three), (b) OCR (about 2 feet tall), (c) a small bank of power factor capacitors, and (d) lightning arrestors/switches in both their closed and opened positions.

SIMULATING MAGNETIC COUPLED CIRCUITS IN PSPICE

To model magnetic coupling in PSpice, we need two things, a way to designate the which inductors are coupled and a way to define the dot orientations. Coupled inductors are designated using the K_linear part in the ANALOG library. The K_linear part defines a single coupling coefficient for up to six coupled inductors and is easy to use. Dot conventions are trickier in that the dotted terminal is pin 1 on the inductor. (Pin numbers are discussed in the PSpice sections in the BECA text.) Here's how you keep it straight, when you get an inductor part, it is placed on *Schematics* horizontally. *For horizontally oriented parts, pin 1 is on the left.* Therefore, the dotted terminal is on the left. If you rotate the inductor, (PSpice always rotates parts counterclockwise) be cognizant of where pin 1 goes. Should you loose track of pin 1, go to netlist. The inductor's node numbers are always listed pin 1 first, pin 2 second.

SIMULATION ONE

As an example, let's simulate the magnetically coupled circuit in Figure 11.5 for the phasor form of the output voltage. The corresponding *Schematics* diagram in Figure 11.6 includes the K_linear part which has been edited as shown in Figure 11.7. To set the coupling, we simply enter the names of the coupled inductors and the coupling coefficient into the appropriate fields. The K_linear part can be placed anywhere on the *Schematics* page - it has no connections. When drawing the circuit, we were careful to orient pin 1, and so the dots, at the top of each inductor.

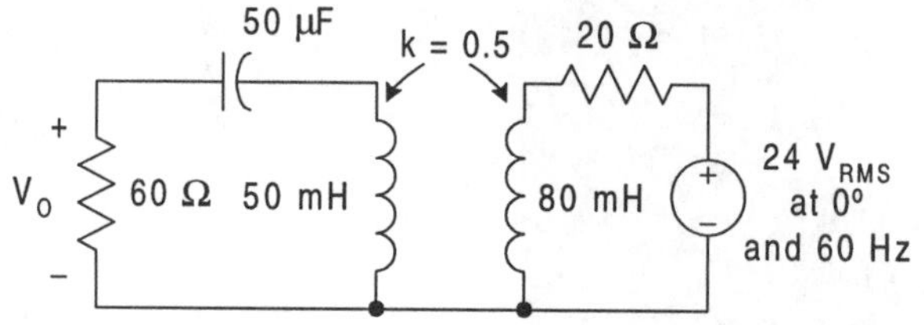

Figure 11.5. A simple magnetically coupled circuit for Simulation One.

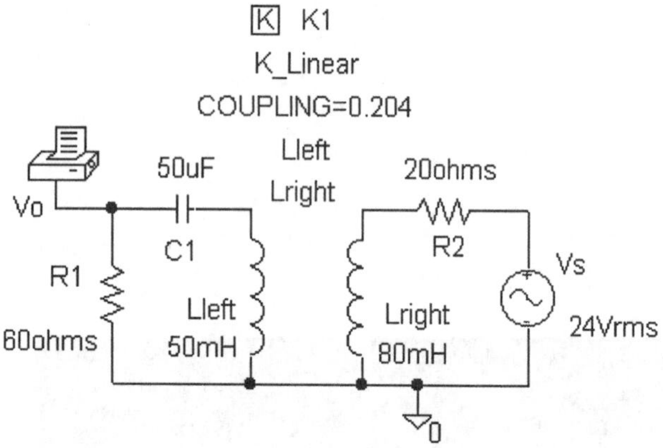

Figure 11.6. Schematics diagram for Simulation One

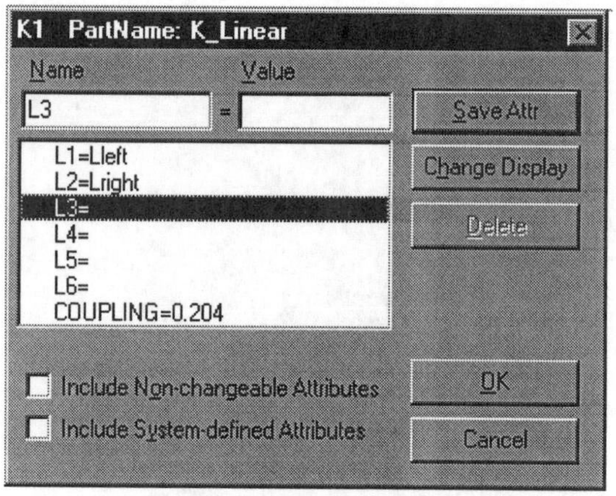

Figure 11.7. The K_Linear Attributes box.

Of course, a single frequency AC Sweep is requested and the phasor data gathered by the VPRINT1 part is in the output file.

```
****     AC ANALYSIS     ****

Vo = 2.779 Volts at 63.48 degrees
```

Simulation Two

It is entirely possible for two circuits to be coupled unintentionally. If the current in one circuit is large, perhaps feeding a motor, a sizable magnetic field can be generated. A nearby circuit can easily be affected. As an example, consider the automated control scenario depicted in Figure 11.8. The motor's speed, rated at 5000 rpm, is read by the tachometer which, in turn, sends a voltage proportional to speed back to a speed control circuit. The tachometer output voltage is 1 V per 1000 rpm. How could the two circuits be magnetically coupled? Think of the conductors of each circuit as forming single loop inductors. If the current to the motor is large enough, the tachometer loop, acting like an antenna, will "pick up" 60 Hz noise. Obviously, reducing the area of the loop is an easy way to decrease the noise susceptibility. A popular technique, called twisted wire pair, twists the conductors (the tachometer's) together creating a long chain of smaller loops. Due to the twisting, adjacent mini-loops have opposite dot positions. Thus, the noise induced in one loop will be cancelled by the neighboring loops.

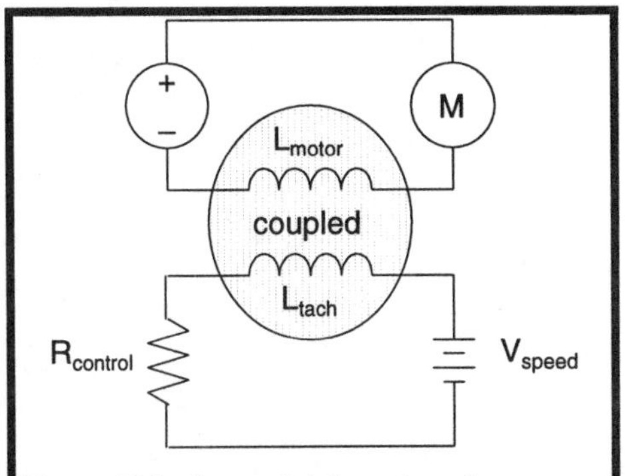

Figure 11.8. A candidate system for magnetic coupling.

To demonstrate the effect of twisted wire pair on reducing magnetic noise pickup, we will perform a transient simulation of our magnetic coupling scenario in using the *Schematics* diagrams in Figure 11.9a and b. In Figure 11.9a, the motor current is sinusoidal at 10 A, the coupling coefficient is 0.25 and twisted wire pair is not used. In Figure 11.9b, the tach wiring has been twisted 9 times. Accordingly, the inductance of each loop has been divided by nine and the coupling by 3. From the PROBE results in Figure 11.10, we see that there is a 0.95-V, 60-Hz sinewave is induced in the tachometer circuit when twisted wire pair is not used. This represent a motor speed error of $\pm$ 950 rpm! When twisted wire pair is used, we see the noise pickup is nine times less. Guess what would happen if we used 1000 twists?

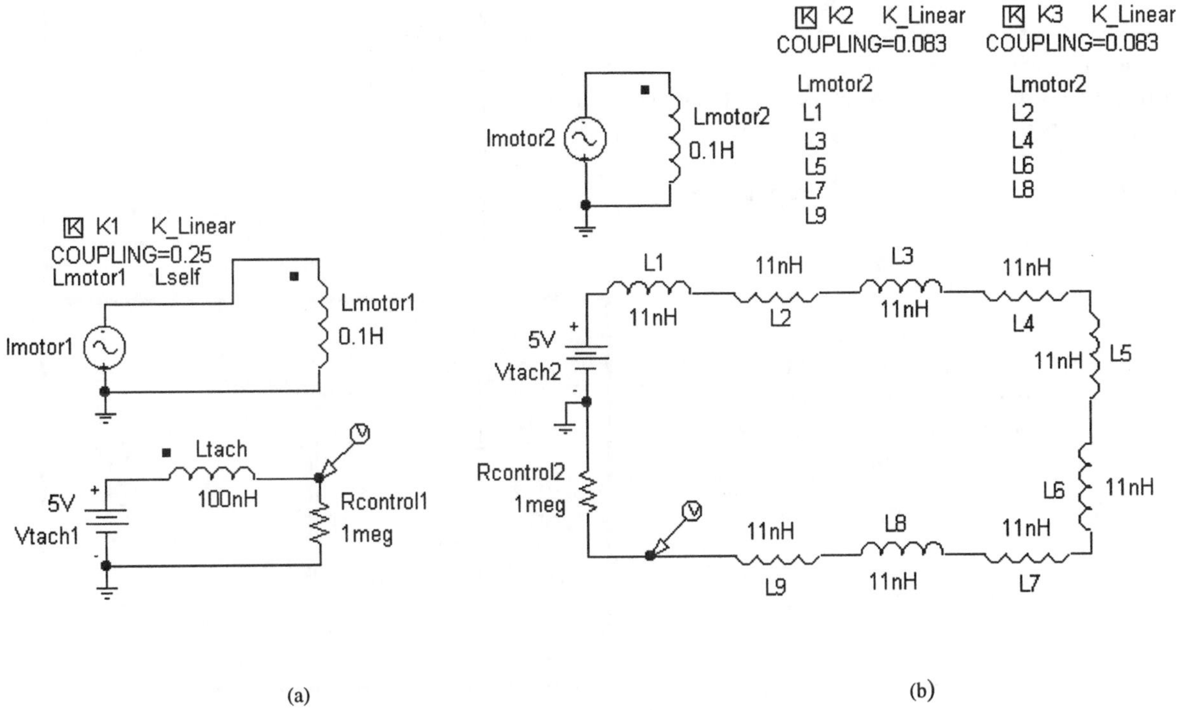

Figure 11.9. Schematics diagrams for Simulation Two both (a) without and (b) with twisted wire pair.

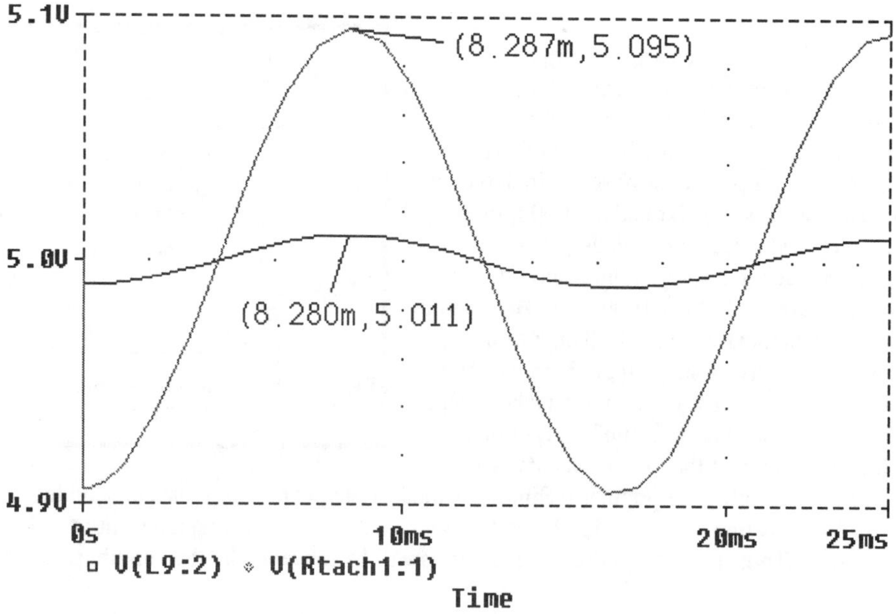

Figure 11.10. PROBE output for Simulation Two demonstrating the advantage of twisted wire pair.

SIMULATION THREE - IDEAL TRANSFORMERS IN THE BECA LIBRARY

In many analyzes, transformers can be treated as ideal without an appreciable loss in accuracy. While there is no ideal transformer in PSpice, one does exist in the BECA library. Actually, there are two: Ideal_XFMR+ with additive windings and Ideal_XFMR- with subtractive windings. Both transformers, shown in Figure 11.11, have but one editable attribute - the turns ratio (secondary/primary). To demonstrate their use, we will simulate the circuit in Figure 11.12 to determine the impedance at the primary terminals.

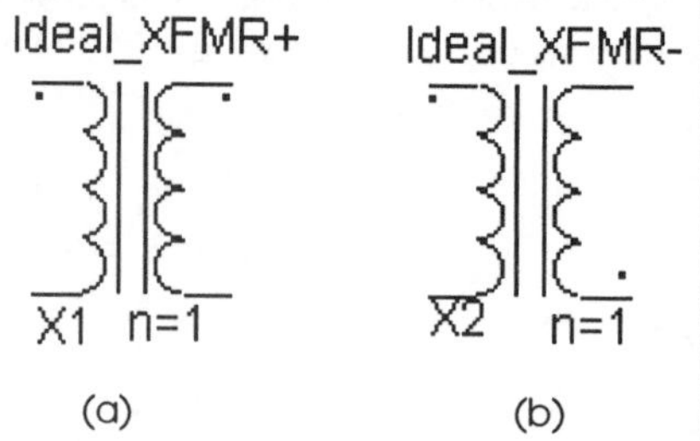

Figure 11.11. The near ideal transformers in the BECA library: (a) additive windings and (b) subtractive windings.

The primary and secondary impedances are related by

$$Z_{PRIM} = \frac{Z_{SEC}}{n^2}$$

Thus, we expect the primary impedance in this case to be 25 Ω. Since the primary current is 1 A, the primary voltage and impedance have the same numeric value. From the output file, the primary impedance is dead on.

```
****     AC ANALYSIS     ****

Zprimary = 25.00 Ohms at 0.0 degrees
```

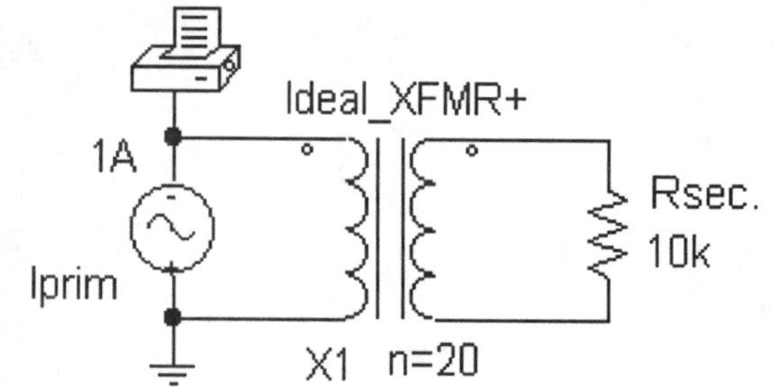

Figure 11.12. *Schematics* diagram for Simulation Three

SIMULATION FOUR

As demonstrated in the Simulation Four of Chapter 10, PSpice can perform single phase power factor simulations. Here we will show a three-phase simulation for a 60 Hz, wye-wye, 7200 V_{RMS} line-to-neutral system where the lines are characterized by a resistance and inductance of 5 Ω and 100 mH respectively. The three phase load is rated at 200 kVA and 0.7 lagging power factor. We will add power factor correction capacitors, C_{pf}, and determine the power factor for C_{pf} = 1 μF as well the capacitance required for unity power factor.

The *Schematics* diagram in Figure 11.13 models the system where the PARAM part defines the variable Cpf. Within the Setup menu, the AC Sweep is set for a 60 Hz, single frequency simulation and the Parametric will sweep the Global Parameter, Cpf, from 10 nF to 10μF. When the simulation is finished and PROBE opens, we plot the power factor by entering the Trace Expression below

Figure 11.14 shows the resulting PROBE plot where unity power factor occurs at C_{pf} = 1.202 μF and when C_{pf} = 1.0 μF, the power factor is 0.984 lagging.

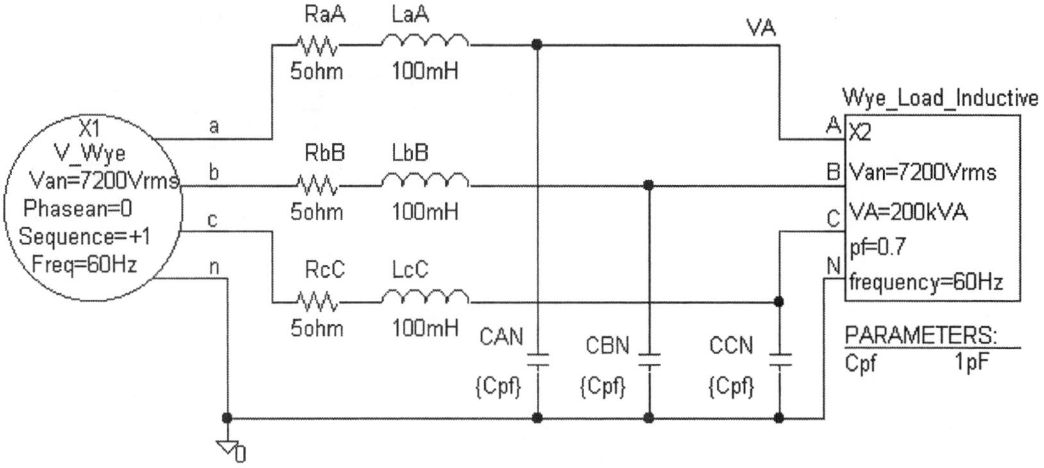

Figure 11.13. *Schematics* diagram for Simulation Four.

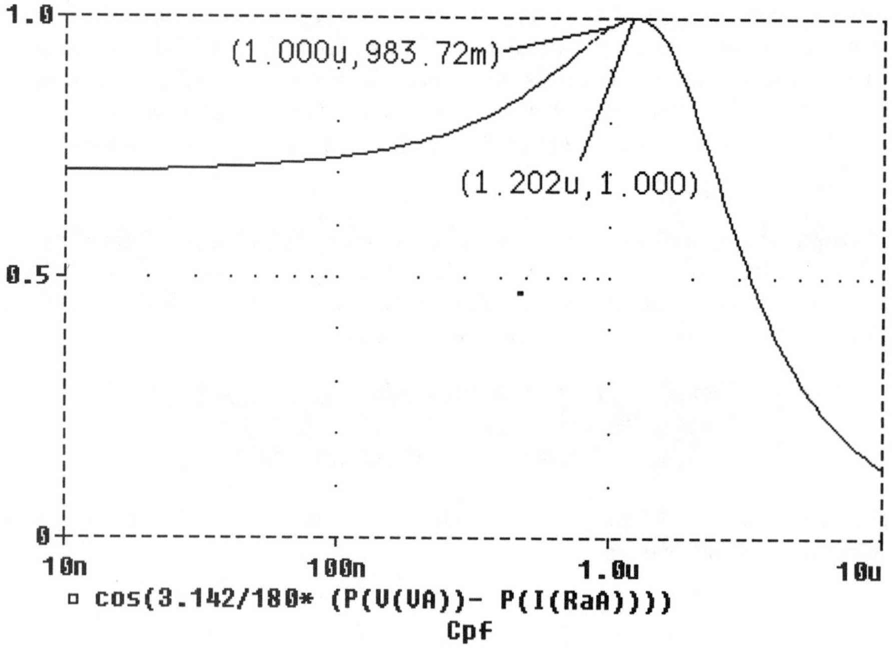

Figure 11.14. PROBE plot of power factor for the three-phase system in Simulation Four.

NOTES

EWB SIMULATION

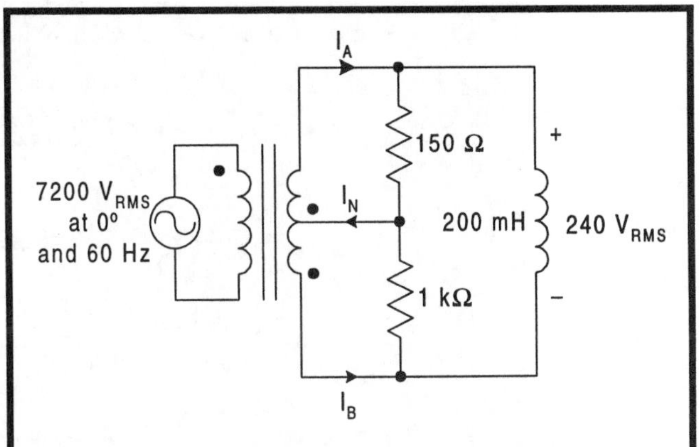

Figure 11.15. Circuit diagram for the EWB simulation.

While Electronics Workbench has no three phases parts, it does have a center-tapped linear transformer part. We will introduce this part by simulating the single-phase, three-wire system in Figure 11.15 to determine the currents I_A, I_B and I_N. The requisite EWB diagram is shown in Figure 11.16 where the dependent sources produce voltages that are numerically equal to the currents of interest. To edit the transformer's characteristics, double-click on the part, select the IDEAL model, and clink on EDIT. The dialog box in Figure 11.17 will open allowing us to edit the transformer. Note that the turns ratio is given as primary/secondary. Thus, in this case, we want 7200/240 = 30/1 for the turns ratio. For an ideal transformer, Leakage inductance, Primary winding resistance and Secondary winding resistance should be zero. Also, the Magnetizing inductance should be very large. The values in Figure 11.17 are sufficient.

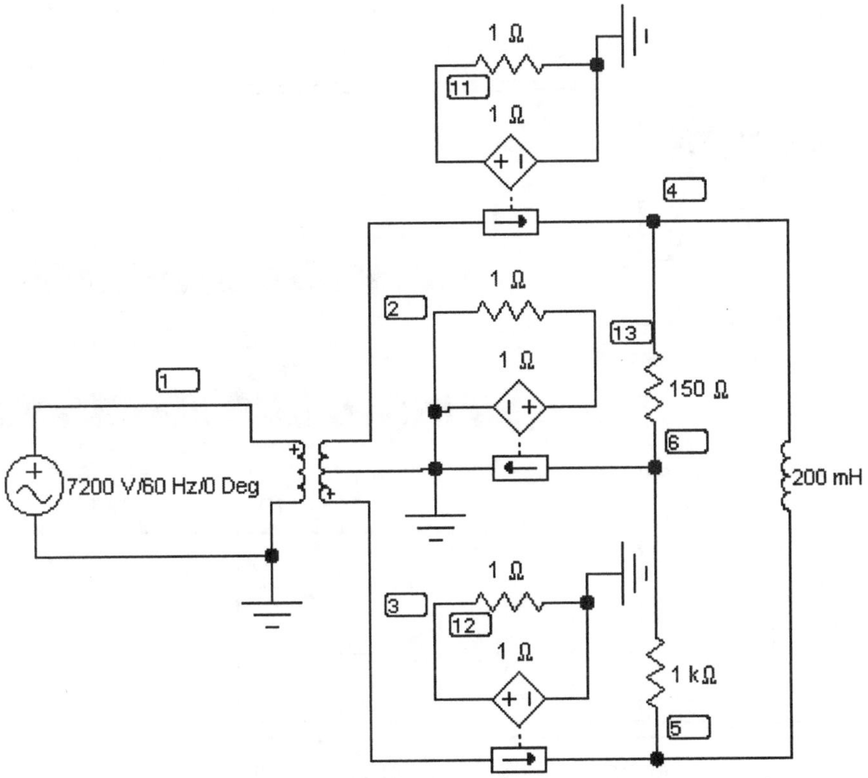

Figure 11.16. EWB diagram for the circuit in Figure 11.13.

AC simulation results are shown in Figure 11.18 where,

$$I_A = 3.28 \underline{/\ 104.11^\circ}\ \ A_{RMS}$$
$$I_B = 3.19 \underline{/\ -87.84^\circ}\ \ A_{RMS}$$
$$I_N = 0.68 \underline{/\ 180^\circ}\ \ A_{RMS}$$

The non-zero current in the neutral wire is due to the resistance mismatch in the two 120-V loops.

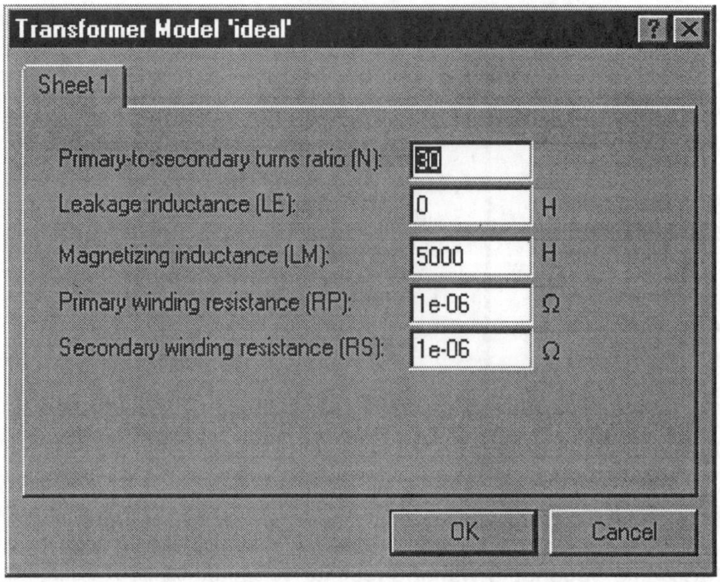

Figure 11.17. The dialog box for editing the linear transformer showing appropriate values for the EWB simulation.

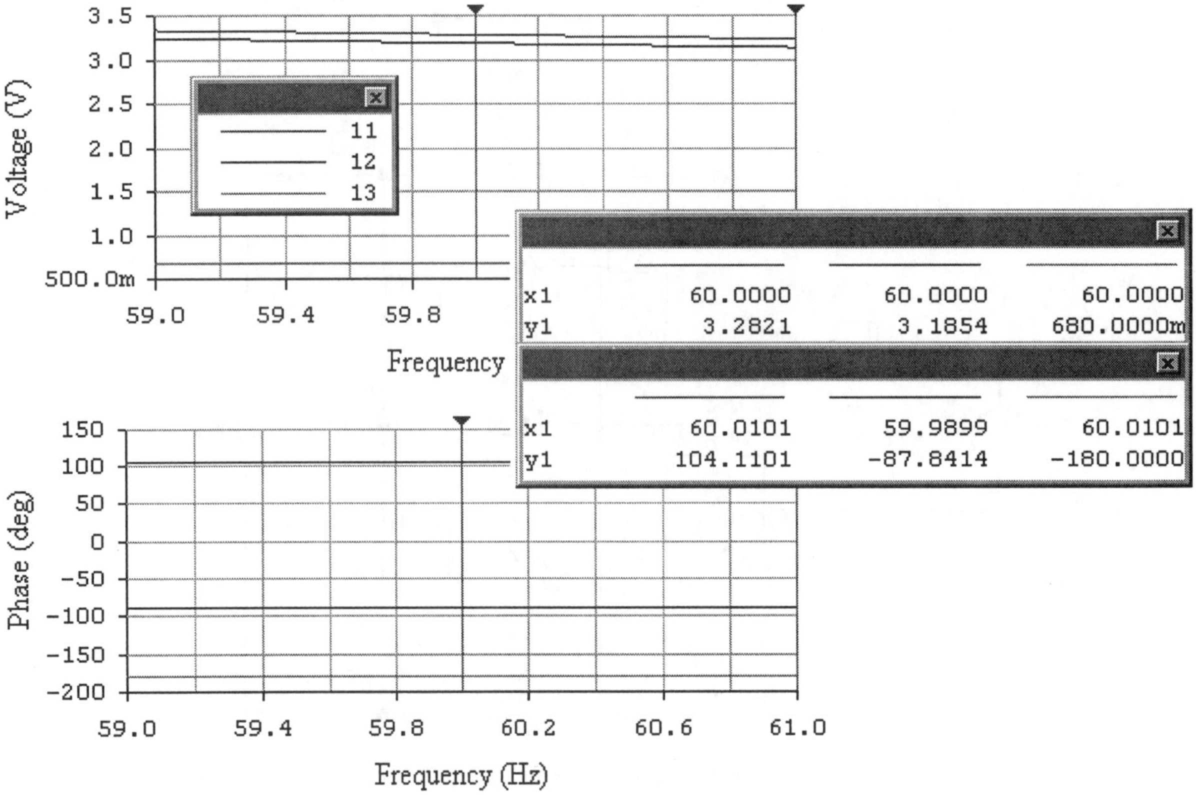

Figure 11.18. The much-edited results of the EWB AC simulation.

EXCEL DEMO

On the CD-ROM, you will find the EXCEL file Power Factor Correction.XLS. This spreadsheet that can calculate the capacitance required to obtain a given power factor. The file was written to accept and output data for a single phase wye-wye equivalent circuit. So, to use the file for three-phase calculations, divide the total power by 3. Also, you will need three capacitors - each equal to the calculated value. An example of the file is shown in Figure 11.19.

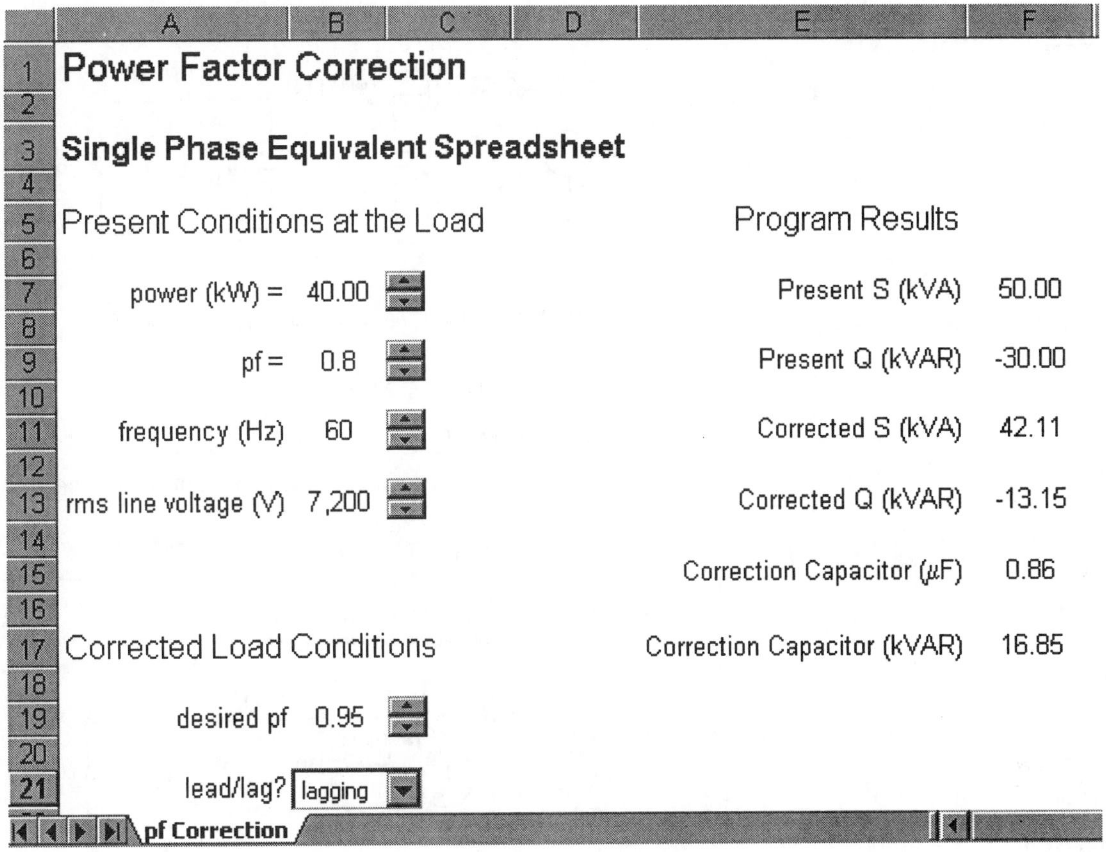

Figure 11.19. An example of the file Power Factor Correction.XLS.

PROBLEM SOLVING EXAMPLES

1. Determine the current I_2 in the network shown below.

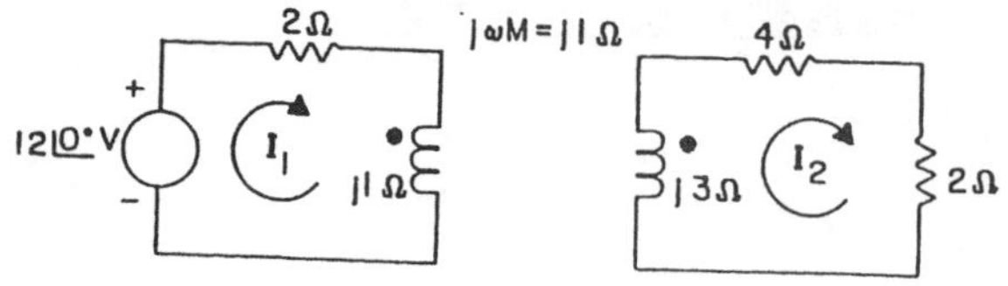

Solution: The network equations are

$(2+j1)I_1-(j1)I_2 = 12\underline{/0^\circ}$

$-(j1)I_1+(6+j3)I_2 = 0$

Solving these equations yields $I_2 = 0.77\underline{/39.81^\circ}A$.

2. Find the current I_1 in the following circuit.

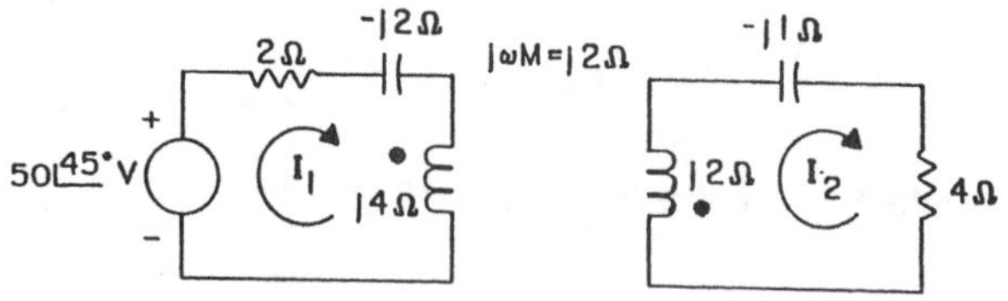

Solution: The loop equations for the network are

$(2+j2)I_1+(j2)I_2 = 50\underline{/45^\circ}$

$(j2)I_1+(4+j1)I_2 = 0$

Solving the equations for I_1 yields $I_1 = 14.57\underline{/14.04^\circ}A$.

3. Determine the impedance seen by the source $V_s = 24\underline{/0^\circ}V$ in the following network.

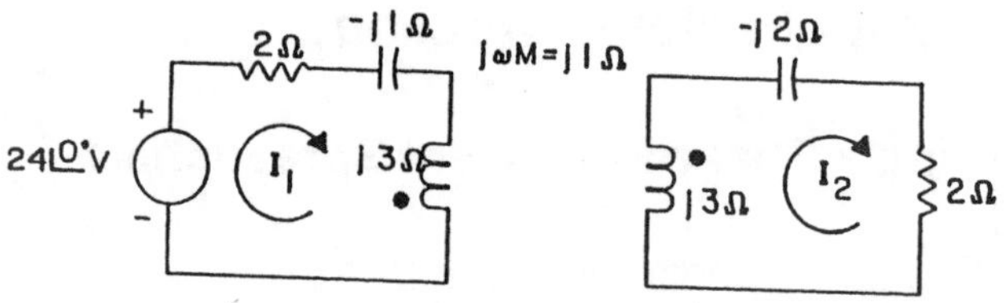

Solution: The loop equations for the network are

$(2+j2)I_1+(j1)I_2 = 24\underline{/0^\circ}$

$(j1)I_1+(2+j1)I_2 = 0$

Solving these equations yields $I_1 = 8.01\underline{/-38.86^\circ}A$

and hence $Z_{in} = \dfrac{24\underline{/0^\circ}}{8.02\underline{/-38.86^\circ}} = 3\underline{/36.86^\circ}\Omega$.

4. Write the circuit equations necessary to find V_o in the network shown below.

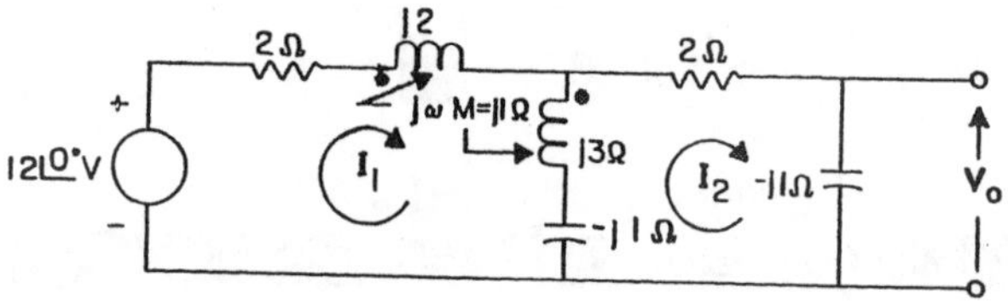

Solution: The network equations are

$-12\underline{/0^\circ}+2I_1+j2I_1+j1(I_1-I_2)+j3(I_1-I_2)+j1I_1$

$\quad -j1(I_1-I_2) = 0$

$-j1(I_2-I_1)+j3(I_2-I_1)-j1I_1+2I_2-j1I_2 = 0$

$V_o = -j1I_2$

5. Write the set of linear independent phasor equations necessary to solve for the currents $i_1(t)$ and $i_2(t)$ in the following network.

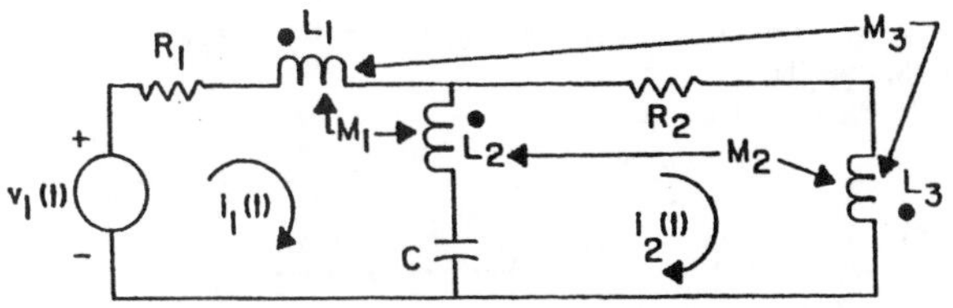

Solution: The loop equations for the network are

$$I_1R_1+j\omega L_1I_1+j\omega M_1(I_1-I_2)-j\omega M_3I_2+j\omega L_2(I_1-I_2)$$

$$+j\omega M_1I_1-j\omega M_2I_2+\frac{1}{j\omega C}(I_1-I_2) = \mathbf{V}_1$$

$$\frac{1}{j\omega C}(I_2-I_1)+j\omega L_2(I_2-I_1)-j\omega M_1I_1+j\omega M_2I_2+R_2I_2$$

$$+j\omega L_3I_2+j\omega M_2(I_2-I_1)-j\omega M_3I_1 = 0$$

6. Determine the expressions for $v_1(t)$ and $v_2(t)$ in the circuit below.

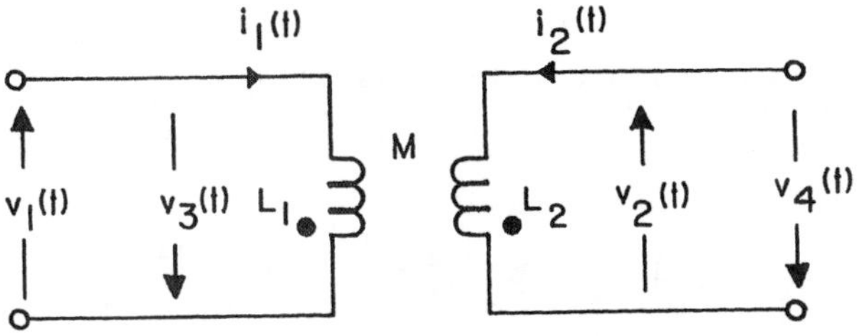

Solution: $v_1(t) = L_1\frac{di_1(t)}{dt} + M\frac{di_2(t)}{dt}$

$v_2(t) = L_2\frac{di_2(t)}{dt} + M\frac{di_1(t)}{dt}$

7. Find the expressions for $v_3(t)$ and $v_4(t)$ in the circuit in Problem 6.

Solution: $v_3(t) = -L_1\frac{di_1(t)}{dt} - M\frac{di_2(t)}{dt}$

$v_4(t) = -L_2\frac{di_2(t)}{dt} - M\frac{di_1(t)}{dt}$

8. Find the expressions for $v_1(t)$ and $v_4(t)$ in the circuit shown in Problem 6.

Solution: $v_1(t) = L_1\frac{di_1(t)}{dt} + M\frac{di_2(t)}{dt}$

$v_4(t) = -L_2\frac{di_2(t)}{dt} - M\frac{di_1(t)}{dt}$

9. Find the expressions for $v_3(t)$ and $v_2(t)$ in the circuit shown in Problem 6.

Soltuion: $v_3(t) = -L_1\frac{di_1(t)}{dt} - M\frac{di_2(t)}{dt}$

$v_2(t) = L_2\frac{di_2(t)}{dt} + M\frac{di_1(t)}{dt}$

10. Determine the output voltage $\mathbf{V}_o$ by converting the linear transformer shown below into its equivalent T network. The coefficient of coupling is .50.

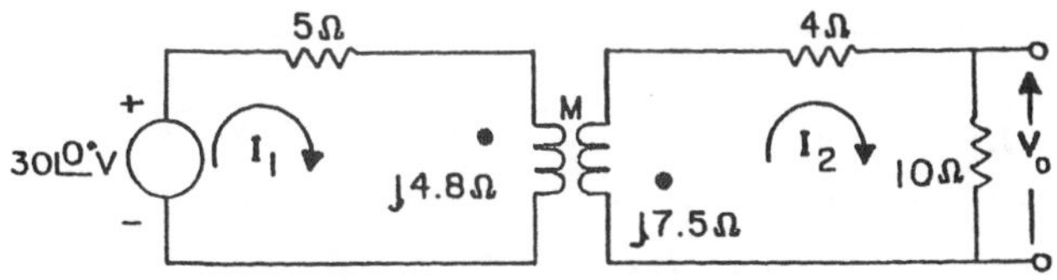

Solution: I_1 and I_2 enter dots and hence M is negative. $M = K\sqrt{L_1L_2} = 3$. Hence the network can be redrawn as

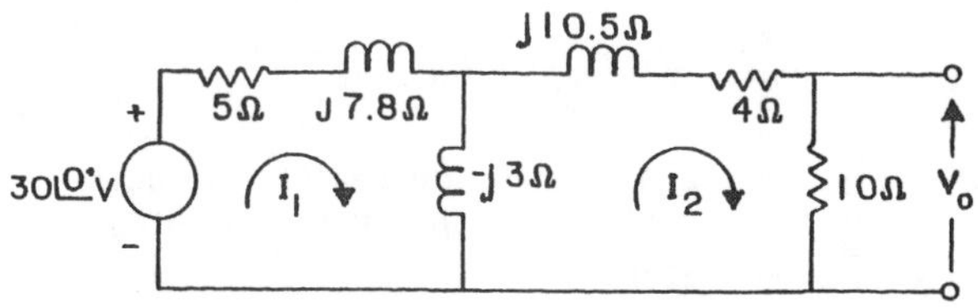

The network equations are

$I_1(5+j7.8-j3)+I_2(j3) = 30\underline{/0°}$

$I_1(j3)+I_2(4+10+j10.5-j3) = 0$

From these equations we find that $I_2 = 0.795\underline{/-157.7°}$A and hence $\mathbf{V}_o = 10I_2 = 7.95\underline{/-157.7°}$V.

11. Determine the impedance at terminals A-B in the following network.

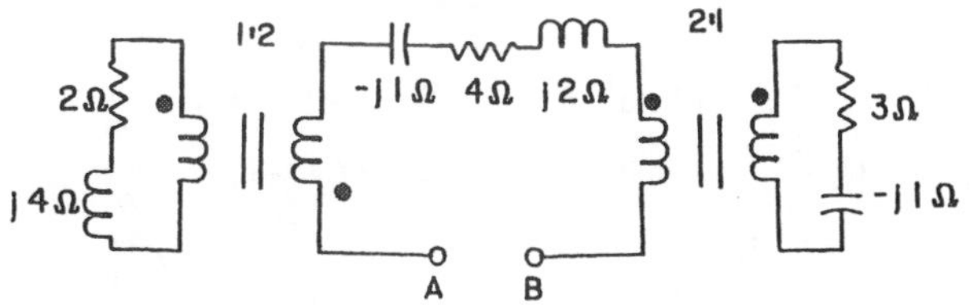

Solution:

$$Z_{left} = \frac{(2+j4)}{(\frac{1}{2})^2} = 8+j16\Omega,\ Z_{right} = \frac{(3+j1)}{(\frac{1}{2})^2} = 12-j4\Omega$$

Therefore $Z_{AB} = 8+j16-j1+4+j2+12-j4 = 24+j13\Omega$.

12. If I_1 is known to be $4\underline{/0°}$A, determine V_o in the following network.

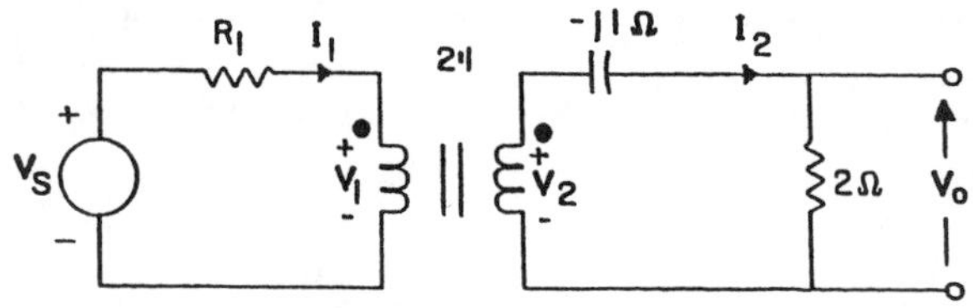

Solution: $I_1 = nI_2$ and $n = \frac{1}{2}$. Hence $I_2 = 2I_1 = 8\underline{/0°}$A and therefore $V_o = 16\underline{/0°}$V.

13. If $V_o = 10\underline{/45°}$V in the circuit of Problem 12, determine the voltage V_1.

Solution: If $V_o = 10\underline{/45°}$V then $I_2 = \frac{10\underline{/45°}}{2} = 5\underline{/45°}$A. $V_2 = I_2(-j1)+10\underline{/45°} = 11.18\underline{/18.41°}$V and since $V_1 = \frac{V_2}{n}$; $V_1 = 22.36\underline{/18.41°}$V.

14. The output voltage is known to be $100\underline{/0°}$V, determine the input voltage and the current I_1 in the network shown below.

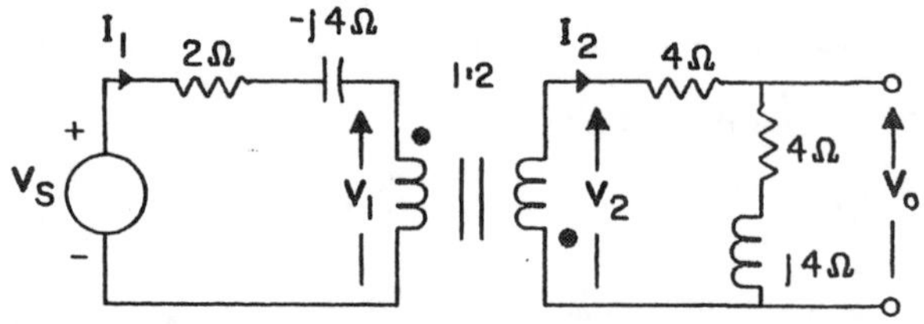

Solution: $I_2 = \frac{100\underline{/0°}}{4+j4} = 17.68\underline{/-45°}$A.

Therefore $V_2 = 100\underline{/0°}+I_2(4) = 158\underline{/-18.43°}$V.

$V_1 = -\frac{V_2}{2} = -79\underline{/-18.43°}$V. $I_1 = -2I_2 = -35.36\underline{/-45°}$A and hence $V_5 = I_1(2-j4)+V_1 = 176.78\underline{/98.13°}$V.

15. Find I_2 in the network shown below.

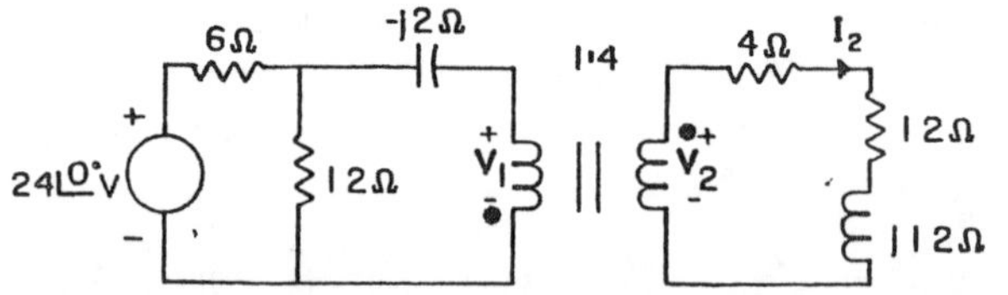

Solution: Using Thevenin's theorem the network is reduced to

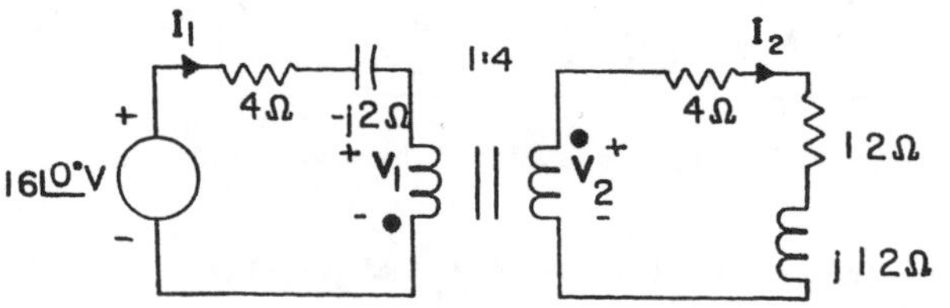

$$I_1 = \frac{16\underline{/0°}}{4-j2+\frac{16+j12}{16}} = 3.10\underline{/14.04°}\text{A}$$

$$I_2 = \frac{I_1}{n} = 0.78\underline{/14.04°}\text{A}$$

16. Determine the value of Z_L for maximum power transfer and the maximum power delivered for the following network.

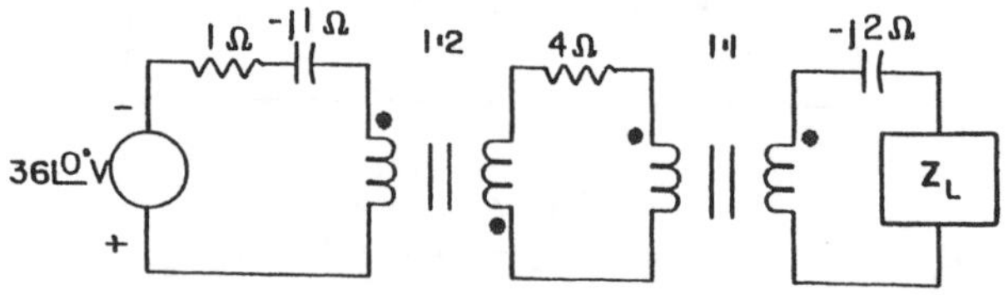

Solution: Reflecting the voltage source and impedance to the output yields

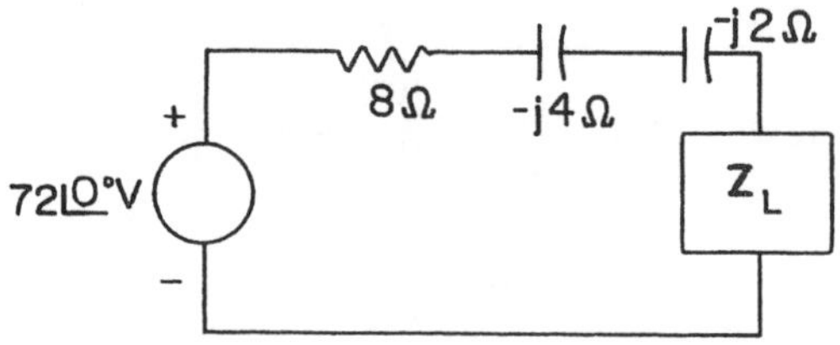

From the network we see that $Z_L = 8+j6\Omega$ and $I_L = \frac{72\underline{/0°}}{16} = 4.5\underline{/0°}$A and $P_{load} = \frac{1}{2}(4.5)^2(8) = 81$W.

17. If V_o in the following network is $120\underline{/0°}$Vrms and the load absorbs 600 watts at 0.8PF lagging, determine the value of the capacitor C so that the current I_1 is in phase with the 60-Hz voltage source.

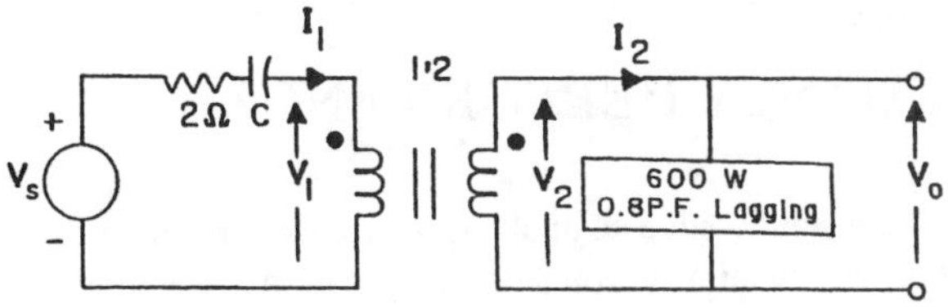

Solution: The magnitude of the load current is

$I_2 = \frac{600}{0.8(200)} = 6.25A$ hence $Z_L = \frac{120}{6.25} \underline{/-36.8°} =$

$15.37+j11.5Ω$. $Z_i = \frac{Z_L}{2^2} = 3.84+j2.875Ω$.

I_1 will be in phase with V_s if $X_C = 2.875$ hence

$\frac{1}{377C} = 2.875$ and $C \simeq 900μF$.

18. Form an equivalent circuit for the transformer and the secondary in the following network and determine the current I_1.

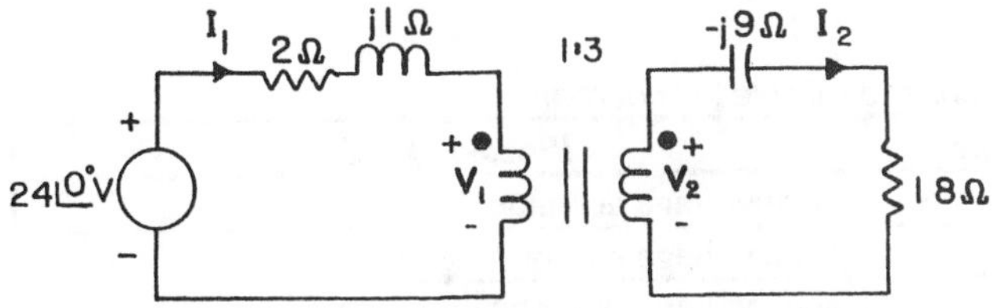

Solution: The network can be changed to

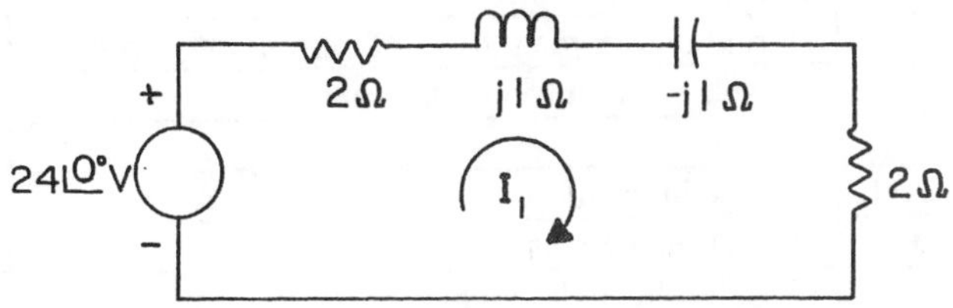

$$I_1 = \frac{24\underline{/0°}}{4} = 6\underline{/0°}A$$

19. Form an equivalent circuit for the transformer and primary of the network in Problem 18 and determine the current I_2.

Solution: Reflecting the primary circuit to the secondary yields

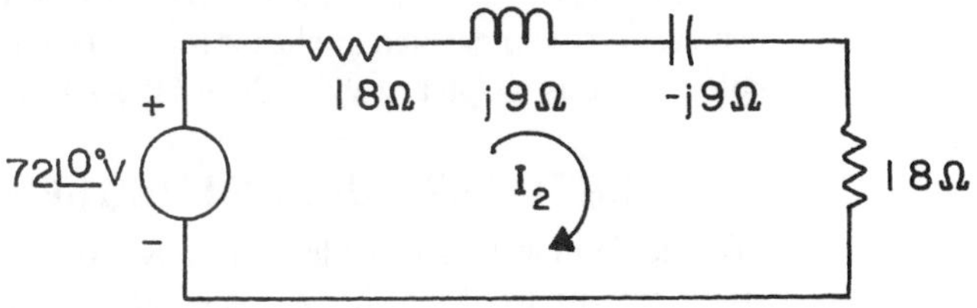

$$I_2 = \frac{72\underline{/0°}}{36} = 2\underline{/0°}A$$

Chapter 12 VARIABLE FREQUENCY PERFORMANCE

Variable frequency network analysis is nothing more than steady-state ac analysis where the frequency of excitation is a variable. As in Chapters 9 - 11, the solution contains no transient information and voltages and currents are represented in phasor form. Typically, since frequency is a variable, solutions are shown in Bode plots. In this chapter, we will demonstrate how to create and utilize Bode plots in both PSpice and EWB.

SIMULATING VARIABLE FREQUENCY CIRCUITS IN PSPICE

The performance of variable frequency circuits, often called filters, is usually depicted in a Bode plot. Bode plots are graphs of the phasor magnitude and phase versus frequency where magnitude is plotted in dB, phase in degrees and frequency on a log axis. It is very easy to create Bode plots in PSpice, as we will show in the following simulations. However, we will take a moment to introduce the **Markers** menu in *Schematics*.

THE ADVANCED MARKERS

A marker is a pseudo-part in *Schematics*. While a marker has no effect on the simulation, it does inform PROBE to plot the marker quality. The primary markers are shown in Figure 12.1 and listed in Table 12.1.

Table 12.1

The principle markers available in *Schematics*

Marker	Simulation Type	Plotted Quality
Voltage/Level	dc	node voltage dc value
	ac	node voltage phasor magnitude
	transient	node voltage instantaneous value
Voltage Differential	dc	dc voltage between two nodes
	ac	magnitude of phasor voltage between two nodes
	transient	instantaneous voltage between two nodes
Current into Pin*	dc	dc current into a pin
	ac	current phasor magnitude of current into a pin
	transient	instantaneous current into a pin
Advanced		
vdb	same and Voltage/Level	voltage in decibels
idb*	same as Current into Pin	current in decibels
vphase	ac only	phase of the node voltage
iphase*	ac only	phase of the pin current
*Must be placed directly on the pin of interest.		

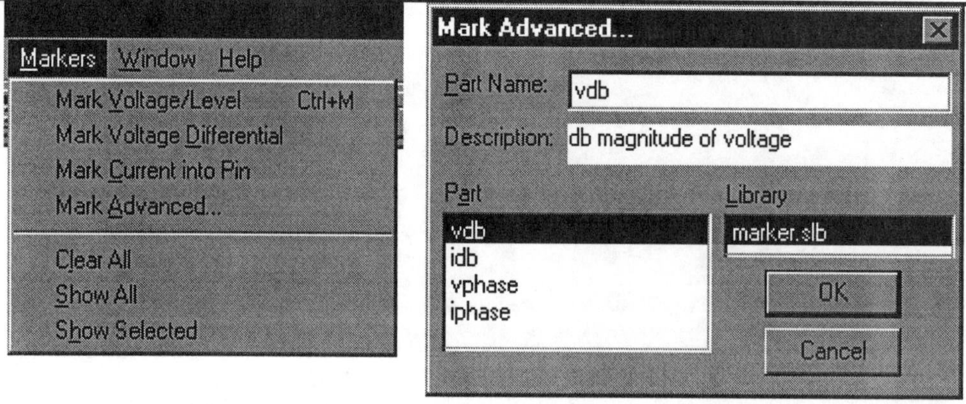

(a) (b)

Figure 12.1. Accessing the markers in Schematics: (a) the **Markers** menu and (b) some of the advanced markers.

Simulation One

Let's demonstrate a variable frequency analysis and the use of the markers by simulating the passive low-pass filter in Figure 12.2. We will also produce a Bode plot of the output voltage and extract the bandwidth of the filter and the phase when the frequency equals the bandwidth.

Note that in Figure 12.2, the vdb and vphase advanced markers have been used to create the Bode plot. The frequency will be varied from 1 Hz to 10 kHz in decades (log scale) as shown in the AC Sweep dialog box in Figure 12.3. The resulting Bode plot shown in Figure 12.4 where the critical data points have been tagged. The bandwidth is found to be about 174 Hz and the phase at 174 Hz is -44.8°.

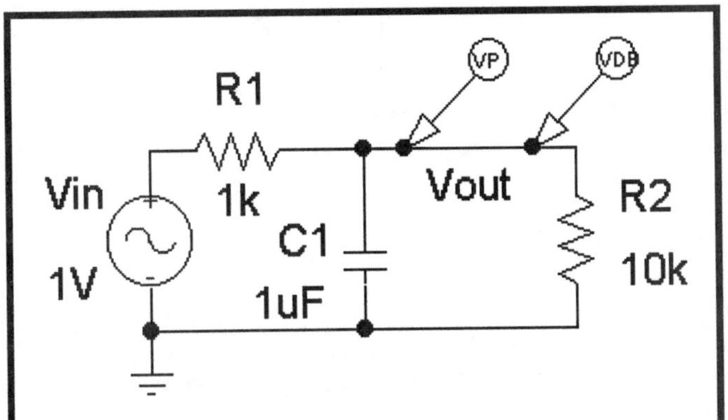

Figure 12.2. *Schematics* diagram for Simulation One.

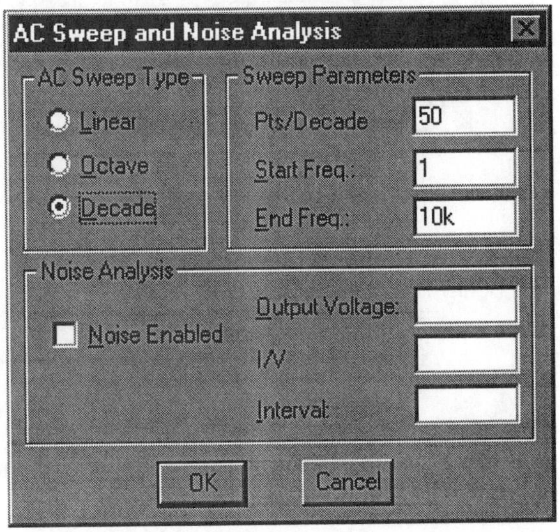

Figure 12.3. The AC Sweep dialog box edited to vary frequency in decades from 1 Hz to 10 kHz.

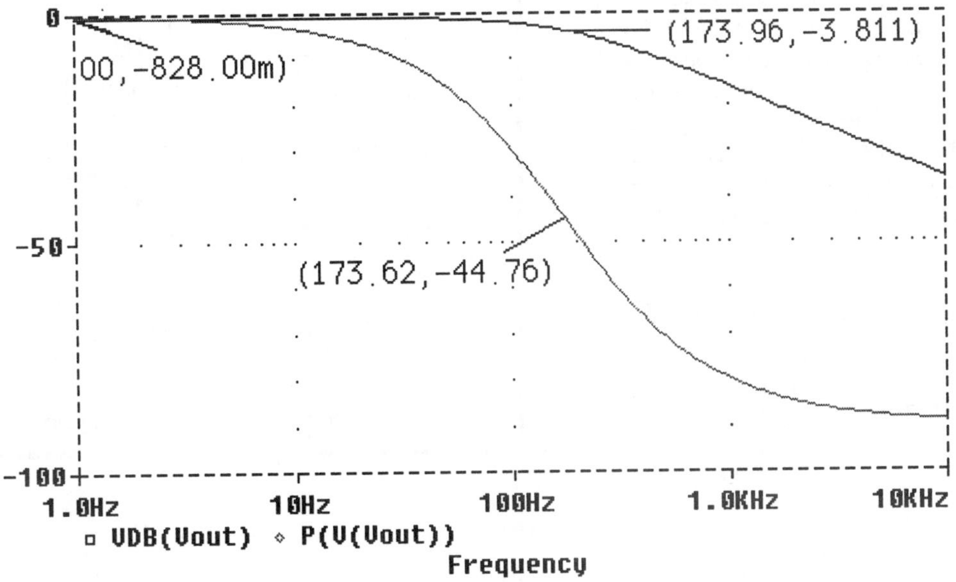

Figure 12.4. The Bode plot for Simulation One showing the data points of interest.

Simulation Two

As a demonstration of resonance, let's simulate the Panasonic ELJ-FC100JF, 10-μH inductor introduced on page 55. Using the inductor's parasitic resistance and capacitance values from page 55, we can find the resonant frequency using the *Schematics* circuit in Figure 12.5. We will vary the frequency of the ac current source until the voltage reaches a maximum. The frequency at which the maximum occurs is the resonant frequency. From the simulation results plotted in Figure 12.6, we find that the resonant frequency of the inductor is 32.36 MHz, which matches the manufacturer's specification of 32 MHz very well. Note that resonance occurs without any other passive elements connected to the inductor. We call this kind of resonance, ***self-resonance***.

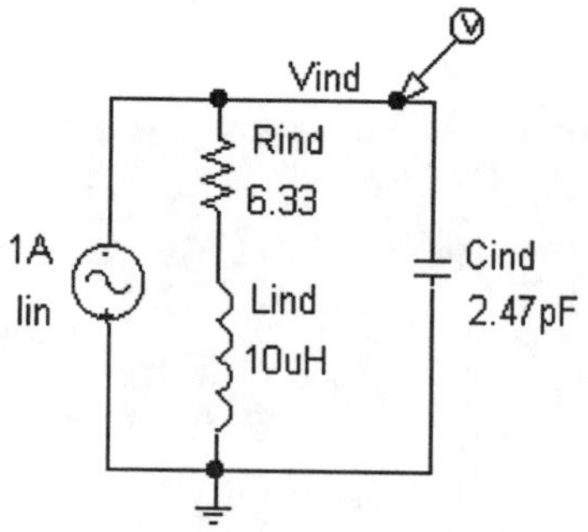

Figure 12.5. *Schematics* diagram for Simulation Two.

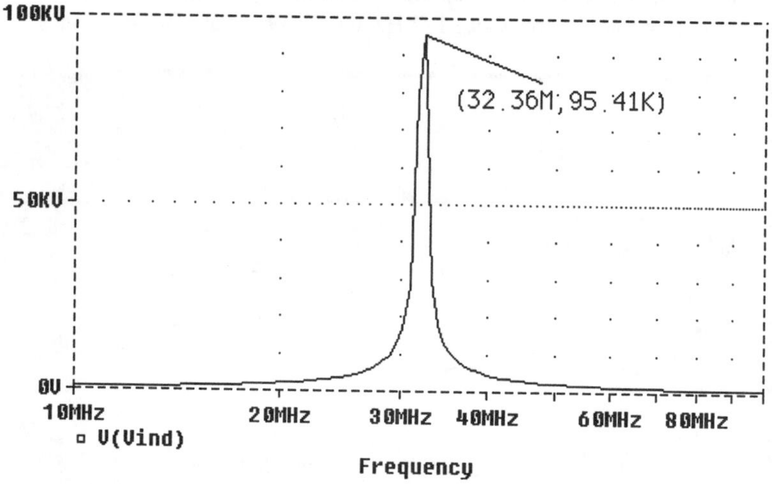

Figure 12.6. PROBE results for Simulation Two indicating inductor self-resonance.

Simulation Three

The *Schematics* circuit diagram in Figure 12.7 is a 2nd order low-pass filter. Let's simulate it, using the quasi-ideal opamp in the BECA library, to determine the center frequency, the bandwidth and the Q factor of the filter. From the resulting PROBE plot, shown in Figure 12.8, we find the center frequency is 501.2 kHz and the 3 dB corner frequencies are 511.8 kHz and 494.8 kHz. Therefore, bandwidth and Q factor are

$$BW = 511.8k - 494.8k = 7.0 \text{ kHz}$$

$$Q = \frac{f_O}{BW} = \frac{501.2k}{7.0k} = 71.6$$

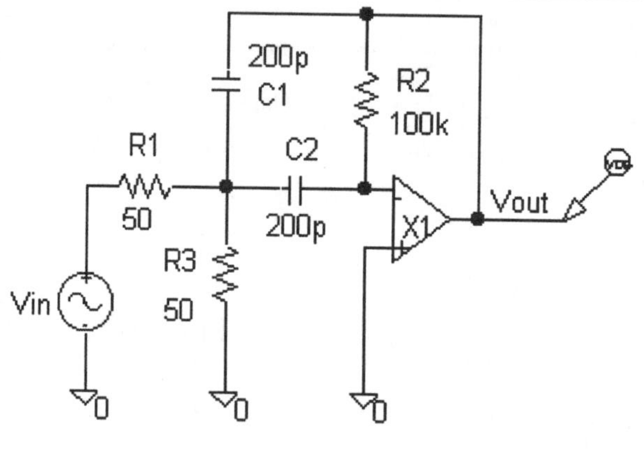

Figure 12.7. *Schematics* diagram for Simulation Three.

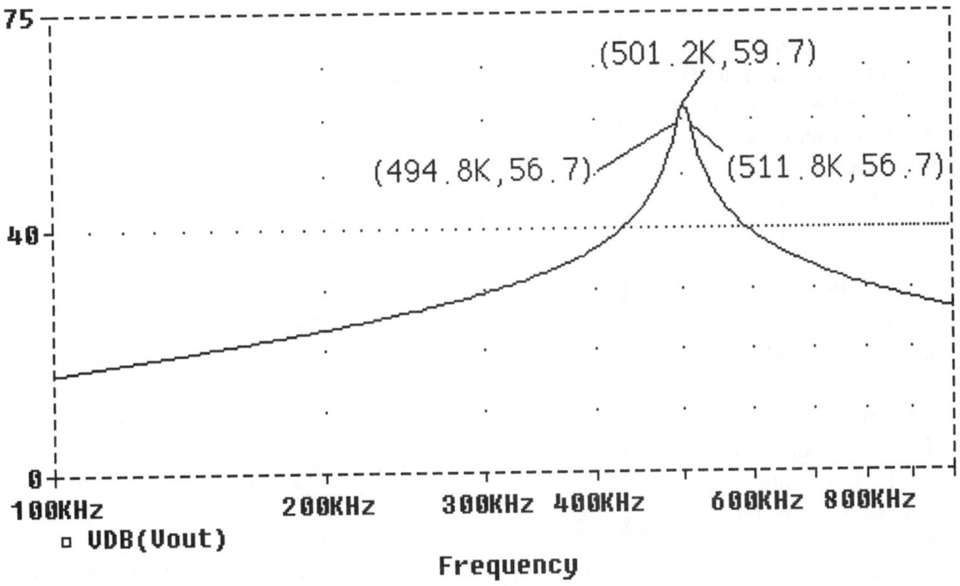

Figure 12.8. PROBE results for Simulation Three.

Simulation Four

For the final PSpice simulation of this chapter, we return to the Performance Analysis simulation. In this case, we will simulate the 2nd order bandpass filter in Figure 12.9 to determine the value of R_B that yields the highest Q value. This requires an AC Sweep across a range of frequencies, a Parametric sweep of R_B and a means of plotting Q. Figure 12.10 shows the corresponding *Schematics* diagram. Using the PARAM part and setting the AC Sweep and the Parametric dialog boxes as shown in Figure 12.11 satisfies the first two requirements. After simulating and opening PROBE, select Performance Analysis from the Trace menu. This causes the variable R to become the *x*-axis variable.

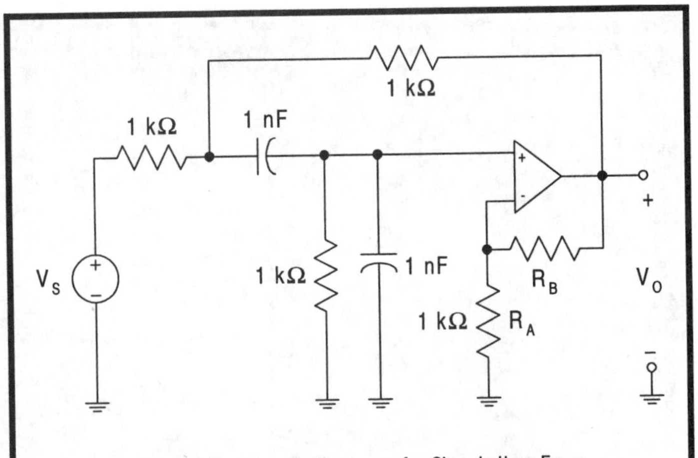

Figure 12.9. Circuit diagram for Simulation Four.

Now we will plot the quality factor, Q, which we know to be

$$Q = \frac{center\ frequency}{bandwidth}$$

Select Add from the Trace menu to open the window in Figure 12.12. On the right side of the window is a list of goal functions. We are interested in two in particular: CenterFreq(1,db_level) and BPBW(1,db_level). Let's examine the arguments of these goal functions. First, for CenterFreq(1,db_level), the number 1 indicates that CenterFreq is a function of only one voltage or current. The remaining variable, db_level, is just a number - we recommend 0.1. To demonstrate how the CenterFreq(1,db_level) goal function works, consider the following example

CenterFreq(V(Vout),0.1)

PROBE finds the maximum value of the voltage Vout. Next, the frequencies that correspond to that maximum minus 0.1 dB are found. The center frequency is then calculated as the average of these two frequencies. The accuracy of this averaging approach improves as the db_level value is decreased.

Arguments for the BPBW(1,db_level) function (bandpass bandwidth) are exactly the same. Since bandwidth is specified at 3 dB below the maximum value, we should set db_level = 3. Plots of center frequency and bandwidth are shown in Figure 12.13a. Notice that the center frequency is fairly constant while the bandwidth depends on R_B. This will cause Q to also depend on R_B as shown in Figure 12.13b. We find that Q has a maximum value of 21.8 when $R_B = 4\ k\Omega$.

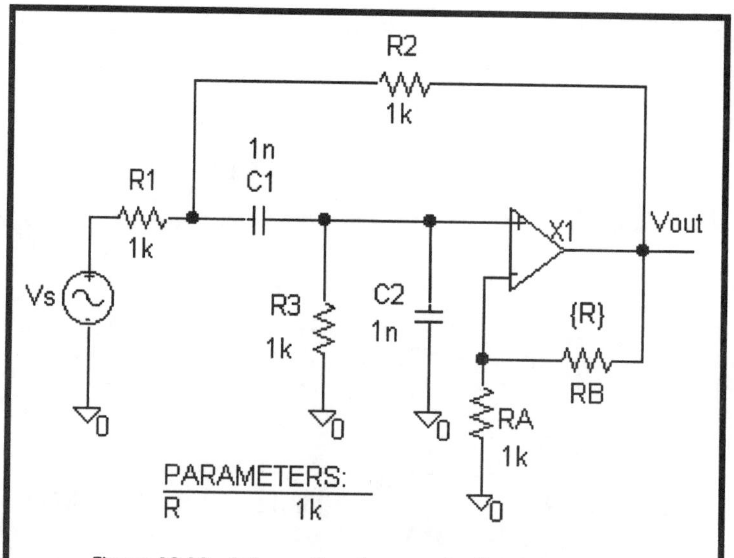

Figure 12.10. *Schematics* diagram for Simulation Four.

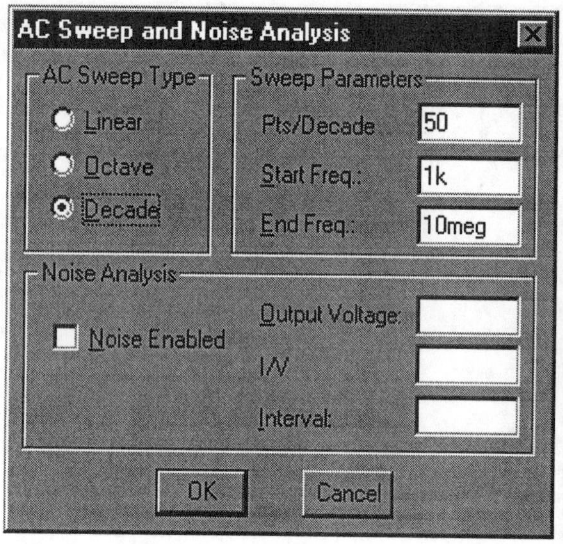

(a)

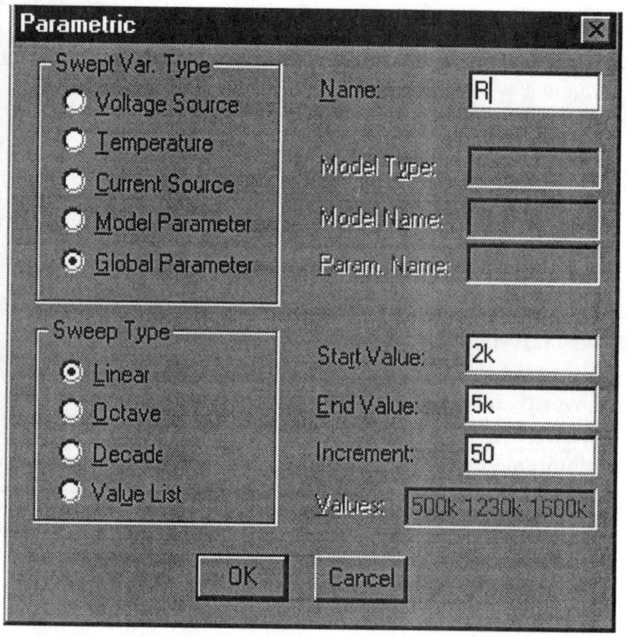

(b)

Figure 12.11. Setup dialog boxes edited for Simulation Four: (a) AC Sweep and (b) Parametric.

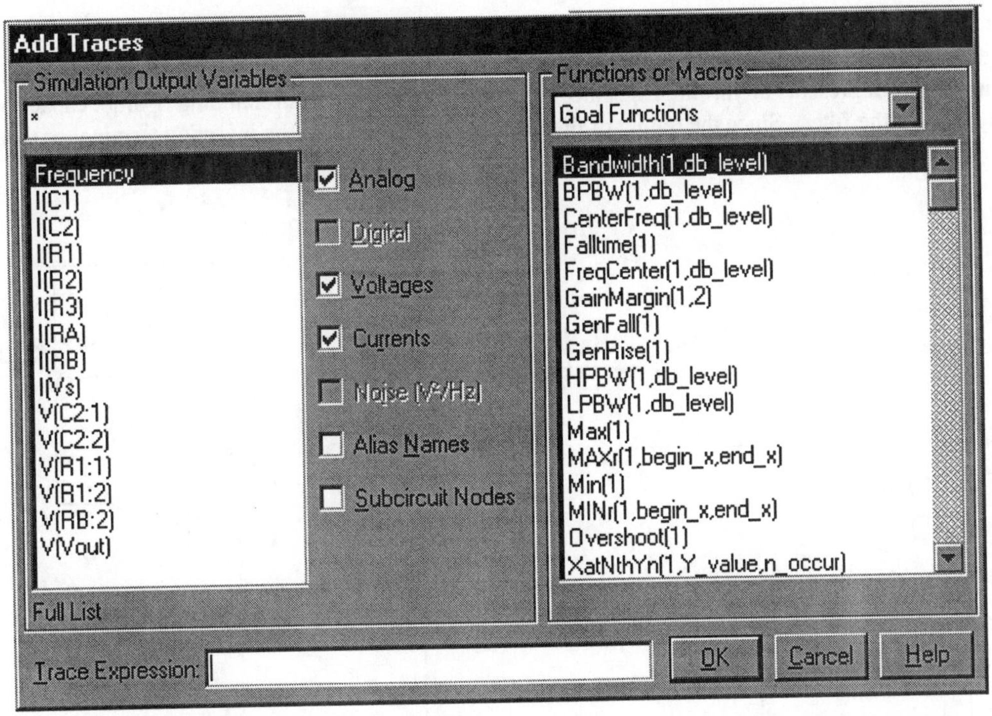

Figure 12.12. The Performance Analysis Add Traces window. Goal functions are on the right while voltages and currents are on the left.

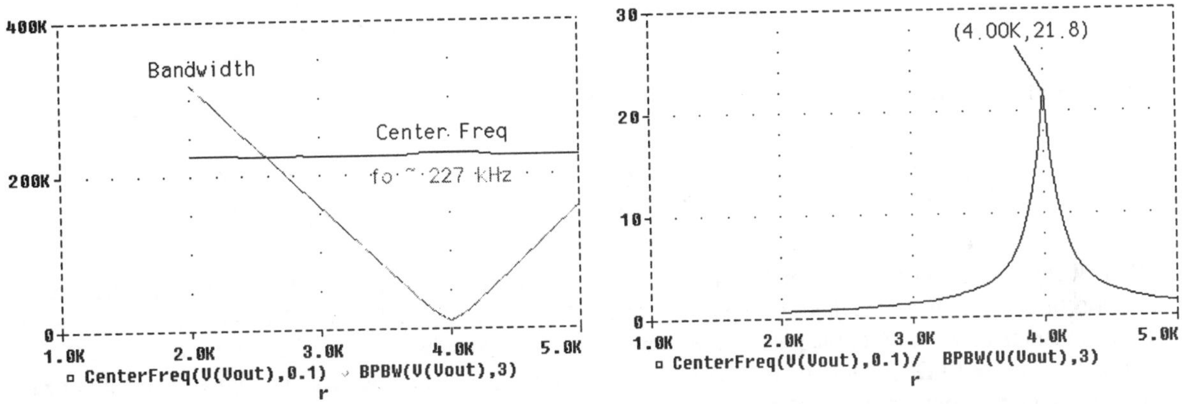

(a) (b)

Figure 12.13. Results for Simulation Four: (a) bandwidth and center frequency and (b) quality factor.

EWB SIMULATION

For our EWB simulation, we will use a lowpass filter to eliminate line noise (60 Hz) from a thermocouple's voltage signal. A thermocouple consists of the two wires made of dissimilar metals that are joined at one end. If there is a temperature difference between the joint and the rest of the thermocouple, then a voltage is developed between the two wires. This phenomenon is called the Seebeck effect. A typical thermocouple sensitivity is about 40 μV/C - not very much. So, for a temperature of 100°C, the thermocouple voltage will be only 4 mV! Since thermocouples are used extensively in factories and industrial plants, susceptibility to 60 Hz noise is a serious concerns. In most cases, the thermocouple voltage is filtered using a lowpass active filter. In this way, the small thermocouple voltage can be amplified while the noise is attenuated.

As an example, consider the EWB diagram in Figure 12.14. The 1-V sinusoidal source models the 60 Hz noise while the battery and series resistor model the thermocouple's voltage and wire resistance. The first opamp is a unity gain buffer. It presents an infinite resistance to the thermocouple. The remaining opamp, capacitors and resistors make up the filter. Since the output of the buffer eventually enters the inverting terminal of the filter opamp, we can expect the overall gain of the network to be negative.

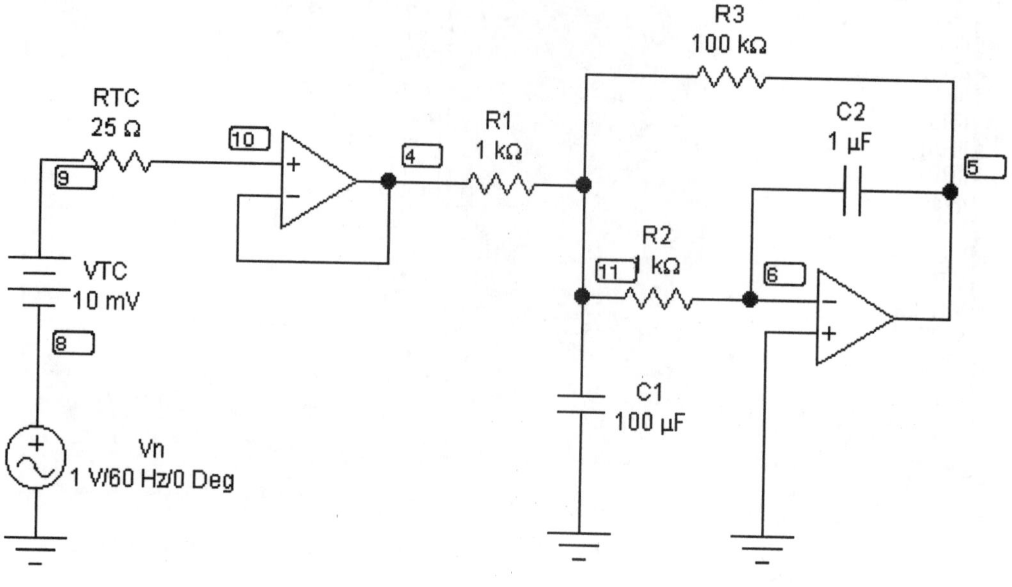

Figure 12.14. EWB diagram for a second order thermocouple filter.

From the BECA text, the characteristics of the filter are

$$A_O = \frac{-R_3}{R_1} \qquad \omega_C^2 = \frac{1}{C_1 C_2 R_2 R_3} \qquad 2\varsigma\omega_C = \frac{1}{C_1}\left(\frac{1}{R_1} + \frac{1}{R_2} + \frac{1}{R_3}\right)$$

We will choose a dc gain of - 100, a damping ratio of unity and ω_C = 10 rad/sec.. These choices ensure three features of interest. First, the thermocouple voltage will be amplified by a factor of 100. Second, the poles will be equal and real. And, third, the pole frequencies will be much less than 60 Hz. We arbitrarily choose R_3 = 100 kΩ and R_2 = 1 kΩ, yielding R_1 = 1 kΩ, C_1 = 100 µF and C_2 = 1 µF. Figure 12.15 shows the resulting gain magnitude plot. From the cursor data, we see that the gain at dc is indeed 100 and drops to 0.071 at 60 Hz. So, the thermocouple voltage is amplified by 100 while the noise is reduced (attenuated) by a factor of 0.071!

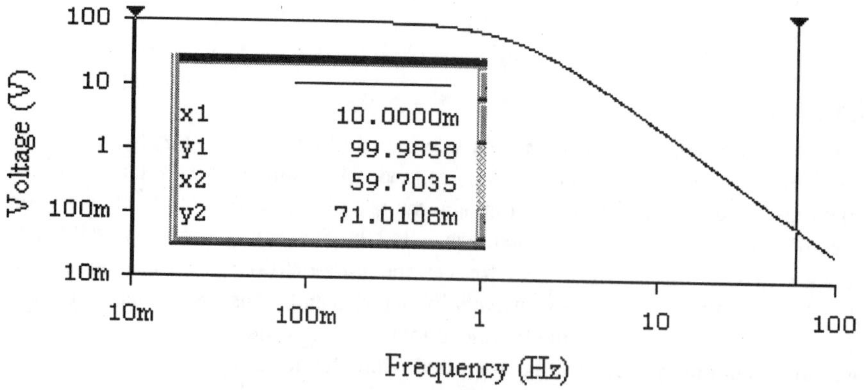

Figure 12.15. EWB AC Frequency Analysis results demonstrating 60 Hz noise filtering.

A transient simulation of the voltage on the thermocouple before (large sinewave) and after filtering is shown in Figure 12.16. We expect the 60 Hz component to be 0.071 times smaller and the dc portion of the output to be -1 V. While extracting exact values from the plot is difficult, we do see the desired results.

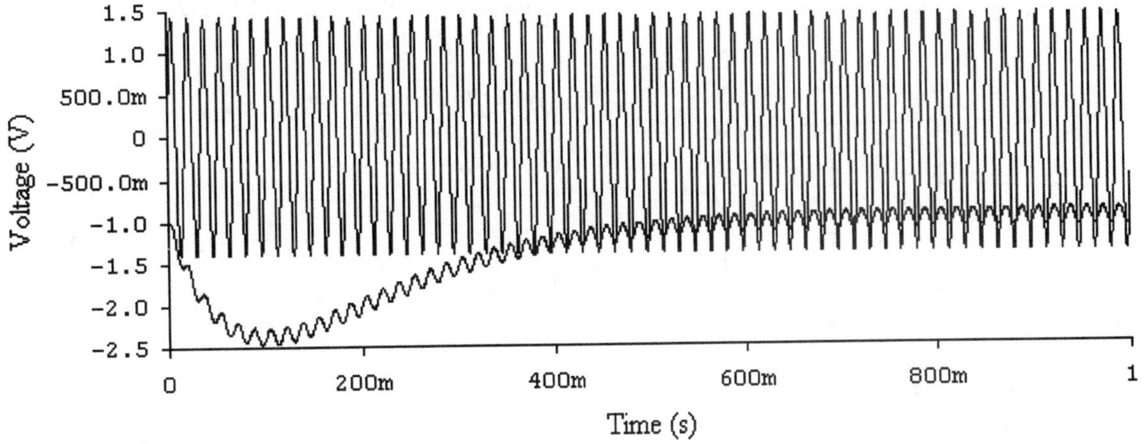

Figure 12.16. EWB Transient simulation of the thermocouple filter.

NOTES

PROBLEM SOLVING EXAMPLES

1. Determine the driving-point impedance at the input terminals of the network shown below as a function of s.

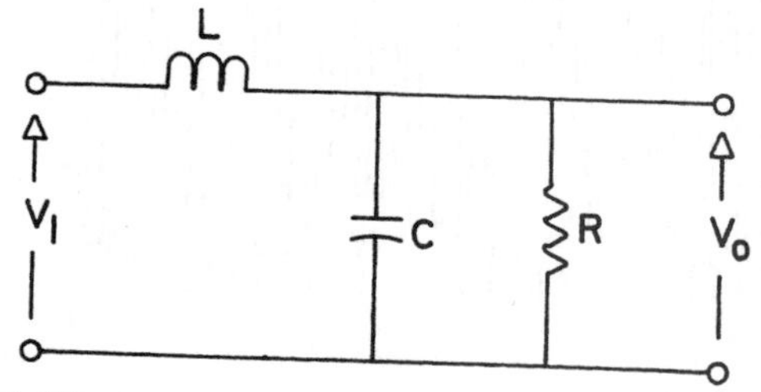

Solution:

$$Z_{in}(s) = sL + \frac{\frac{R}{sC}}{R+\frac{1}{sC}} = \frac{s^2LCR+sL+R}{sCR+1}$$

2. Compute the voltage transfer function as a function of s for the network in Problem 1.

Solution: From the solution to the previous problem

$$V_o(s) = \frac{V_i(s)}{\frac{s^2LCR+sL+R}{sCR+1}}\left(\frac{\frac{R}{sC}}{R+\frac{1}{sC}}\right)$$

Hence

$$G_v(s) = \frac{R}{s^2LCR+sL+R}$$

3. Sketch the magnitude and phase characteristics as a function of frequency of the network functions defined by each transfer function.

a) $\frac{1}{j\omega+\alpha}$ b) $\frac{j\omega+\alpha}{j\omega+\beta}$, $\alpha > \beta$ c) $\frac{j\omega}{j\omega+\alpha}$

Solution:

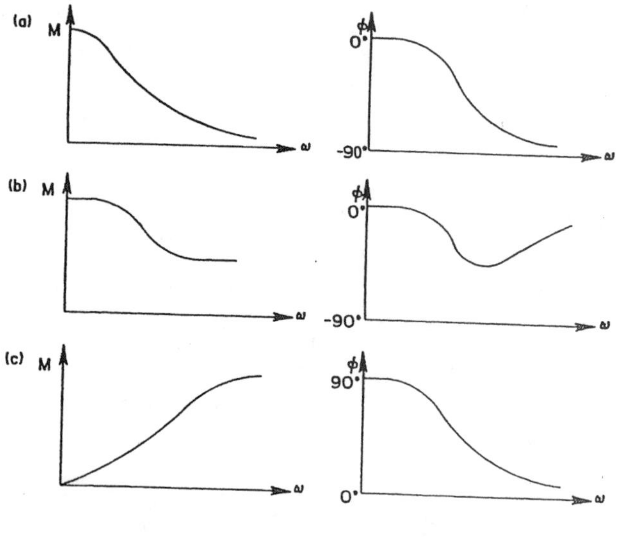

4. Given the network functions defined by each transfer function, sketch their magnitude and phase characteristics as a function of frequency.

a) $\frac{j\omega-\alpha}{j\omega+\alpha}$ b) $\frac{j\omega+\alpha}{j\omega+\beta}$, $\beta > \alpha$

Solution:

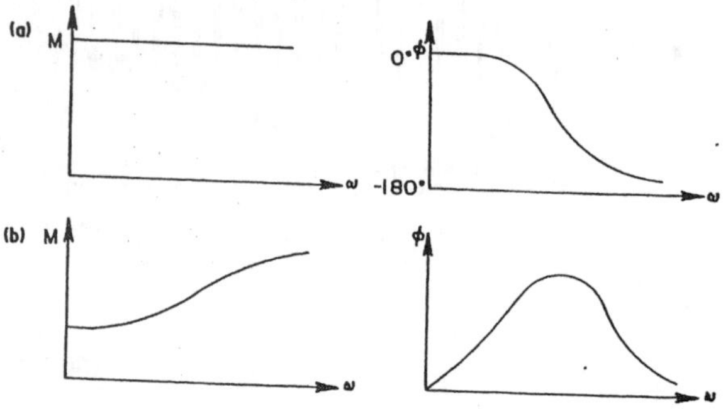

5. Draw the Bode plot for the network function

$$H(j\omega) = \frac{1}{(j\omega+5)(j\omega+10)}$$

Solution:

$$H(j\omega) = \frac{\frac{1}{50}}{(0.2j\omega+1)(0.1j\omega+1)} \text{ and } 20\log_{10}\left(\frac{1}{50}\right) =$$

-34dB

Therefore the Bode plot is

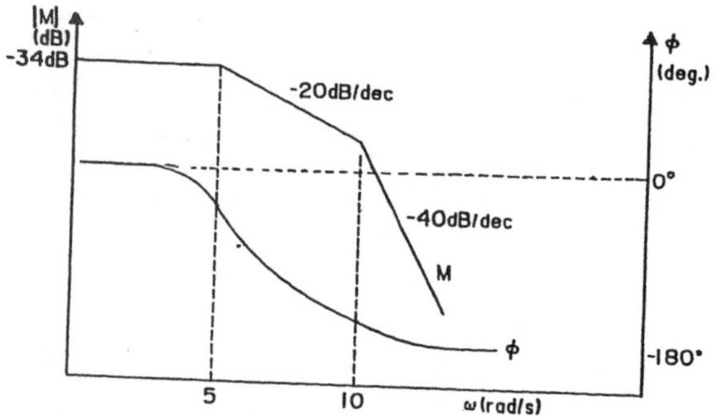

6. Draw the Bode plot for the network function

$$H(j\omega) = \frac{(j\omega+40)}{(j\omega+1)(j\omega+2)}$$

Solution: $H(j\omega)$ can be written as

$$H(j\omega) = \frac{20(j\omega/40+1)}{(j\omega+1)(j\omega/2+1)}$$

Therefore

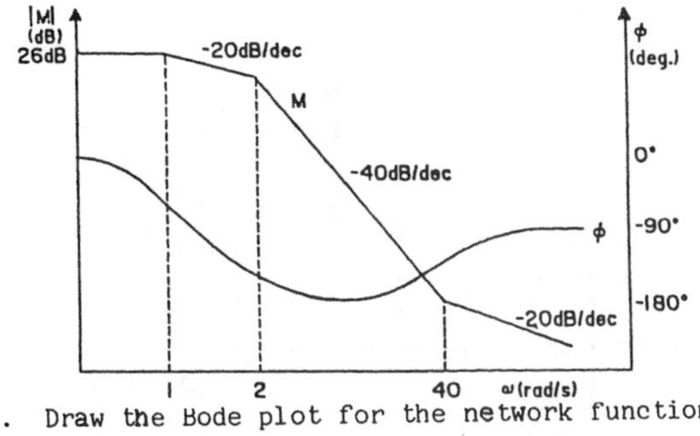

7. Draw the Bode plot for the network function

$$H(j\omega) = \frac{(j\omega+8)(j\omega+2)}{-\omega^2}$$

Solution: The network function can be written as

$$H(j\omega) = 16\,\frac{(0.125j\omega+1)(0.5j\omega+1)}{(j\omega)^2}$$

The Bode plot is then

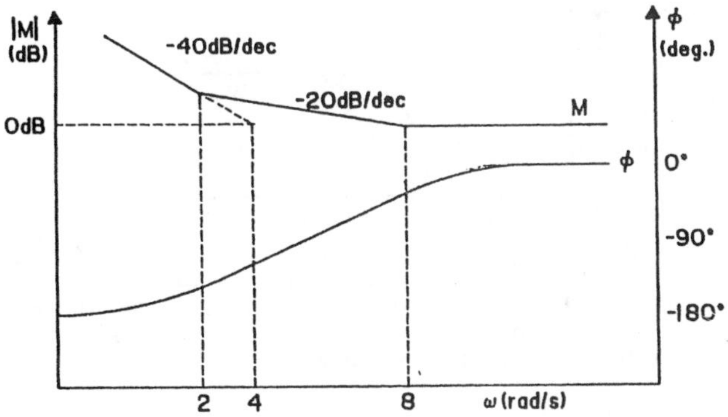

8. Sketch the magnitude characteristic of the Bode plot for the transfer function

$$G(j\omega) = \frac{(j\omega+4)^2}{-\omega^2(\frac{j\omega}{10}+1)(\frac{j\omega}{20}+1)}$$

Solution: The transfer function can be written as

$$G(j\omega) = \frac{16(\frac{j\omega}{4}+1)^2}{(j\omega)^2(\frac{j\omega}{10}+1)(\frac{j\omega}{20}+1)}$$

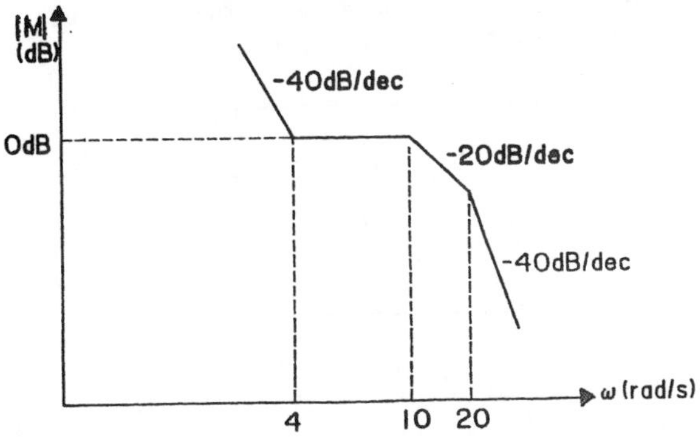

9. Find $H(j\omega)$ for the magnitude characteristic shown below.

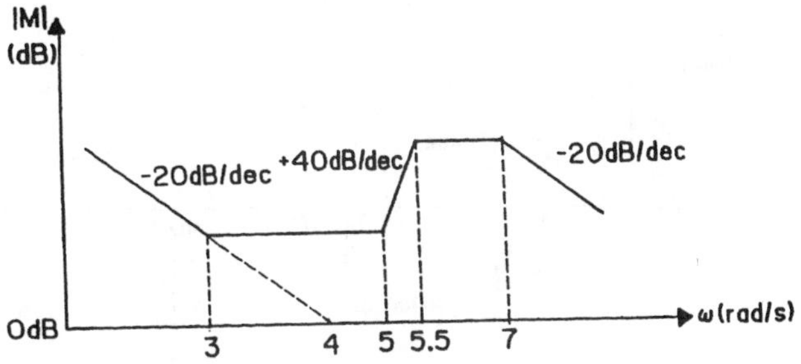

Solution:

$$H(j\omega) = \frac{4(\frac{j\omega}{3}+1)(\frac{j\omega}{5}+1)^2}{(j\omega)(\frac{j\omega}{5.5}+1)^2(\frac{j\omega}{7}+1)}$$

10. Find $H(j\omega)$ for the following magnitude characteristic.

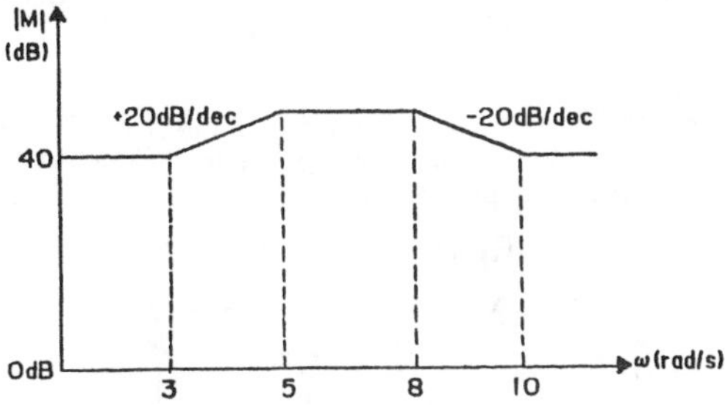

Solution:

$$H(j\omega) = \frac{100(\frac{j\omega}{3}+1)(\frac{j\omega}{10}+1)}{(\frac{j\omega}{5}+1)(\frac{j\omega}{8}+1)}$$

11. Determine $H(j\omega)$ for the following magnitude characteristic.

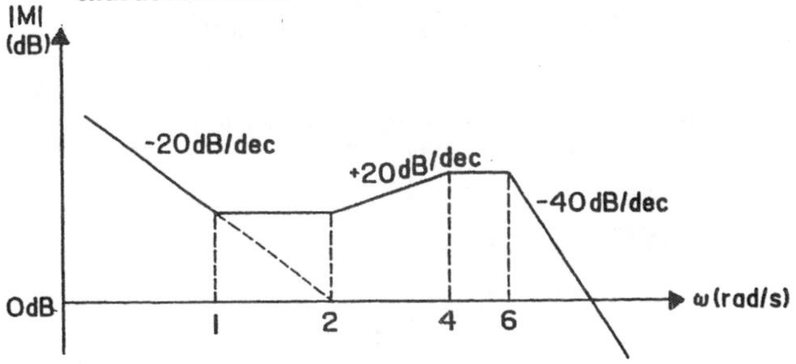

Solution:

$$H(j\omega) = \frac{2(j\omega+1)(\frac{j\omega}{2}+1)}{(j\omega)(\frac{j\omega}{4}+1)(\frac{j\omega}{6}+1)^2}$$

12. Given the following magnitude characteristic for $G(j\omega)$ determine the transfer function $G(j\omega)$.

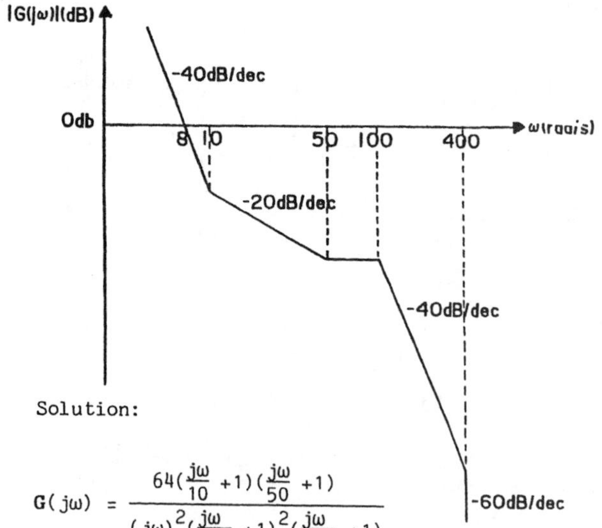

Solution:

$$G(j\omega) = \frac{64(\frac{j\omega}{10}+1)(\frac{j\omega}{50}+1)}{(j\omega)^2(\frac{j\omega}{100}+1)^2(\frac{j\omega}{400}+1)}$$

13. Compute the voltage transfer function for the network shown below and tell what type of filter the network represents.

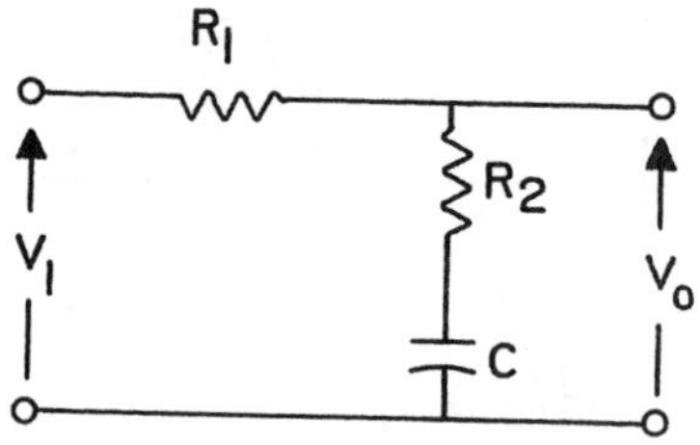

Solution:

$$\frac{V_o}{V_1}(j\omega) = \frac{R_2+\frac{1}{j\omega C}}{R_1+R_2+\frac{1}{j\omega C}} = \frac{j\omega CR_2+1}{j\omega C(R_1+R_2)+1}$$

The magnitude characteristic as a function of frequency is of the form

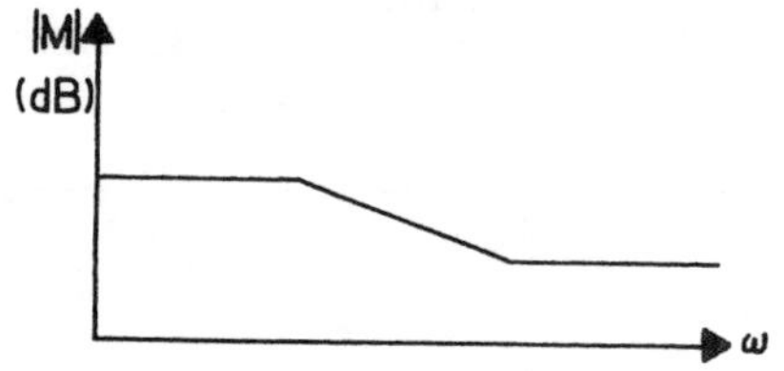

Therefore this is a low-pass filter.

14. Given the following network, sketch the magnitude characteristic for the voltage transfer function and determine what type of filter the network represents.

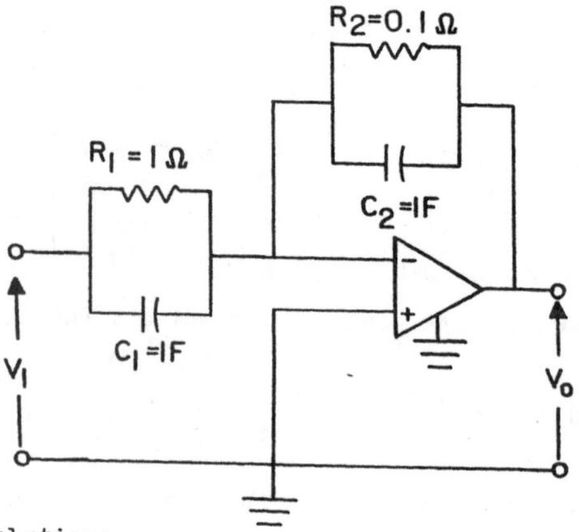

Solution:

$$\frac{V_o}{V_1}(j\omega) = -\frac{Z_2(j\omega)}{Z_1(j\omega)} = -\frac{.1(j\omega+1)}{.1j\omega+1}$$

The magnitude characteristic is of the form

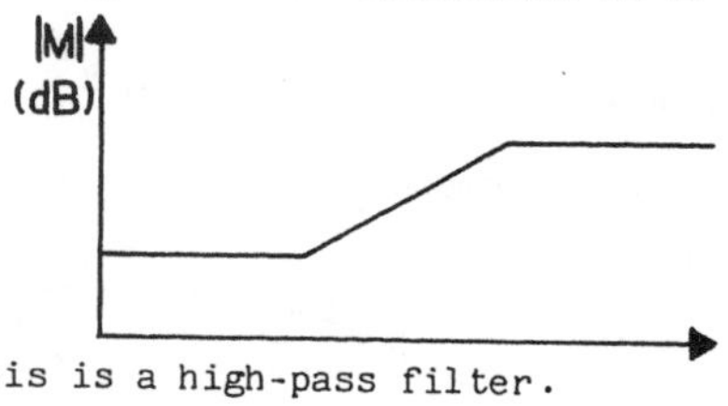

This is a high-pass filter.

15. A series RLC circuit is required to have a resonant frequency of 1 MHz and a bandwidth that is 2% of the resonant frequency. If R = 100Ω, find the values for L and C.

Solution:

$$BW = (0.02)(2\pi)(10^6) = (4\pi)(10^4)\text{rad/s}$$

$$Q = \frac{\omega_o}{BW} = \frac{(2\pi)(10^6)}{(4\pi)(10^4)} = 50$$

$$L = \frac{QR}{\omega_o} = \frac{(50)(100)}{(2\pi)(10^6)} = 0.8\text{mH}$$

$$C = \frac{1}{\omega_o^2 L} = \frac{1}{[(2\pi)(10^6)]^2(0.8)(10^{-3})} = 31.7\text{pF}$$

16. Determine the expression for the frequency at which the input impedance of the following network is real.

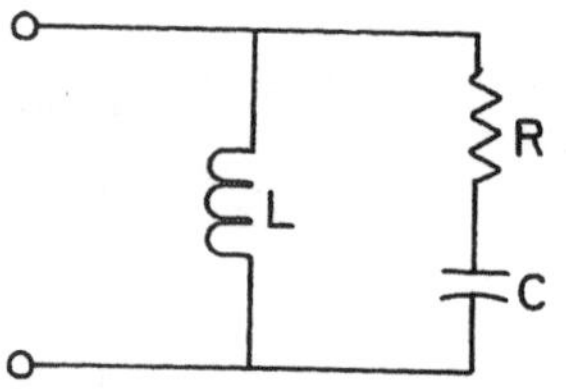

Solution: The input impedance is

$$Z(j\omega) = \frac{j\omega L\left(R+\frac{1}{j\omega C}\right)}{j\omega L+R+\frac{1}{j\omega C}} = \frac{j\omega L-\omega^2 LCR}{(1-\omega^2 LC)+j\omega CR}$$

$$= \frac{(j\omega L-\omega^2 LCR)[(1-\omega^2 LC)-j\omega CR]}{[1-\omega^2 LC]^2+(\omega CR)^2}$$

Setting the j term in the numerator equal to zero yields

$$\omega L(1-\omega^2 LC)+\omega CR(\omega^2 LCR) = 0$$

or

$$\omega = \frac{1}{\sqrt{LC-C^2R^2}}$$

17. Given the following network, find the frequency ω_M at which the magnitude of impedance $\mathbf{Z}(j\omega)$ is a maximum. In addition, determine the phase of the impedance at the frequency ω_M.

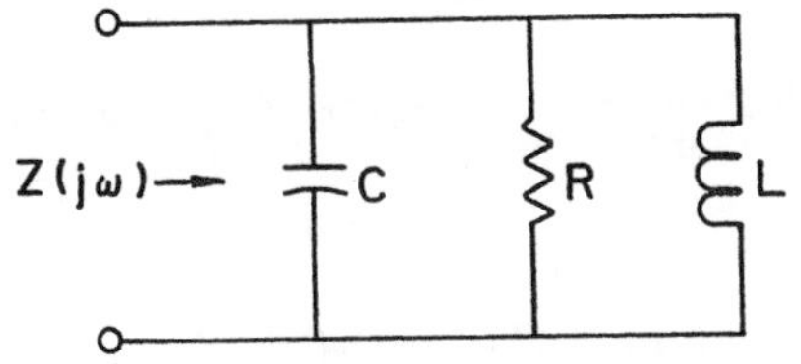

Solution:

$$\mathbf{Z}(j\omega) = \frac{1}{\frac{1}{R}+j\omega C+\frac{1}{j\omega L}} = \frac{1}{\frac{1}{R}+j\left(\omega C-\frac{1}{\omega L}\right)}$$

Therefore $|\mathbf{Z}|_{max}$ occurs when $\omega C-\frac{1}{\omega L} = 0$ or

$$\omega_M = \frac{1}{\sqrt{LC}} \text{ and } |\mathbf{Z}|_{max} = R$$

$$\phi(\omega) = \tan^{-1}\left[\frac{\omega C-\frac{1}{\omega L}}{\frac{1}{R}}\right]$$

Hence

$$\phi(\omega_M) = 0$$

18. Consider the following network. The source is a sinusoidal voltage source. The frequency is 60Hz and the rms value is 120 volts. We wish to deliver maximum power to the 5Ω resistor. Determine the value of a capacitor which, when placed in series with the source, will provide maximum power transfer to the 5Ω load. In addition, find the value of the voltage across the capacitor and inductor and the maximum power delivered to the load. Comment on the magnitude of the capacitor and inductor voltages.

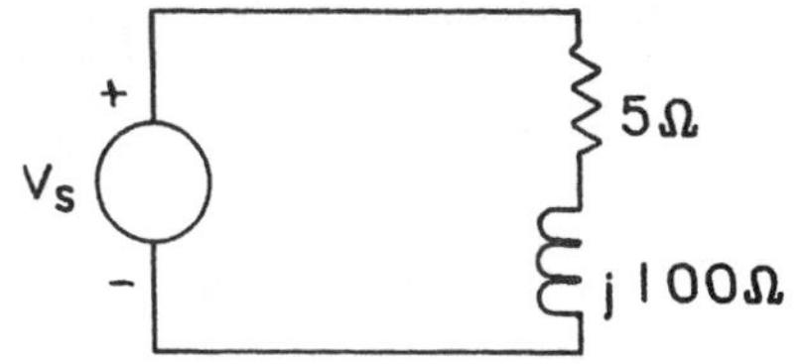

Solution: To deliver maximum power to the 5Ω load we establish series resonance.

$$\omega L = 100 \text{ and hence } L = \frac{100}{377} = 0.2653\text{H}$$

$$\text{Then } C = \frac{1}{\omega_o^2 L} = 26.52\mu\text{F}$$

$I_{rms} = \frac{120}{5} = 24\text{A}$ and therefore $P_{Load} = (24)^2(5) = 2.88\text{KW}$

$$|\mathbf{V}_L|_{max} = |\mathbf{V}_C|_{max} = (100)(24)\sqrt{2} = 3394\text{V}$$

This magnitude is very high due to resonance condition and may be damaging to the network components.

Chapter 13 THE LAPLACE TRANSFORM

PROBLEM SOLVING EXAMPLES

1. Use the results of Theorem-3 and the fact that if $f(t) = e^{-t}\sin t$, then $F(s) = \frac{1}{(s+1)^2+1}$ to find the Laplace transform of $f(t) = e^{-2t}\sin 2t$.

Solution: Since $\mathcal{L}[f(at)] = \frac{1}{a}F(\frac{s}{a})$, then

$$\mathcal{L}[e^{-2t}\sin 2t] = (\frac{1}{2})\frac{1}{(\frac{s}{2}+1)^2+1} = \frac{2}{(s+2)^2+4}$$

2. Use Theorem-8 to find $\mathcal{L}[f(t)]$ if $f(t) = te^{-at}u(t-1)$.

Solution:

$$\mathcal{L}[e^{-at}u(t-1)] = e^{-s}\mathcal{L}[e^{-a(t+1)}]$$

$$= e^{-(s+a)}\mathcal{L}[e^{-at}] = \frac{e^{-(s+a)}}{s+a}\text{. Then } \mathcal{L}[te^{-at}u(t-1)]$$

$$= -\frac{d}{ds}\left[\frac{e^{-(s+a)}}{s+a}\right] = \frac{e^{-(s+a)}}{s+a} + \frac{e^{-(s+a)}}{(s+a)^2}$$

3. If $f(t)$ is given by the waveform shown below, find $F(s)$.

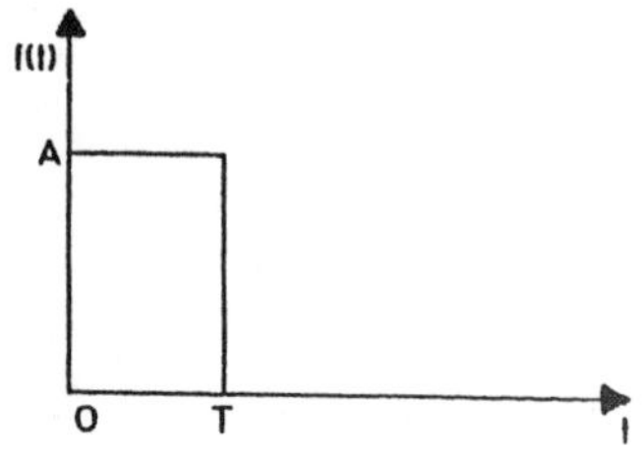

Solution:

$$f(t) = A[u(t)-u(t-T)]$$

Therefore

$$F(s) = \frac{A}{s} - \frac{A}{s}e^{-s} = \frac{A}{s}[1-e^{-s}]$$

4. Determine the Laplace transform of the following waveform.

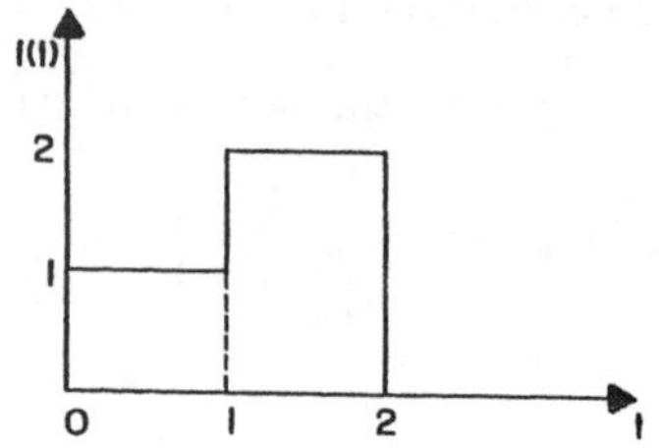

Solution:

$$f(t) = 1[u(t)-u(t-1)]+2[u(t-1)-u(t-2)]$$

$$= u(t)+u(t-1)-2u(t-2)$$

Therefore

$$F(s) = \frac{1}{s}[1+e^{-s}-2e^{-2s}]$$

5. Determine the Laplace transform of the following periodic waveform.

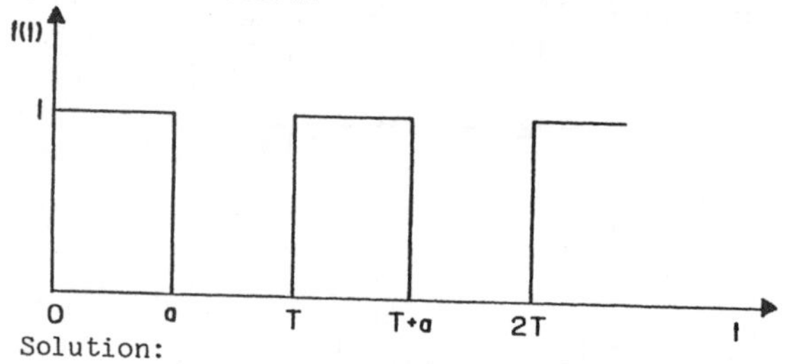

Solution:

$$f_1(t) = u(t)-u(t-a)$$

Hence

$$F_1(s) = \frac{1}{s}(1-e^{-as})$$

Therefore **F**(s) for the periodic function is

$$\mathbf{F}(s) = \frac{1}{s}\left[\frac{1-e^{-as}}{1-e^{-Ts}}\right]$$

6. Find the Laplace transform of the following waveform.

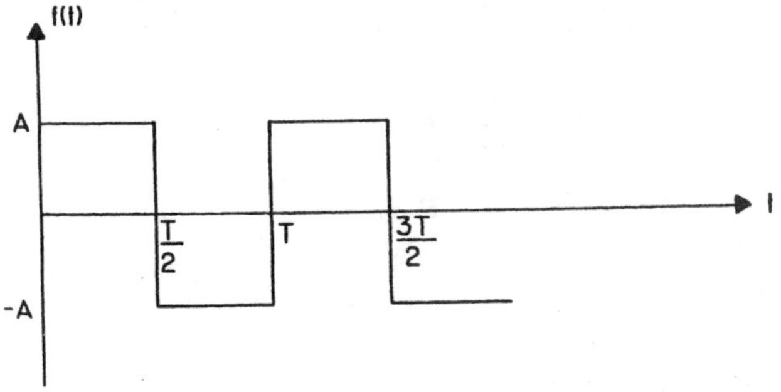

Solution:

$$f_1(t) = A[u(t)-u(t-\tfrac{T}{2})]-A[u(t-\tfrac{T}{2})-u(t-T)]$$

Hence

$$\mathbf{F}_1(s) = A\left[\frac{1-2e^{-\frac{T}{2}s}+e^{-Ts}}{s}\right]$$

Therefore **F**(s) for the periodic function is

$$\mathbf{F}(s) = A\left[\frac{1-2e^{-\frac{T}{2}s}+e^{-Ts}}{s(1-e^{-Ts})}\right]$$

7. Given the following functions **F**(s), find the inverse Laplace transform of each function.

(a) $\mathbf{F}(s) = \dfrac{2(s+1)}{(s+2)(s+3)}$

(b) $\mathbf{F}(s) = \dfrac{10(s+2)}{(s+1)(s+4)}$

(c) $\mathbf{F}(s) = \dfrac{s^2+2s+3}{s(s+1)(s+2)}$

Solution:

(a) $\dfrac{2(s+1)}{(s+2)(s+3)} = \dfrac{-2}{s+2} + \dfrac{4}{s+3}$

Hence $f(t) = [-2e^{-2t}+4e^{-3t}]u(t)$

(b) $\dfrac{10(s+2)}{(s+1)(s+4)} = \dfrac{\frac{10}{3}}{s+1} + \dfrac{\frac{20}{3}}{s+4}$

Hence $f(t) = [\frac{10}{3}e^{-t}+\frac{20}{3}e^{-4t}]u(t)$

(c) $\dfrac{s^2+2s+3}{s(s+1)(s+2)} = \dfrac{\frac{3}{2}}{s} - \dfrac{2}{s+1} + \dfrac{\frac{3}{2}}{s+2}$

Hence $f(t) = [\frac{3}{2} -2e^{-t}+ \frac{3}{2}e^{-2t}]u(t)$

8. Find f(t) if **F**(s) is given by the expression

$$\mathbf{F}(s) = \frac{(s+1)^2}{s(s+2)(s^2+2s+2)}$$

Solution:

$$\mathbf{F}(s) = \frac{k_0}{s} + \frac{k_1}{s+2} + \frac{k_2}{s+1-j1} + \frac{k_2^*}{s+1+j1}$$

$$k_0 = \frac{1}{4}, \quad k_1 = -\frac{1}{4}$$

$$k_2 = \left.\frac{(s+1)^2}{s(s+2)(s+1+j1)}\right|_{s=-1+j1} = -\frac{j}{4}$$

Hence

$$f(t) = [\tfrac{1}{4} - \tfrac{1}{4}e^{-2t}+ \tfrac{1}{2}e^{-t}\cos(t-90°)]u(t)$$

9. Find the inverse Laplace transform of **F**(s) where

$$\mathbf{F}(s) = \frac{e^{-s}}{s^2+2s+2}$$

Solution:

$$\frac{1}{s^2+2s+2} = \frac{1}{(s+1+j1)(s+1-j1)} = \frac{\frac{1}{2}\angle{-90°}}{s+1-j1} + \frac{\frac{1}{2}\angle{90°}}{s+1+j1}$$

Hence

$$f(t) = e^{-(t-1)}\cos[(t-1)-90°]u(t-1)$$

10. Find the inverse Laplace transform of the function

$$\mathbf{F}(s) = \frac{e^{-s}}{s^2(s^2+1)(s^2+4)}$$

Solution:

$$\frac{1}{s^2(s^2+1)(s^2+4)} = \frac{k_{01}}{s^2} + \frac{k_{02}}{s} + \frac{k_1}{s-j} + \frac{k_1^*}{s+j} + \frac{k_2}{s-2j} + \frac{k_2^*}{s+2j}$$

$$k_{01} = \frac{1}{4}, \quad k_{02} = \frac{d}{ds}\left[\frac{1}{(s^2+1)(s^2+4)}\right]_{s=0} = 0$$

$$k_1 = \frac{1}{(j)^2(4+j)^2} = \frac{1}{3}\,\underline{/180^\circ}$$

$$k_2 = \frac{1}{(2j)^2[1+(2j)^2]} = \frac{1}{12}$$

Therefore

$$f(t) = [\tfrac{1}{4}(t-1)+\tfrac{2}{3}\cos[(t-1)+180^\circ]+\tfrac{1}{6}\cos(t-1)]u(t-1)$$

11. Find the inverse Laplace transform of the function $\mathbf{F}(s)$ using the convolution integral

$$\mathbf{F}(s) = \frac{1}{s(s+2)}$$

Solution:

If $\mathbf{F}_1(s) = \frac{1}{s}$ and $\mathbf{F}_2(s) = \frac{1}{s+2}$, then

$f_1(t) = u(t)$ and $f_2(t) = e^{-2t}$

Hence

$$f(t) = \int_0^t e^{-2\lambda}u(t-\lambda)d\lambda = \int_0^t e^{-2\lambda}d\lambda, \quad t>0$$

$$= \tfrac{1}{2}[1-e^{-2t}]u(t)$$

12. Find $f(t)$ using the convolution integral if

$$\mathbf{F}(s) = \frac{s+2}{s^2(s+1)}$$

Solution: If $\mathbf{F}_1(s) = \frac{1}{s^2}$ and $\mathbf{F}_2(s) = \frac{s+2}{s+1} = 1+\frac{1}{s+1}$

Then $f_1(t) = t$ and $f_2(t) = \delta(t)+e^{-t}$

Therefore

$$f(t) = \int_0^t (t-\lambda)(\delta(\lambda)+e^{-\lambda})d\lambda$$

$$= \int_0^t t\delta(\lambda)d\lambda+\int_0^t te^{-\lambda}d\lambda-\int_0^t\lambda\delta(\lambda)d\lambda-\int_0^t\lambda e^{-\lambda}d\lambda$$

$$= [2t+e^{-t}-1]u(t)$$

13. Determine the initial and final values of the time function $f(t)$ if $\mathbf{F}(s)$ is given by the expression

(a) $\mathbf{F}(s) = \dfrac{2(s+2)}{s(s+1)}$

(b) $\mathbf{F}(s) = \dfrac{2(s^2+2s+6)}{(s+1)(s+2)(s+3)}$

(c) $\mathbf{F}(s) = \dfrac{2s^2}{(s+1)(s^2+2s+2)}$

Solution:

(a) $f(0) = \lim_{s\to\infty} s\mathbf{F}(s) = 2$

$f(\infty) = \lim_{s\to 0} s\mathbf{F}(s) = 4$

(b) $f(0) = \lim_{s\to\infty} s\mathbf{F}(s) = 2$

$f(\infty) = \lim_{s\to 0} s\mathbf{F}(s) = 0$

(c) $f(0) = \lim_{s\to\infty} s\mathbf{F}(s) = 2$

$f(\infty) = \lim_{s\to 0} s\mathbf{F}(s) = 0$

14. Use Laplace transforms to solve the following differential equations.

(a) $\dfrac{dy(t)}{dt} +3y(t) = e^{-t}, \quad y(0) = 1$

(b) $\dfrac{dy(t)}{dt} +4y(t) = 2u(t), \quad y(0) = 2$

Solution:

(a) $s\mathbf{Y}(s)-y(0)+3\mathbf{Y}(s) = \dfrac{1}{s+1}$

$(s+2)\mathbf{Y}(s) = \dfrac{1}{s+1} +1$

$\mathbf{Y}(s) = \dfrac{s+2}{(s+1)(s+3)}$

Hence $y(t) = [\dfrac{e^{-t}}{2} + \dfrac{e^{-3t}}{2}]u(t)$

(b) $s\mathbf{Y}(s)-y(0)+4\mathbf{Y}(s) = \dfrac{2}{s}$

$\mathbf{Y}(s) = \dfrac{1}{s+4}\left[\dfrac{2}{s} +2\right]$

$= \dfrac{2(s+1)}{s(s+4)}$

Hence $y(t) = [\frac{1}{2} + \frac{3}{2}e^{-4t}]u(t)$

Chapter 14 APPLICATION OF THE LAPLACE TRANSFORM

Using the Laplace transform in circuit analysis is the culmination of our work. It converts differential equations to algebraic equations. Of course, the frequency domain (Chapter 8) did the same thing. But the frequency domain is limited to steady-state ac systems while the s-domain is applicable to dc, transient and ac analyzes. Also, as you become more comfortable with the s-domain, you will find that the mathematics readily surrenders information about the circuit under inspection.

PSPICE SIMULATIONS

SIMULATION ONE

For our first simulation, we will demonstrate the generic relationship between the time and *s*-domains for systems described by simple differential equations. A time-honored application for electronic circuitry, both digital and analog, has been calculating the trajectories of artillery fired projectiles. In fact, ENIAC, the world's first true digital computer, was constructed for the sole purpose of calculating such trajectories. We will instead use an analog approach to plot range versus altitude and range versus time for projectiles fired from the artillery piece depicted in Figure 14.1. We assume that at $t = 0$, the projectile is launched with particular horizontal and vertical velocities. Also, the projectile has no means of self propulsion. From our freshman physics classes, we know that the range $x(t)$ and altitude $y(t)$ are given by

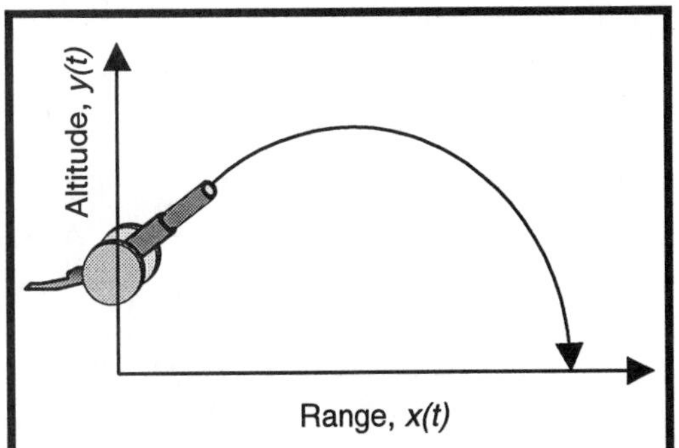

Figure 14.1. Depiction of the artillery piece described in Simulation One.

$$x(t) = \int \dot{x}(0)\, dt \qquad\qquad y(t) = y(0) + \int \dot{y}(0)\, dt - \int\int g\, dt^2 \qquad (14.1)$$

where g is the gravitational constant and the dotted terms represent velocities in the x and y directions.

To develop a circuit that can model the system described in (14.1), we will make use of the inverting integrator shown in Figure 14.2. Remember that for a particular simulation, the initial velocities, altitude and g are constants. The circuit in Figure 14.3 models the equations in (14.1) very well. Each op-amp subcircuit is either an integrator, a summer or a summing integrator. You should be able to verify that,

$$V_X(t) = \int V_{VXO}\, dt \qquad\qquad V_Y(t) = V_{YO} + \int V_{VYO}\, dt - \int\int V_G\, dt^2 \qquad (14.2)$$

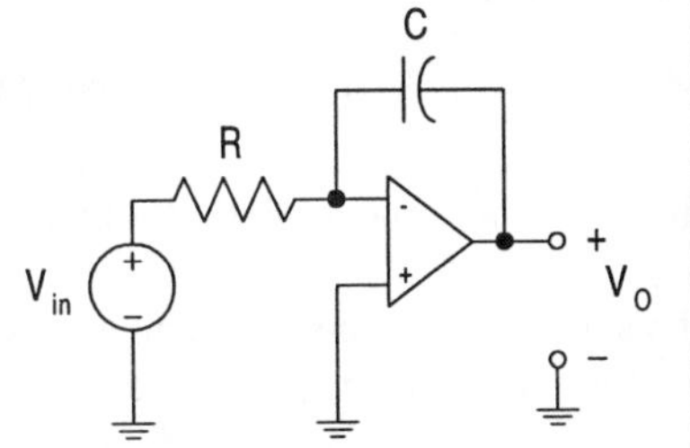

$$v_o(t) = \frac{-1}{RC} \int v_{in}(t)\, dt$$

Figure 14.2. The inverting integrator.

Comparing (14.1) to (14.2), we see the following analogies

Equation 14.1	Equation 14.2
$x(t)$	$V_X(t)$
$\dot{x}(0)$	V_{VXO}
$y(t)$	$V_Y(t)$
$y(0)$	V_{YO}
$\dot{y}(0)$	V_{VYO}
g	$V_G(t)$

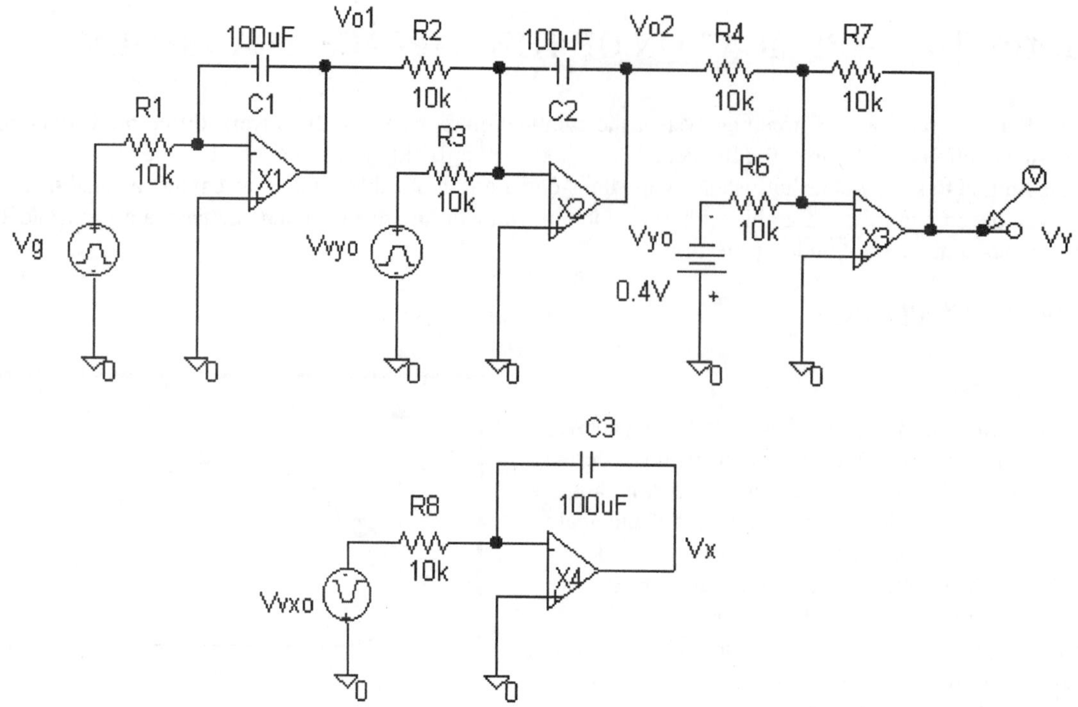

Figure 14.3. Schematics diagram for the analog computer in Simulation One.

To complete the analogy, we must decide how many kilometers on the firing range correspond to one volt in the circuit. Here, we have chosen to let one volt represent one km. Although it is an option, we have decided not to scale the simulation time axis.

Now for a particular simulation. The projectile is launched at a 45° angle with an initial velocity of 500 m/s. Also, the artillery piece itself is at 400 m above ground level. We will find the range versus time and the trajectory. Since $g = 9.8$ m/s^2, and 1 volt corresponds to 1 km, we find $V_G = 9.8$ mV. Also, due to the 45° launch angle, the initial x and y velocities are equal, and

$$\dot{x}(0) = \dot{y}(0) = 500\cos(45) = 353.5 \text{ m/s} \qquad V_{VXO} = V_{VYO} = 0.3535 \text{ V}$$

Similarly, an initial altitude of 400 m corresponds to $V_{YO} = 0.4$ V.

The simulation results for range versus time are shown in Figure 14.4. Using the cursor, the maximum range and time-to-target are 25.9 km and 73.2 seconds. Plotting the trajectory is a two step process. First, delete Vx and add the trace, Vy. Second, change the x-axis variable from time to Vx. Do this by selecting **X-Axis Settings** from the **Plot** menu in PROBE. The dialog box in Figure 14.5 opens. Select **Axis Variable.** When the **X Axis Variable** dialog box appears, simply choose **V(Vx)**. The resulting trajectory plot is shown in Figure 14.6, where we find the maximum altitude is 6.77 km.

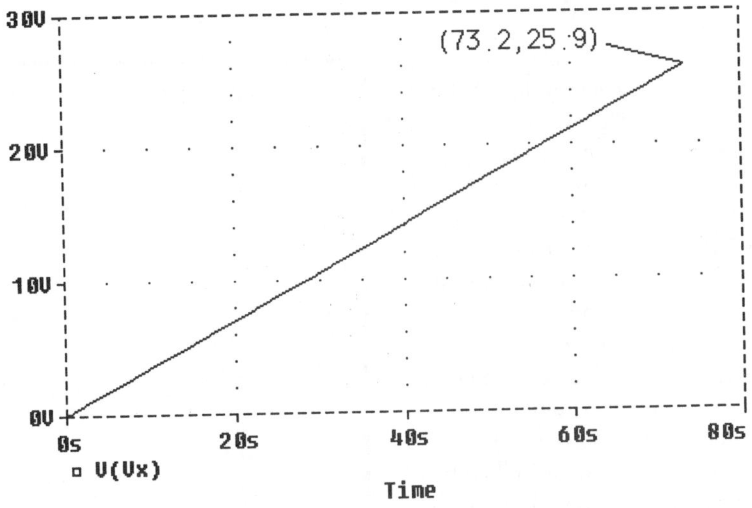

Figure 14.4. PROBE results for range and time-to-target for the artillery piece in Simulation One.

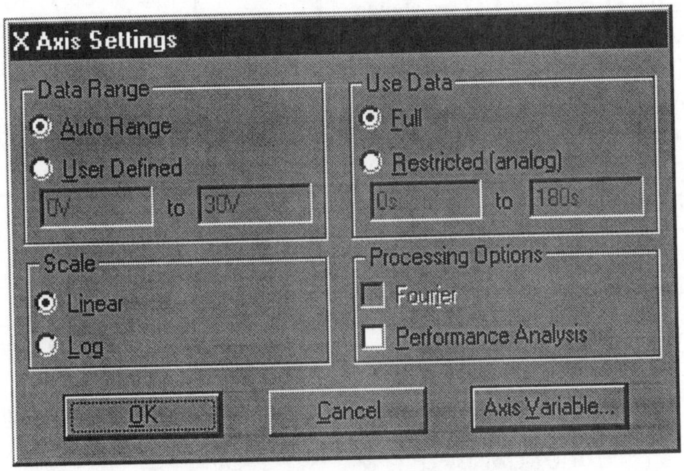

Figure 14.5. The X Axis Settings dialog box.

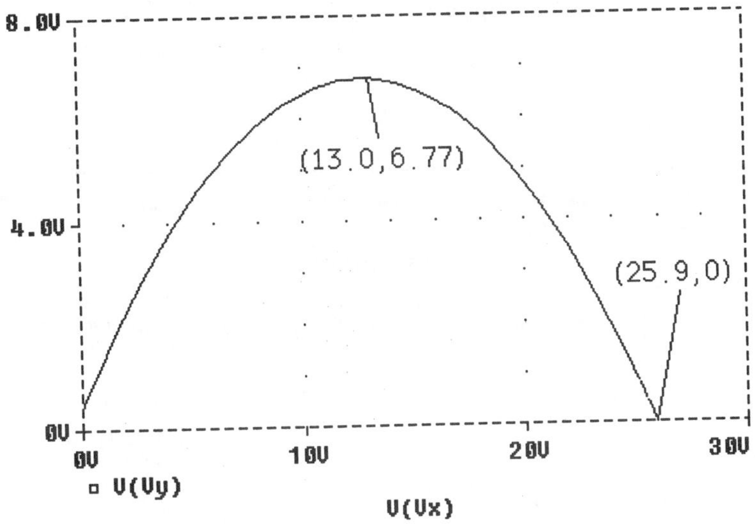

Figure 14.6. Projectile trajectory for the artillery piece in Simulation One.

SIMULATION TWO

In this simulation, we will examine the relationship between the Bode plot and the *s*-plane for the 2nd order network in Figure 14.7. In particular, we are interested in the pole locations of the transfer function for over-, under- and critical damping. By varying the resistor, we can produce each of these cases. Here, R has been stepped from 0.5 Ω to 5 Ω in 1.5 Ω increments. Critical damping occurs at 2 Ω. Figure 14.8 shows the resulting Bode plots (magnitude only) which were created in PSpice using the PROBE utility. The conspicuous "hump" in the Bode plot makes the underdamped case stand out. As for the critically damped and the two overdamped cases, they are difficult to distinguish because the poles are close to one another. When R = 5 Ω, the poles are farthest apart at 4.79 and 0.21 rad/s. But on a Bode plot, this quite close.

The *s*-plane plots for R = 0.5 Ω, 2 Ω and 5 Ω are shown in Figure 14.9. When the network is underdamped, R = 0.5 Ω, we see the poles are complex conjugates of one another, as expected. For critical damping, the poles are real and equal. And, as the damping increases and the system is more and more overdamped, the poles are real both move further and further apart.

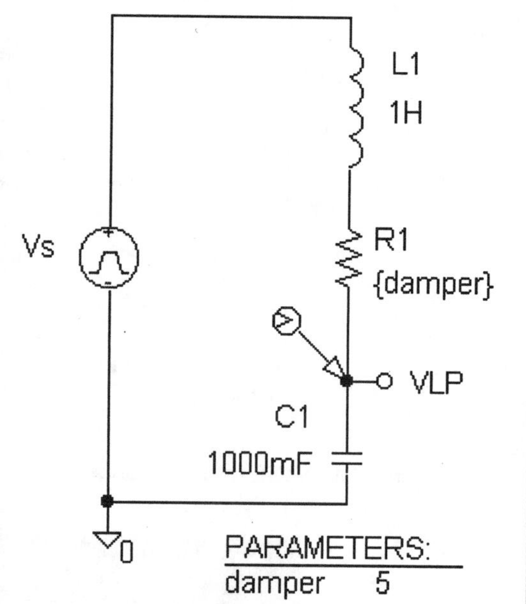

Figure 14.7. *Schematics* diagram for Simulation Two.

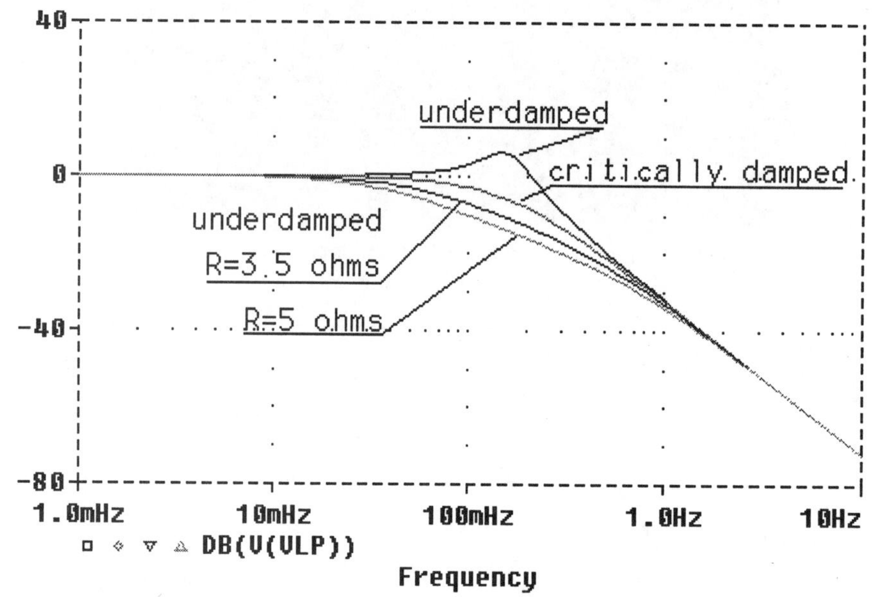

Figure 14.8. Effect of damping on the Bode plot (magnitude only) for the network in Simulation One.

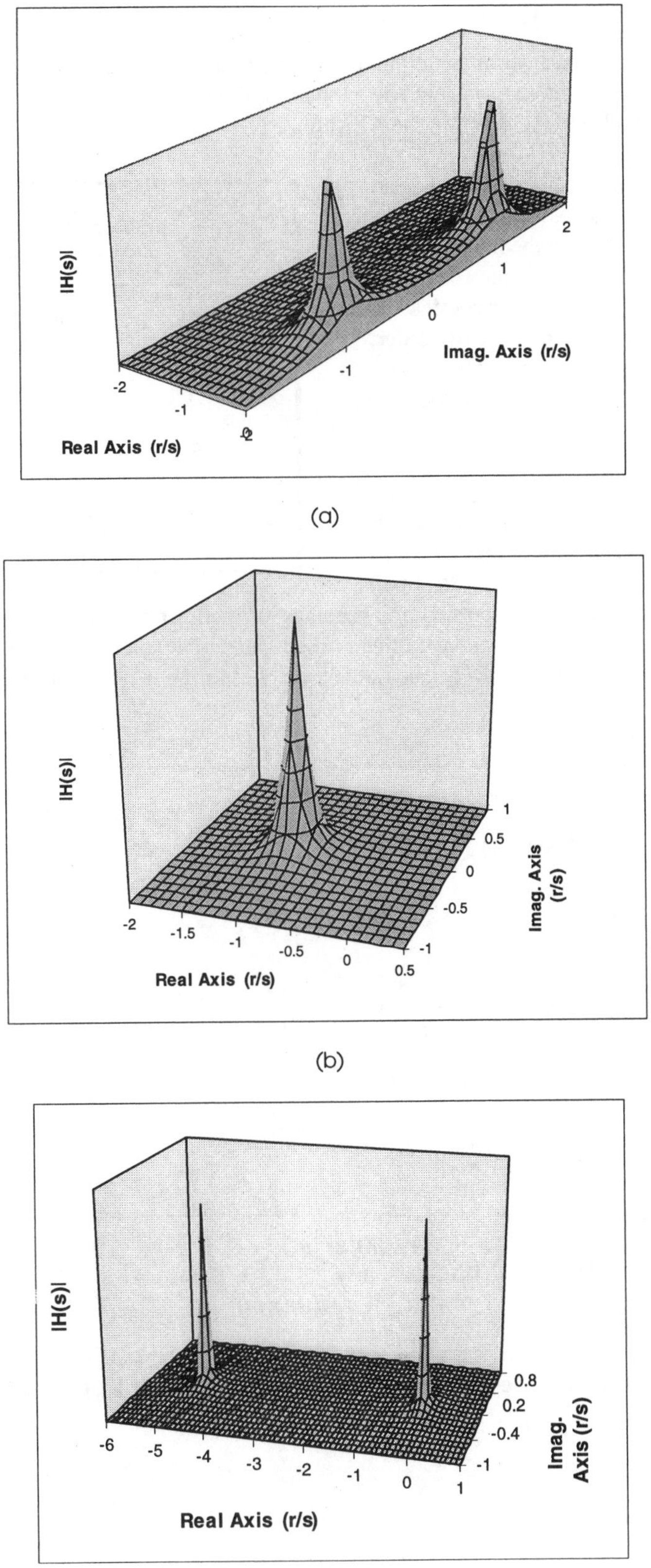

Figure 14.9. *s*-plane plots for (a) R = 0.5 Ω, (b) R = 2 Ω and (c) R = 5 Ω.

SIMULATION THREE

Sometimes, we know a circuit's transfer function in the s-domain rather than its topology (components and connections). For simulation purposes, we could develop an equivalent network that has the same transfer function., or we can use the LAPLACE part in the ABM library.

The LAPLACE part, shown in Figure 14.10, has an input, an output and two attributes: the transfer function numerator and denominator. Both the numerator and denominator can be functions of *s* but they cannot be functions of time or any other voltage or current. As an example, the part in Figure 14.10 models a 1st order lowpass filter with a pole frequency of 1.59 kHz (*s* is in rad./s) and a dc gain of 100 (40 dB). The Bode plot in Figure 14.11 verifies the model.

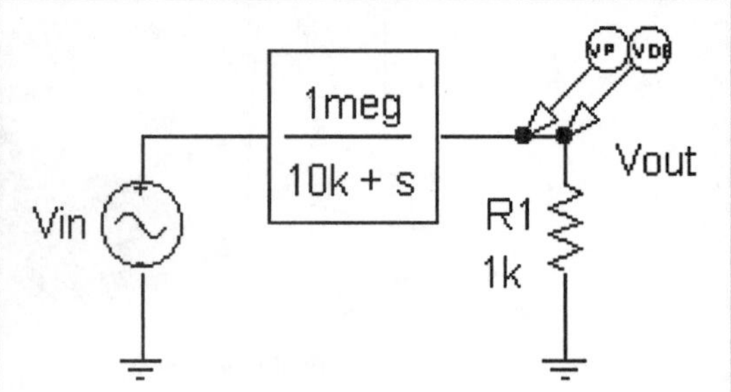

Figure 14.10. The LAPLACE part discussed in Simulation Three.

The LAPLACE part is modeled differently in each of the three PSpice analysis types - dc, ac and transient. In a dc analysis, the output is the dc input times transfer function evaluated at $s = 0$. For an ac analysis, the variable *s* is set equal to jω and the transfer function becomes just another complex number. Finally, in a transient analysis, the output signal is the convolution of the input and the part's impulse respond. Ouch. A forewarning, performing these calculations can be rather time consuming in a network with multiple LAPLACE parts.

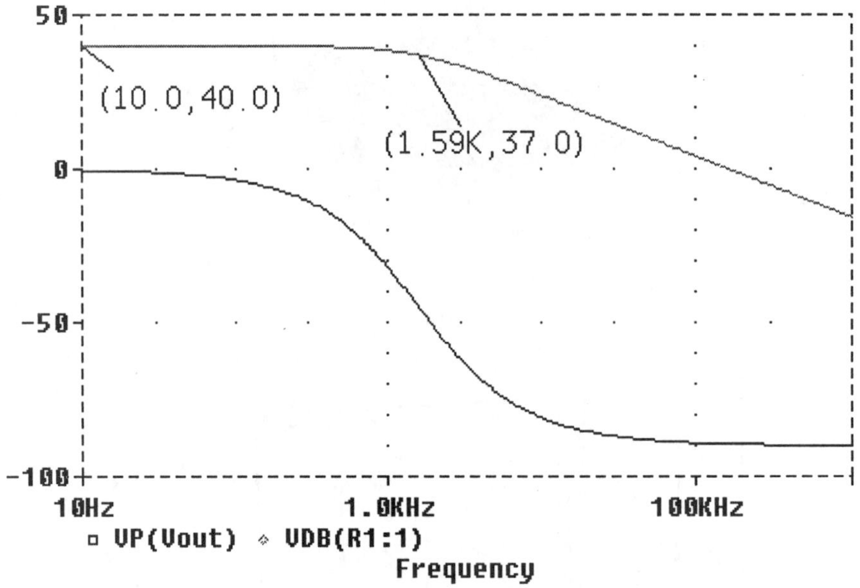

Figure 14.11. Bode plot for the lowpass filter modeled by the LAPLACE part in Simulation Three.

SIMULATION FOUR

To demonstrate the utility of the LAPLACE part, we will simulate an AM radio tuner filter and amplifier. First, a little background on AM radio details. AM stands for *amplitude modulation*. The information signal (voice, music, etc.) is used to modulate the amplitude of a second signal, called the carrier, which is then broadcast by the station. Each station is assigned a specific carrier frequency by the FCC. The assigned frequency can be any value between 530 kHz to 1710 kHz in 10 kHz increments. So, our filter must be able to tune in a station at say 1230 kHz, while rejecting stations at 1220 and 1240 kHz. Based on these specifications, the quality factor must be greater than 100!

We will not get into the details of tuner design. Rather, we will specify general characteristics and use the LAPLACE part to model those specifications. In this example, there are two fictitious stations in our vicinity, KOOL - 1590 and KORN - 1610. Although KOOL has a weaker signal, we prefer it over KORN. In the circuit, KOOL is modeled by the 1-µV source, V_{KOOL}, while the 10-µV source, V_{KORN}, models KORN. The two source voltages are added together using the SUM part (ABM library). The SUM part output is aptly called $V_{ANTENNA}$.

As for the filter itself, a transfer function similar to that sketched in Figure 14.12 would work well. It has the required selectivity and is flat for about 2 kHz on either side of the center frequency. To achieve the needed selectivity, we will use 6th order bandpass filters. To obtain a flat response around the center frequency, we will use two 6th order filters with slightly different center frequencies, then sum the filter outputs, V_{HI} and V_{LO}. As seen in Figure 14.13, each 6th order filter consists of three identical 2nd order stages. Furthermore, the center frequencies of the two filters are different by a factor of 0.9917. The Bode plot for V_{TUNED}, V_{HI} and V_{LO} shown in Figure 14.14, demonstrates the selectivity of the filter. We will return to this example in Chapter 15 for a more thorough investigation.

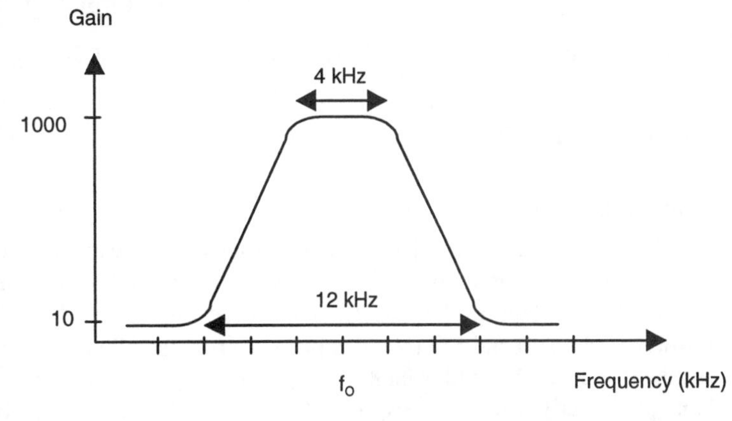

Figure 14.12. Sketch of a adequate AM tuning filter.

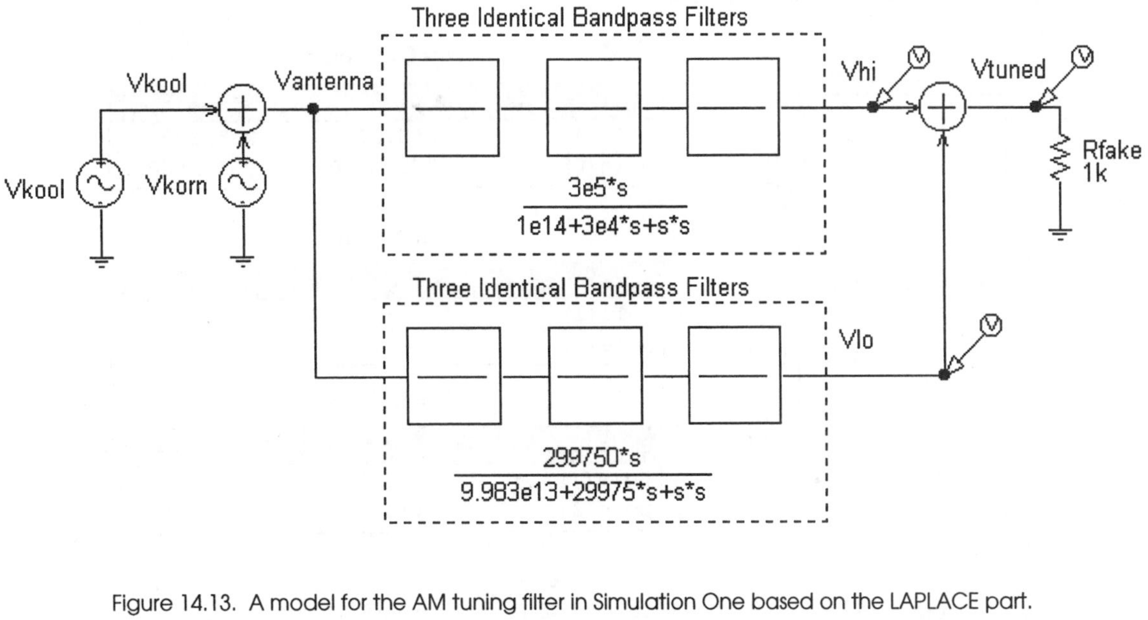

Figure 14.13. A model for the AM tuning filter in Simulation One based on the LAPLACE part.

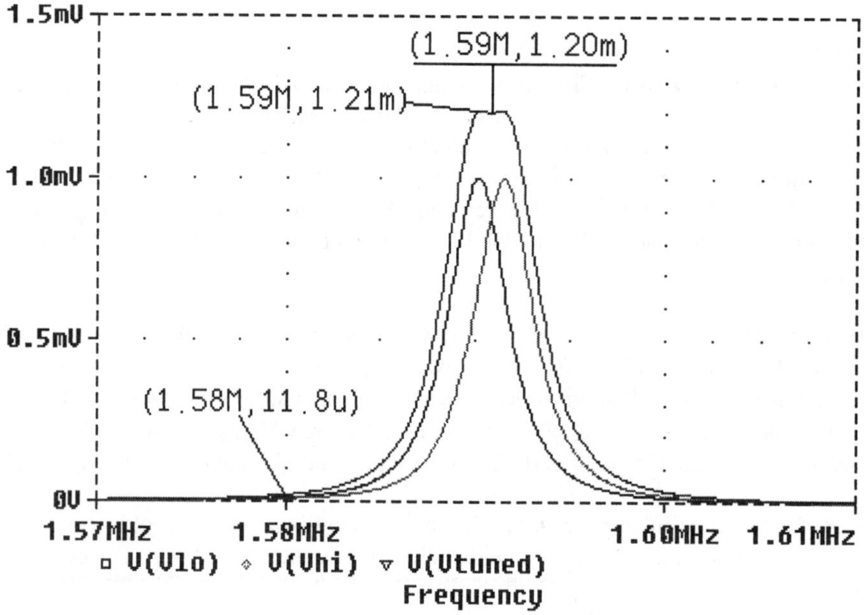

Figure 14.14. PROBE results for AM tuning filter Bode plot showing each 6th order bandpass output and their sum.

EWB SIMULATIONS

SIMULATION ONE

In *s*-domain analyzes, EWB has one advantage over the evaluation version of PSpice. EWB can find the poles and zeros of a network. To demonstrate, we will use EWB to find the poles and zeros of the simple RLC lowpass filter in Figure 14.7, conveniently redrawn in Figure 14.15, for R = 0.5 Ω (underdamped), 2 Ω (critically damped) and 5 Ω (overdamped). To request a Pole/Zero simulation, select **Pole/Zero** from the **Analysis** menu. This will open the dialog box in Figure 14.16 which has already been edited for this example. Note particularly that we have specified a voltage gain transfer function where the node 1 is the output and node 3 is the input. Results of the three simulations, one for each resistor value, are included in Figure 14.15. (Since the transfer function is a lowpass filter, there are no zeros listed.) Take a moment to compare the results here to those in Figure 14.9 - you'll find they are the same.

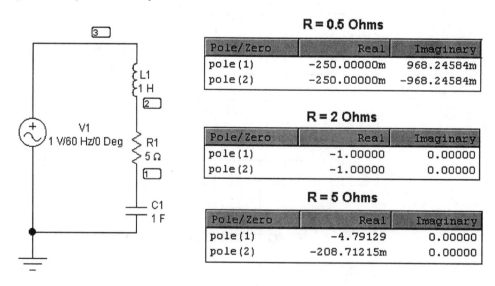

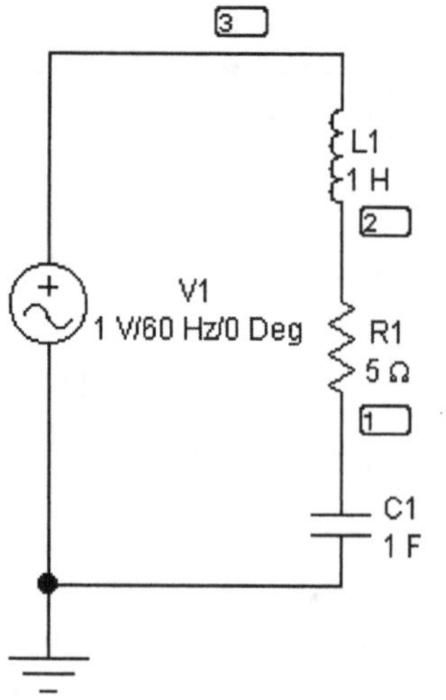

R = 0.5 Ohms

Pole/Zero	Real	Imaginary
pole(1)	-250.00000m	968.24584m
pole(2)	-250.00000m	-968.24584m

R = 2 Ohms

Pole/Zero	Real	Imaginary
pole(1)	-1.00000	0.00000
pole(2)	-1.00000	0.00000

R = 5 Ohms

Pole/Zero	Real	Imaginary
pole(1)	-4.79129	0.00000
pole(2)	-208.71215m	0.00000

Figure 14.15. Circuit diagram and Pole/Zero results for EWB Simulation One.

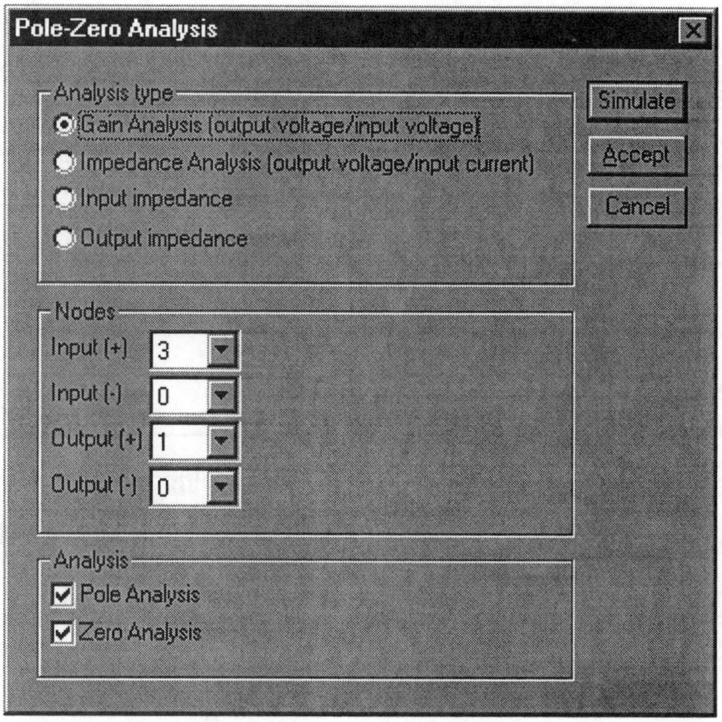

Figure 14.16. The EWB Pole/Zero Analysis dialog box edited for EWB Simulation One.

SIMULATION TWO

In this simulation, we will further explore EWB's Pole/Zero analysis. Consider the circuit in Figure 14.17 which has both poles and zeros. The input and output are at nodes 1 and 2 respectively. We will find the poles and zeros of the voltage gain transfer function, the output impedance and the input impedance. Figure 14.18a shows the Pole/Zero Analysis dialog box properly edited for the voltage gain transfer function. The pole/zero results in Figure 14.18b reveal two complex conjugate poles right on the *s*-plane imaginary axis and two zeros - one at dc and one at 10 rad/s. To obtain the pole/zero locations for the output impedance, simply select the **Output Impedance** radio button in Figure 14.18a. From Figure 14.18b, we see there is a pole and a zero right at - 5 rad./s. These will effectively cancel one another, leaving a zero at dc and a high frequency pole at - 667 krad./s. Finally, the input impedance has complex conjugate zeros and two real poles. Feel free to check these results by-hand.

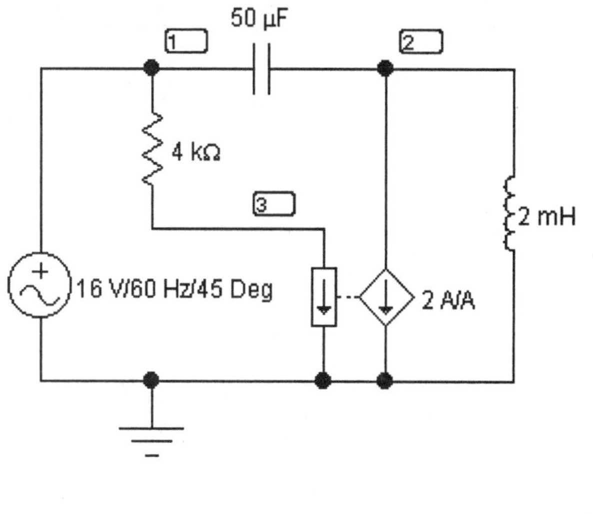

Figure 14.17. Circuit diagram for the Pole/Zero investigation in EWB Simulation Two.

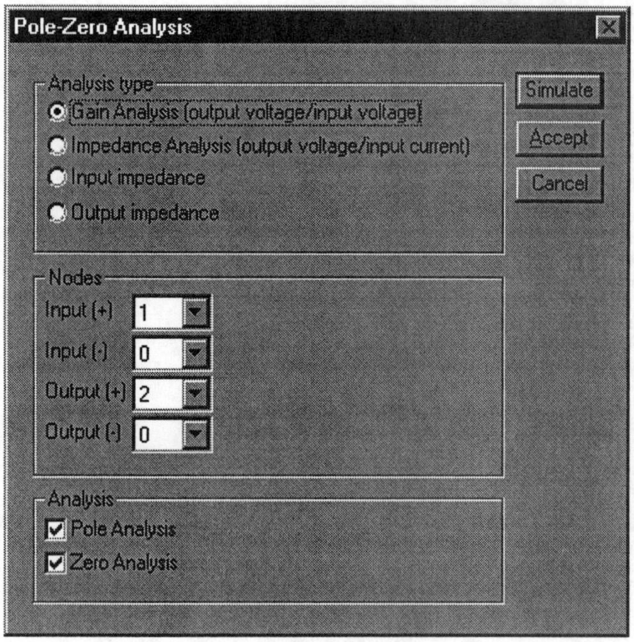

Voltage Gain Results

Pole/Zero	Real	Imaginary
pole(1)	0.00000	3.16228K
pole(2)	0.00000	-3.16228K
zero(1)	0.00000	0.00000
zero(2)	10.00000	0.00000

Output Impedance Results

Pole/Zero	Real	Imaginary
pole(1)	-666.66167K	0.00000
pole(2)	-5.00004	0.00000
zero(1)	-5.00000	0.00000
zero(2)	0.00000	0.00000

Input Impedance Results

Pole/Zero	Real	Imaginary
pole(1)	-666.66167K	0.00000
pole(2)	-5.00004	0.00000
zero(1)	0.00000	3.16228K
zero(2)	0.00000	-3.16228K

(a) (b)

Figure 14.18. For EWB Simulation Two: (a) the Pole/Zero Analysis dialog box and (b) the simulation results for the voltage gain, the output impedance and the input impedance.

NOTES

PROBLEM SOLVING EXAMPLES

1. Find the input impedance **Z**(s) in the following network.

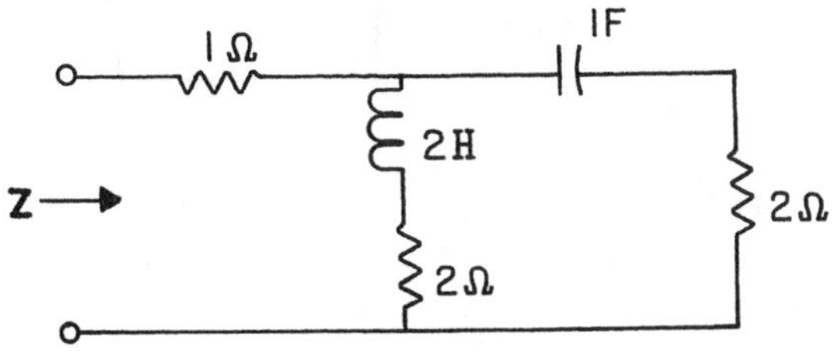

Solution:

$$Z(s) = 1 + \frac{(2s+2)(2+\frac{1}{s})}{2s+2+2+\frac{1}{s}} = \frac{6s^2+10s+3}{2s^2+4s+1}$$

2. Determine the value of the output voltage in the following network as $t \to \infty$.

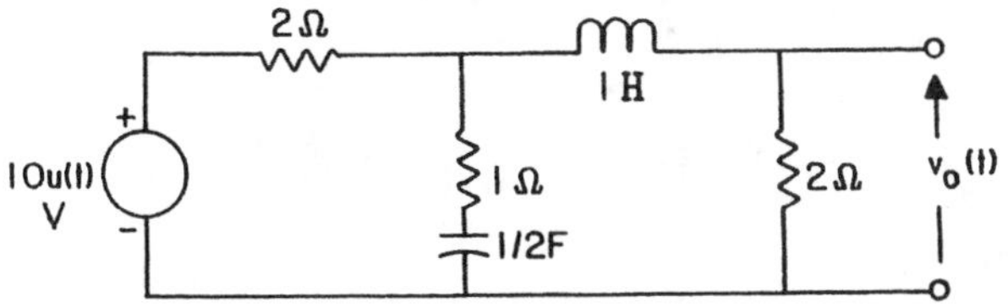

Solution: $v_o(t)\Big|_{t\to\infty} = 5V$

Note that as $t\to\infty$ the capacitor acts like an open circuit and the inductor acts like a short circuit. Hence the answer is obtained using a simple voltage divider.

3. Use Laplace transforms and mesh analysis to find $v_o(t)$ for $t>0$ in the network shown below. Assume zero initial conditions.

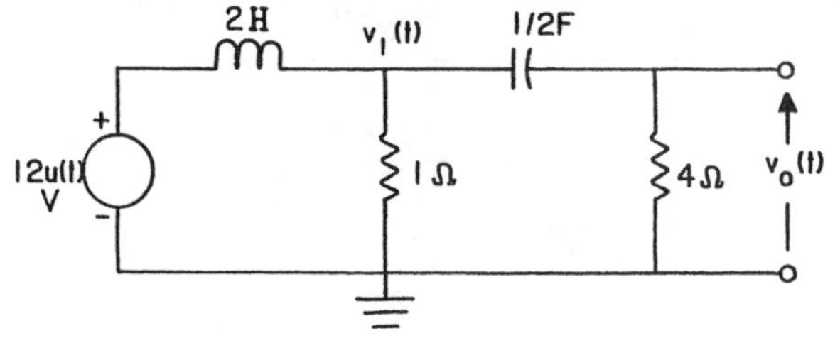

Solution: Since all initial conditions are zero, the transformed circuit is

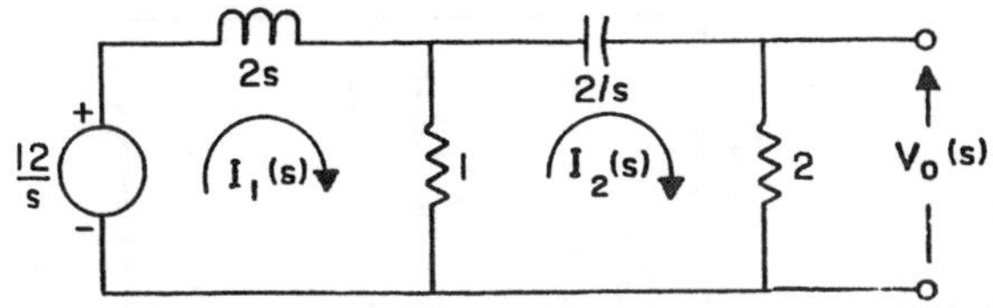

The mesh equations are

$$(2s+1)I_1(s) - I_2(s) = \frac{12}{s}$$

$$-I_1(s) + (1+\frac{2}{s}+2)I_2(s) = 0$$

$$I_2(s) = \frac{(4\sqrt{12})\frac{1}{\sqrt{12}}}{(s+\frac{1}{2})^2+(\frac{1}{\sqrt{12}})^2}$$

Therefore

$$v_o(t) = 4\sqrt{12}e^{-t/2}\sin\frac{t}{\sqrt{12}} \text{ V}$$

4. Solve Problem 3 using Laplace transforms and node equations.

Solution: The nodal equations are

$$\mathbf{V}_1(s)[\frac{1}{2s}+1+\frac{s}{2}] - \mathbf{V}_o(s)[\frac{s}{2}] = \frac{6}{s^2}$$

$$-\mathbf{V}_1(s)[\frac{s}{2}] + \mathbf{V}_o(s)[\frac{s}{2}+\frac{1}{2}] = 0$$

Solving the equations for $\mathbf{V}_o(s)$ yields

$$\mathbf{V}_o(s) = \frac{4}{s^2+s+\frac{1}{3}} = \frac{(4\sqrt{12})\frac{1}{\sqrt{12}}}{(s+\frac{1}{2})^2+(\frac{1}{\sqrt{12}})^2}$$

which is the same answer obtained in Problem 3.

5. Solve Problem 3 using Laplace transforms and source transformation.

Solution: Using the transformed circuit in Problem 3 we can make the following transformations

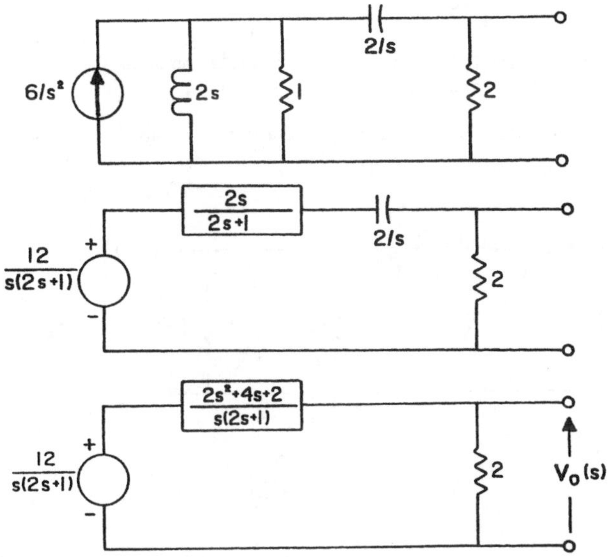

$$V_o(s) = \frac{\frac{12}{s(2s+1)}(2)}{\frac{2s^2+4s+2}{s(2s+1)}+2} = \frac{4}{s^2+s+\frac{1}{3}}$$

which is identical to the result obtained in the previous problem.

6. Solve Problem 3, using Laplace transforms and Thevenin's theorem.

Solution: Using the transformed network in Problem 3 the open circuit voltage is obtained as follows

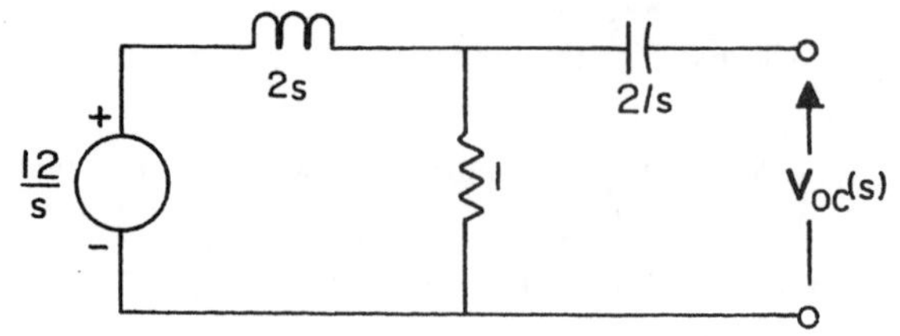

$$V_{oc}(s) = \left(\frac{12}{s}\right)\left(\frac{1}{2s+1}\right) = \frac{12}{2(2s+1)}$$

Looking back into the open terminals shorting out the voltage source we obtain

$$Z_{TH}(s) = \frac{2}{s} + \frac{2s}{2s+1} = \frac{2s^2+4s+2}{s(2s+1)}$$

Therefore

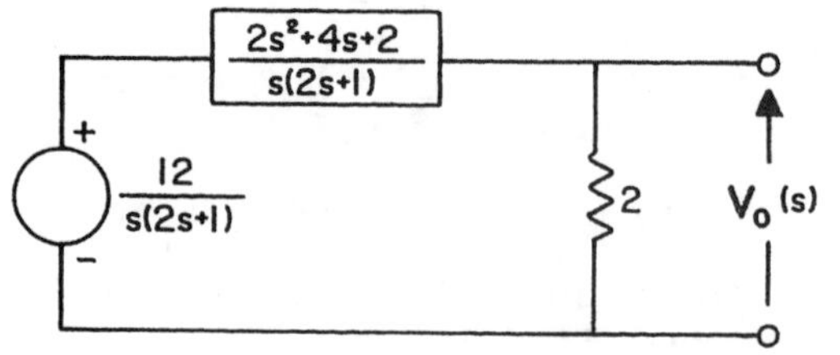

Hence

$$V_o(s) = \frac{4}{s^2+s+\frac{1}{3}}$$

which is identical to the results obtained in previous problems.

7. Use node equations to find $i_o(t)$ in the following network.

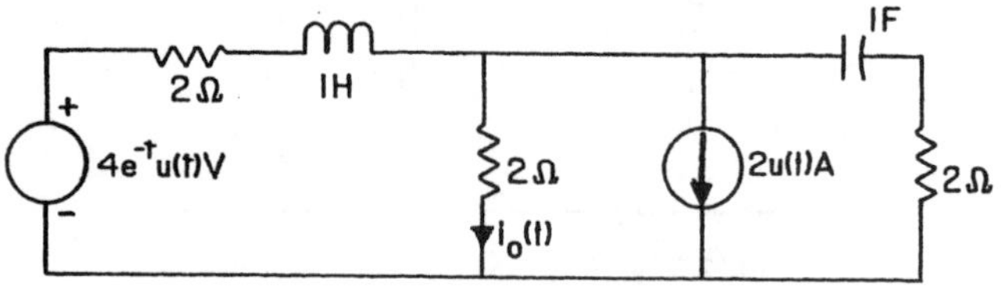

Solution: The transformed network is

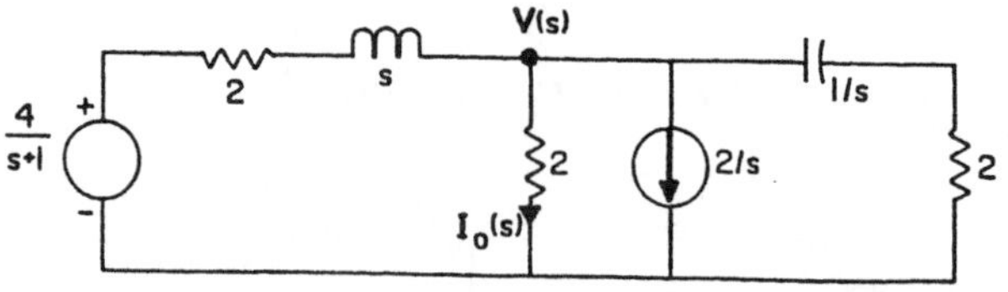

KCL yields

$$V(s)\left[\frac{1}{s+2} + \frac{1}{2} + \frac{1}{2+\frac{1}{s}}\right] = \frac{4}{(s+1)(s+2)} - \frac{2}{s}$$

Hence

$$V(s) = \frac{-4(s^2+s+2)(2s+1)}{s(4s^2+13s+4)(s+1)}$$

and $I_o(s) = \frac{V(s)}{2}$. Expanding $I_o(s)$ in partial fractions yields

$$I_o(s) = \frac{-1}{s} + \frac{0.798}{s+1} + \frac{1.807}{s+2.91} + \frac{0.478}{s+0.344}$$

Therefore

$$i_o(t) = [-1+0.798e^{-t}+1.807e^{-2.91t}+0.478e^{-0.344t}]$$
$$u(t)\,A$$

8. Use superposition to solve Problem 7.

Solution: The two networks to be analyzed are

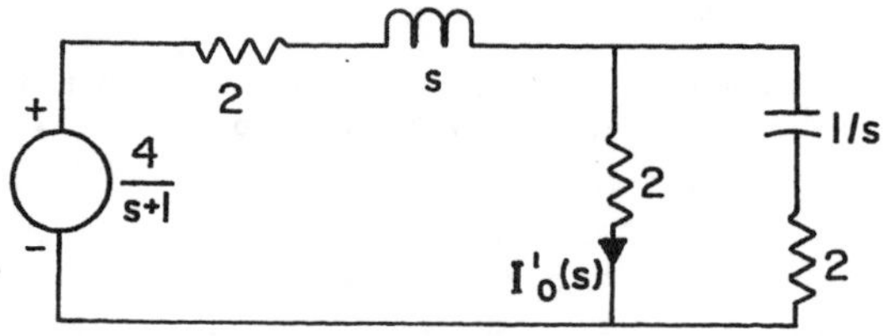

$$I_o^1(s) = \frac{\frac{4}{s+1}}{s+2+\frac{2(\frac{1}{s}+2)}{2+\frac{1}{s}+2}}\left[\frac{2(\frac{1}{s}+2)}{2+\frac{1}{s}+2}\right]\left(\frac{1}{2}\right)$$

$$= \frac{4(2s+1)}{(s+1)(4s^2+13s+4)}$$

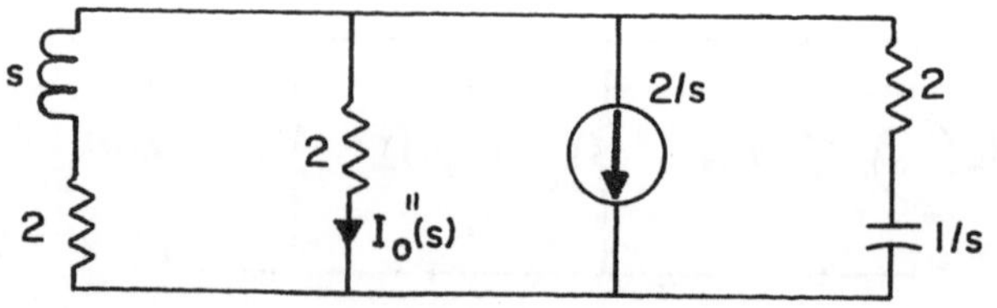

$$I_o^{11}(s) = \frac{-\frac{2}{s}\left[\frac{2s^2+5s+2}{s^2+4s+1}\right]}{2+\frac{2s^2+5s+2}{s^2+4s+1}} = \frac{-(4s^2+10s+4)}{s(4s^2+13s+4)}$$

$$I_o(s) = I_o^1(s)+I_o^{11}(s) = -\frac{1}{2}\,\frac{(2s+1)(s^2+s+2)}{s(s+1)(s^2+\frac{13}{4}s+1)}$$

This is the same answer as that obtained previously.

9. Use source exchange to solve Problem 7.

Solution: The first transformation yields the following

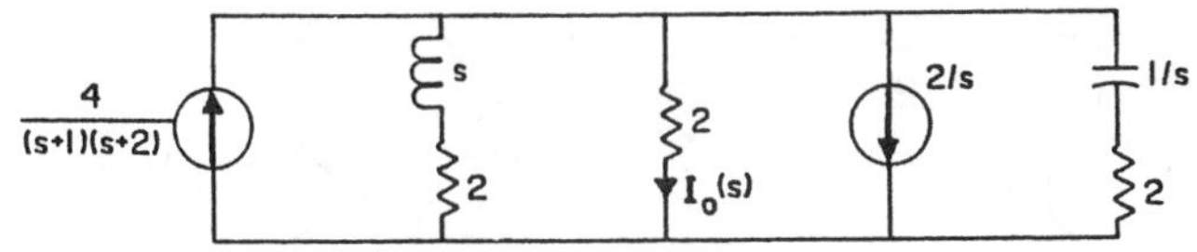

This network is equivalent to

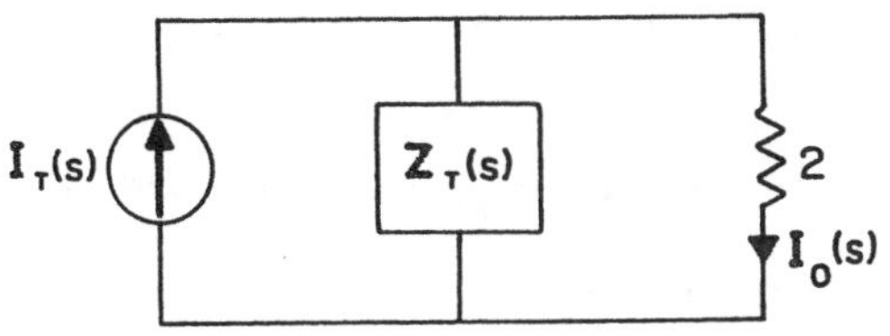

where

$$I_T(s) = \frac{4}{(s+1)(s+2)} - \frac{2}{s} = \frac{-(2s^2+2s+4)}{s(s+1)(s+2)}$$

$$Z_T(s) = \frac{(s+2)(2+\frac{1}{s})}{s+2+2+\frac{1}{s}} = \frac{2s^2+5s+2}{s^2+4s+1}$$

$$I_o(s) = \frac{I_T(s)Z_T(s)}{Z_T(s)+2} = -\frac{1}{2}\,\frac{(s^2+s+2)(2s+1)}{s(s+1)(s^2+\frac{13}{4}s+1)}$$

which is the same as that obtained in the previous problems.

10. Use Norton's Theorem to solve Problem 7.

Solution: $I_{sc}(s)$ is obtained from

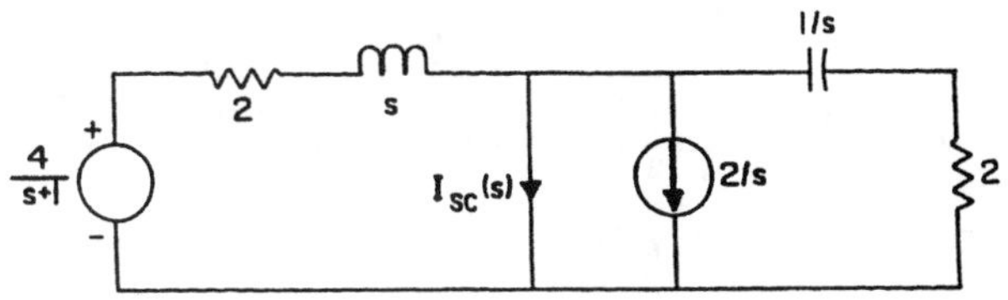

$$I_{sc}(s) = \frac{4}{(s+1)(s+2)} - \frac{2}{s} = \frac{-(2)(s^2+s+2)}{s(s+1)(s+2)}$$

$Z_{TH}(s)$ is derived from

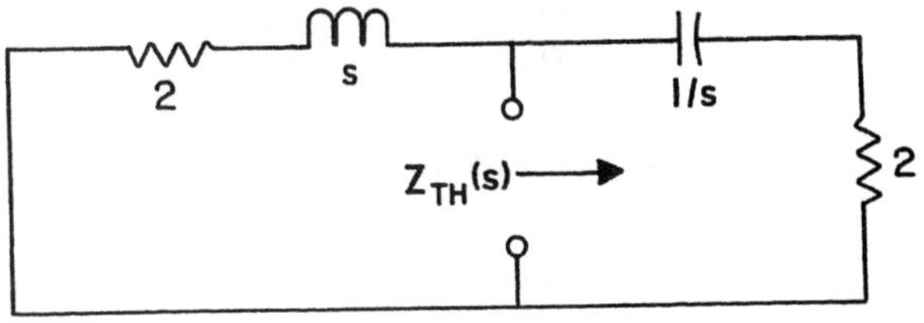

$$Z_{TH}(s) = \frac{(s+2)(s+\frac{1}{2})}{s+2+2+\frac{1}{s}} = \frac{2s^2+5s+2}{s^2+4s+1}$$

Then

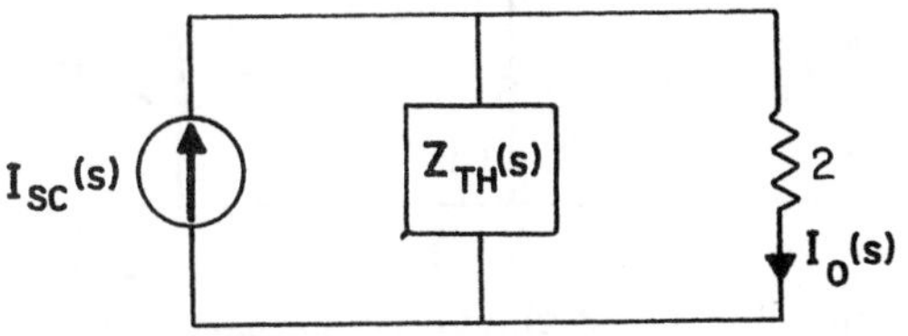

$$I_o(s) = \frac{-\frac{1}{2}(s^2+s+2)(2s+1)}{s(s+1)(s^2+\frac{13}{4}s+1)}$$

This is the same answer obtained earlier.

11. Use nodal equations to find $i_o(t)$ in the following network.

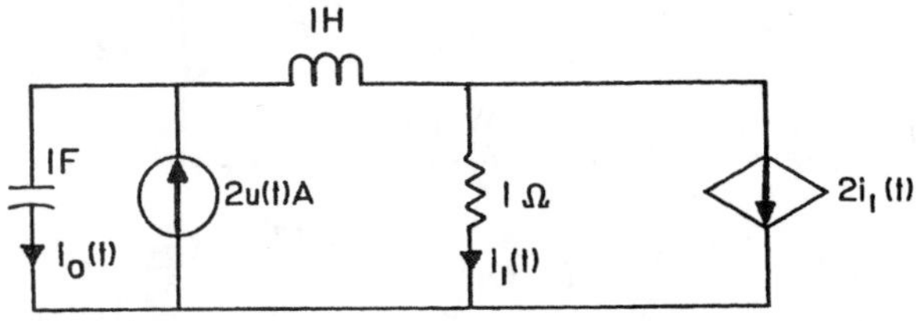

Solution: The transformed network is

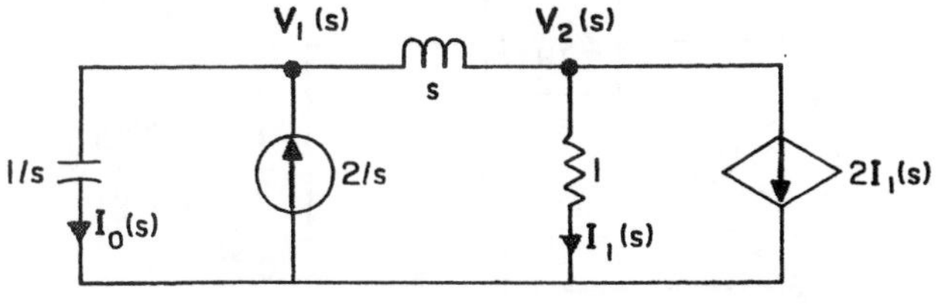

$$V_1(s)[s+\frac{1}{s}]-V_2(s)[\frac{1}{s}] = \frac{2}{s}$$

$$-V_1(s)[\frac{1}{s}]+V_2(s)(\frac{1}{s}+1) = -2I_1(s) = -2(\frac{V_2(s)}{1})$$

These equations yield

$$V_1(s) = \frac{2(3s+1)}{s(3s^2+s+3)} \text{ and } I_o(s) = \frac{V_1(s)}{\frac{1}{s}} = \frac{\frac{2}{3}(3s+1)}{s^2+\frac{1}{3}s+1}$$

$$I_o(s) = \frac{k_1}{s+\frac{1}{6}-j\frac{\sqrt{35}}{6}} + \frac{k_1^*}{s+\frac{1}{6}+j\frac{\sqrt{35}}{6}}$$

$k_1 = 0.37\underline{/-26.6°}$ and hence

$$i_o(t) = [0.74e^{-t/6}\cos(\frac{\sqrt{35}}{6}t-26.6°)]u(t)A$$

12. Use Thevenin's Theorem to solve Problem 11.

Solution: The open-circuit voltage is obtained as follows

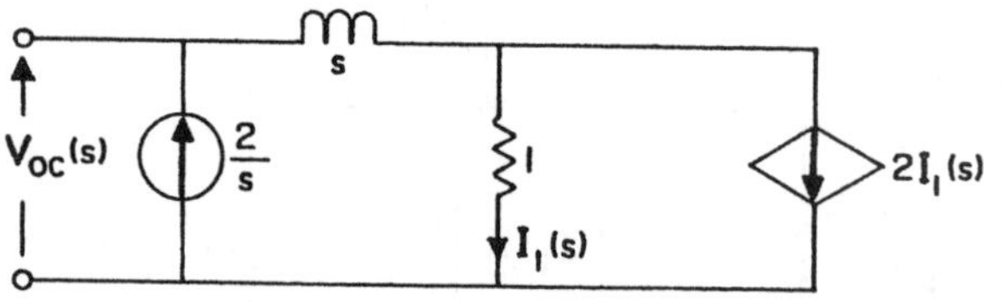

Note that $\frac{2}{s} = I_1(s)+2I_1(s)$ and hence $I_1(s) = \frac{2}{3s}$. Then

$$V_{oc}(s) = \frac{2}{s}(s)+(1)(\frac{2}{3s}) = \frac{6s+2}{3s}$$

$I_{sc}(s)$ is obtained from the circuit

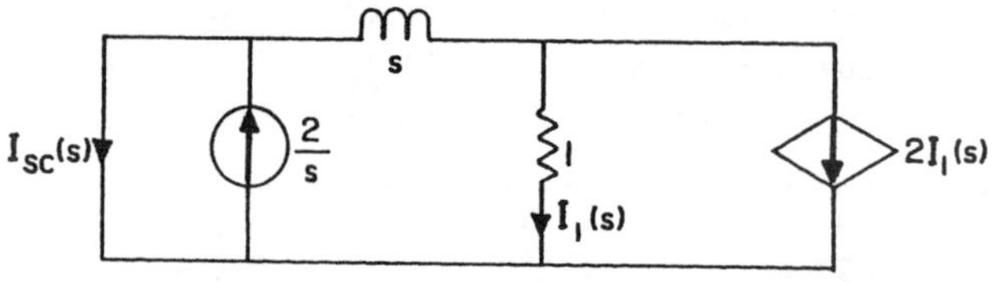

$$I_{sc}(s) = \frac{2}{s} \text{ and hence } Z_{TH}(s) = \frac{V_{oc}(s)}{I_{sc}(s)} = \frac{6s+2}{6}$$

Hence

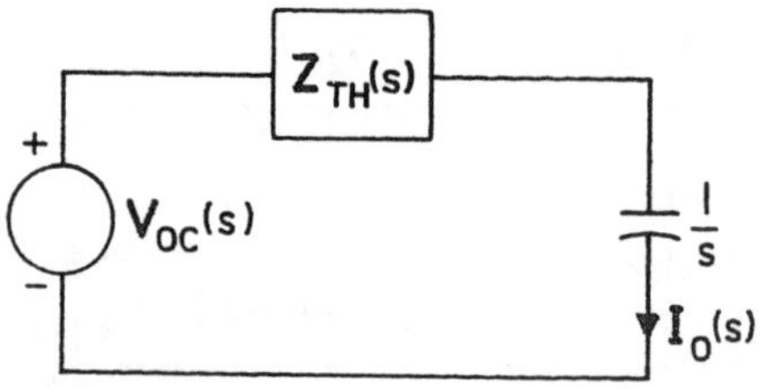

$$I_o(s) = \frac{\frac{6s+2}{3s}}{\frac{6s+2}{6}+\frac{1}{s}} = \frac{2(3s+1)}{3s^2+s+3}$$

which is the same as that obtained in the previous problem.

13. Use loop equations to find $i_1(t)$ in the following network.

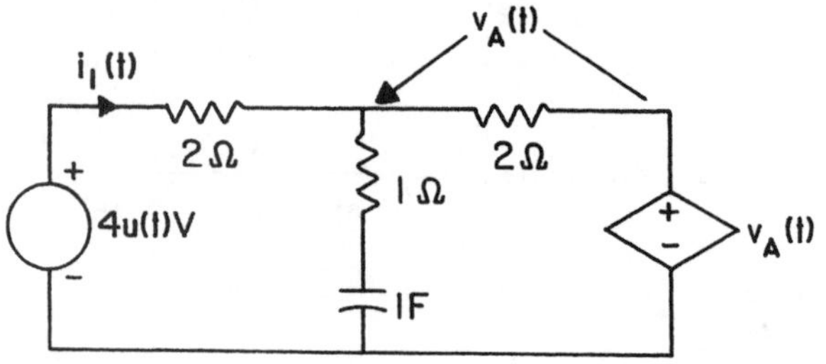

Solution:

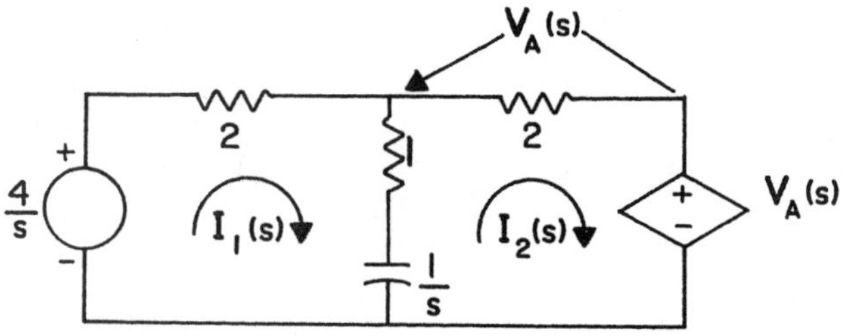

The loop equations are

$$I_1(s)(3+\frac{1}{s})-I_2(s)(1+\frac{1}{s}) = \frac{4}{s}$$

$$-I_1(s)(1+\frac{1}{s})+I_2(s)(3+\frac{1}{s}) = -V_A(s) = -2I_2(s)$$

The equations yield

$$I_1(s) = \frac{\frac{2}{7}(5s+1)}{s(s+\frac{3}{7})} = \frac{k_0}{s} + \frac{k_1}{s+\frac{3}{7}}$$

$k_0 = \frac{2}{3}$ and $k_1 = \frac{2}{7}$ therefore

$$i_1(t) = [\frac{2}{3}+\frac{2}{7}e^{-3/7t}]u(t)A$$

14. Use Thevenin's Theorem to solve Problem 13.

Solution: From the following network we see that $V_{oc}(s) = \frac{4}{s}$

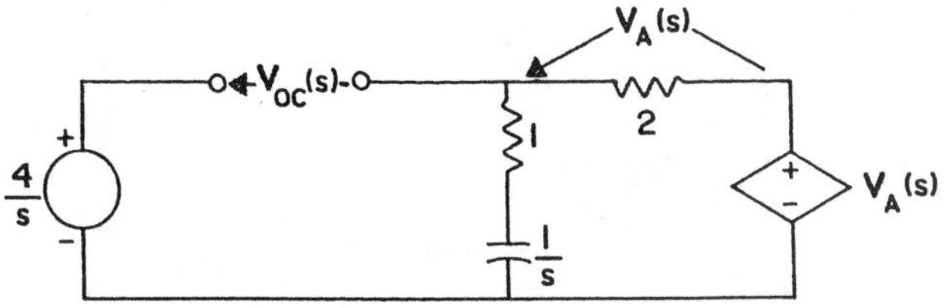

$I_{sc}(s)$ is obtained from the network

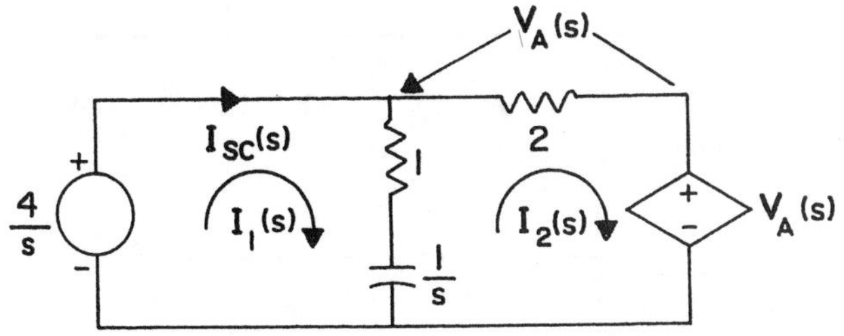

$$I_1(s)(1+\frac{1}{s})-I_2(s)(1+\frac{1}{s}) = \frac{4}{s}$$

$$-I_1(s)(1+\frac{1}{s})+I_2(s)(3+\frac{1}{s}) = -V_A(s) = -2I_2(s)$$

Hence $I_1(s) = I_{sc}(s) = \dfrac{5s+1}{s(s+1)}$ and $Z_{TH}(s) = \dfrac{V_{oc}(s)}{I_{sc}(s)}$

$= \dfrac{4(s+1)}{5s+1}$.

Therefore

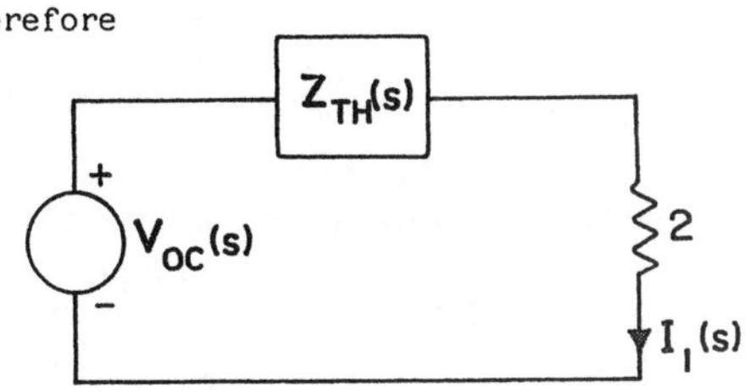

$$I_1(s) = \frac{\frac{4}{s}}{\frac{4(s+1)}{5s+1}+2} = \frac{\frac{2}{7}(5s+1)}{s(s+\frac{3}{7})}$$

which is the same as that obtained in the previous problem.

15. Find the steady-state response $i_o(t)$ in the following network.

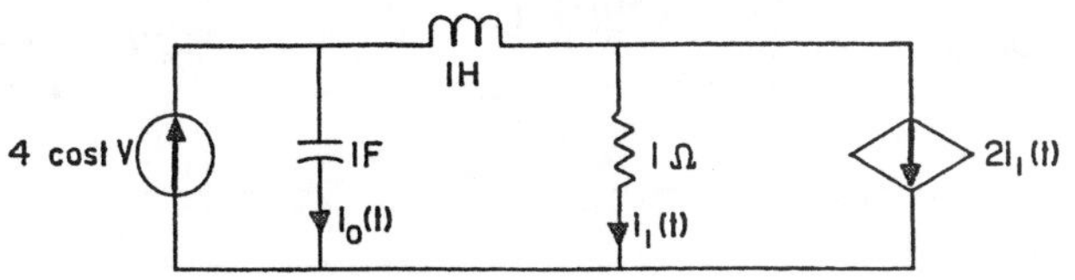

Solution: The transformed network is

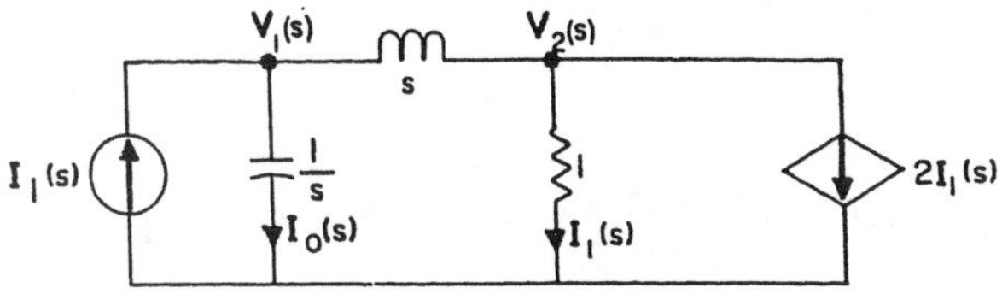

$$V_1(s)(s+\frac{1}{s})-V_2(s)(\frac{1}{s}) = I_i(s)$$

$$-V_1(s)(\frac{1}{s})+V_2(s)(\frac{1}{s}+1) = -2I_1(s) = -2V_2(s)$$

Therefore

$$V_1(s) = [\frac{3s+1}{3s^2+s+3}]I_i(s)$$

and

$$I_o(s) = \frac{V_1(s)}{\frac{1}{s}} = [\frac{s(3s+1)}{3s^2+s+3}]I_i(s)$$

Then

$$H(j1) = \frac{j1(1+j3)}{j1} = 3.16\underline{/71.6°}$$

Therefore

$$i_o(t) = 12.64\cos(t+71.6°)A$$

16. Find the steady-state response $v_o(t)$ in the following network.

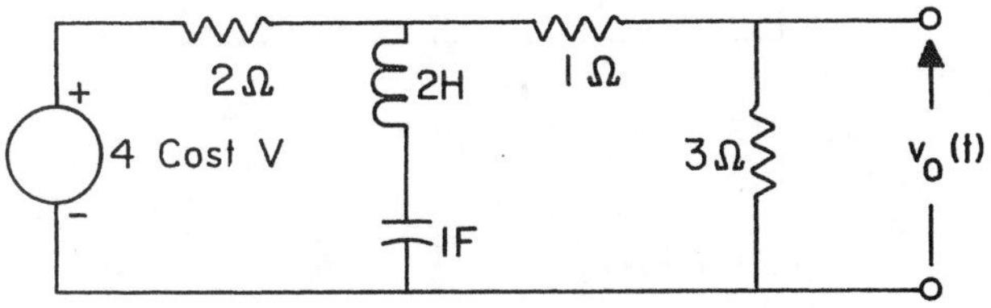

Solution: The transformed circuit is shown below

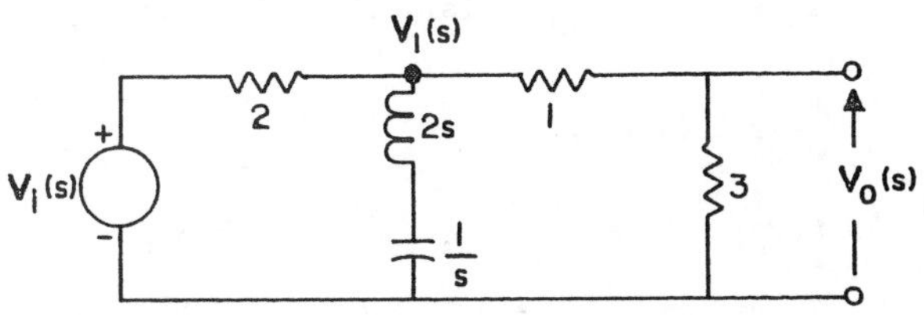

The node equation for $\mathbf{V}_1(s)$ is

$$\mathbf{V}_1(s)\left[\frac{1}{2} + \frac{1}{2s + \frac{1}{s}} + \frac{1}{4}\right] = \frac{\mathbf{V}_i(s)}{2}$$

$$\mathbf{V}_1(s) = \frac{2(2s^2+1)}{6s^2+4s+3}\,\mathbf{V}_i(s) \text{ and } \mathbf{V}_o(s) = \frac{3}{4}\,\mathbf{V}_i(s) =$$

$$\left[\frac{3(s^2+\frac{1}{2})}{6s^2+4s+3}\right]\mathbf{V}_i(s)$$

$$\mathbf{H}(j1) = \frac{3(-1+\frac{1}{2})}{-3+j4} = 0.3\underline{/53.1°}$$

Hence

$$v_o(t) = 1.2\cos(t+53.1°)\text{V}$$

17. Find the steady-state response $v_o(t)$ in the network below.

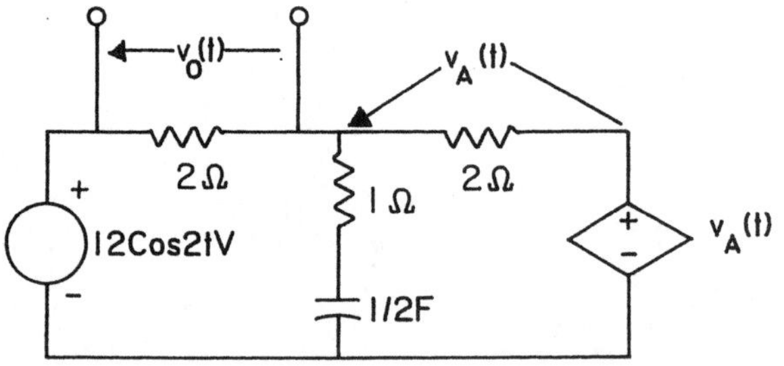

Solution: The transformed network is shown below

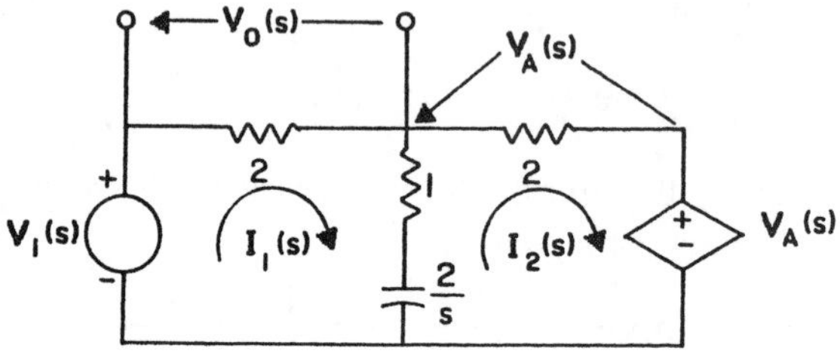

$$\mathbf{I}_1(s)(3+\frac{2}{s}) - \mathbf{I}_2(s)(1+\frac{2}{s}) = \mathbf{V}_i(s)$$

$$-\mathbf{I}_1(s)(1+\frac{2}{s}) + \mathbf{I}_2(s)(3+\frac{2}{s}) = -\mathbf{V}_A(s) = -2\mathbf{I}_2(s)$$

These equations yield

$$\mathbf{I}_1(s) = \left[\frac{5s+2}{14s+12}\right]\mathbf{V}_i(s) \text{ and } \mathbf{V}_o(s) = 2\mathbf{I}_1(s) =$$

$$\left[\frac{2(5s+2)}{14s+12}\right]\mathbf{V}_i(s)$$

$$\mathbf{H}(j2) = 0.66\underline{/11.9°}$$

Therefore $v_o(t) = 8.03\cos(2t+11.9°)\text{V}$

18. Find $v_o(t)$ for $t>0$ in the network below.

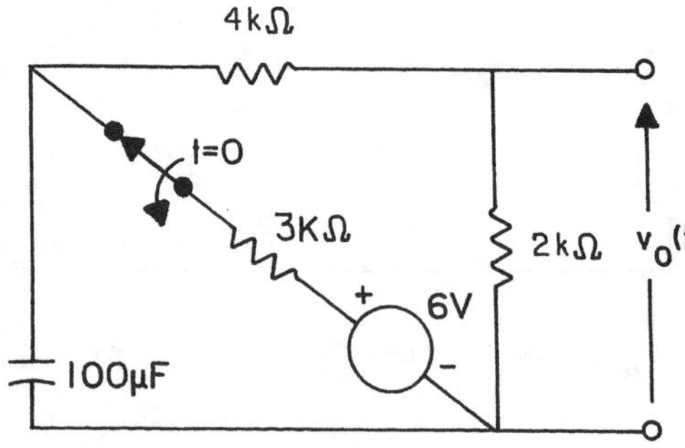

Solution: Note that $v_C(0-) = (6)(\frac{6k}{9k}) = 4\text{V}$. Then the network for $t>0$ is

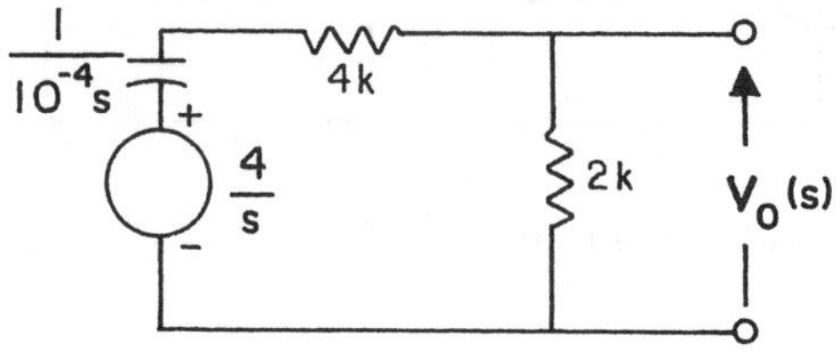

Therefore

$$\mathbf{V}_o(s) = \frac{\frac{4}{s}}{\frac{1}{10^{-4}s} + (6)(10^3)}(2)(10^3) = \frac{\frac{4}{3}}{s+\frac{1}{0.6}}$$

Hence

$$v_o(t) = \frac{4}{3}\,e^{-t/0.6}u(t)\text{V}$$

19. Find $v_o(t)$ for $t>0$ in the network below.

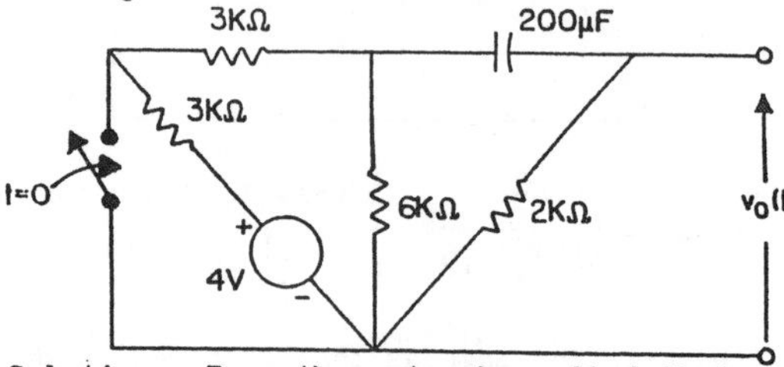

Solution: From the network we find that $v_C(0-) = (\frac{4}{12k})(6k) = 2\text{V}$. The network then for $t>0$ is

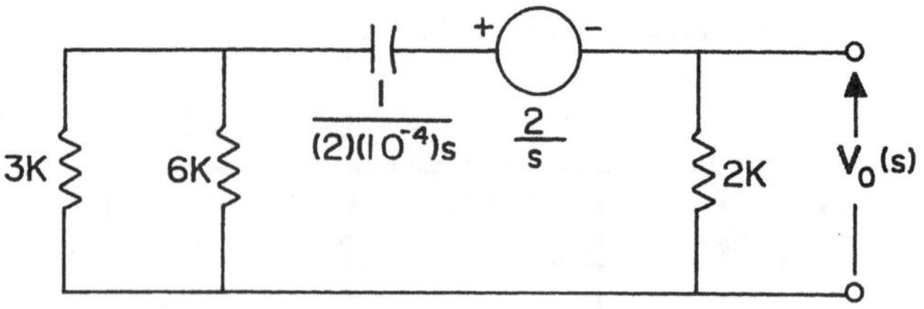

Then

$$\mathbf{V}_o(s) = \frac{-\frac{2}{s}}{\frac{1}{(2)(10^{-4})s} + (4)(10^3)}(2)(10^3) = \frac{-1}{s+1.25}$$

Therefore

$$v_o(t) = -1e^{-1.25t}u(t)\text{V}$$

20. Find $i_o(t)$ for $t>0$ in the following network.

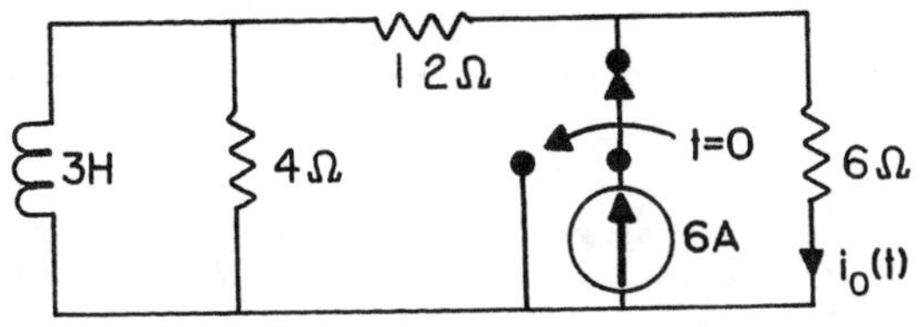

Solution: Note that $i_L(0-) = \frac{(6)(6)}{6+12} = 2\text{A}$. Hence the network for $t>0$ is

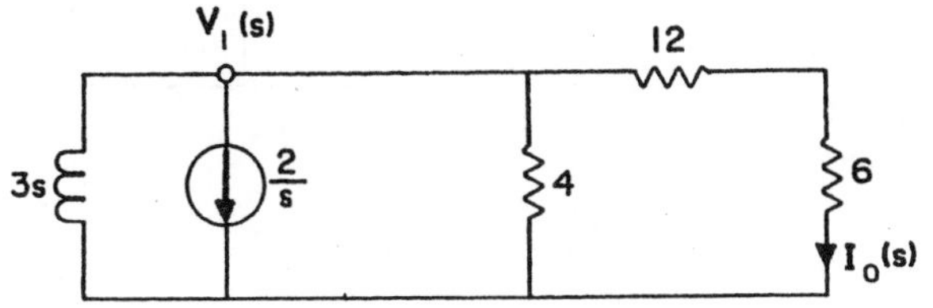

$$\mathbf{V}_1(s)\left[\frac{1}{3s} + \frac{1}{4} + \frac{1}{18}\right] = \frac{-2}{s}$$

$$\mathbf{V}_1(s) = \frac{-72}{11s+12}$$

and $\mathbf{I}_o(s) = \frac{\mathbf{V}_1(s)}{18} = \frac{-\frac{4}{11}}{s+\frac{12}{11}}$. Therefore

$$i_o(t) = \frac{-4}{11}e^{-\frac{12}{11}t}u(t)\text{V}$$

21. Find $v_o(t)$ for $t>0$ in the network below.

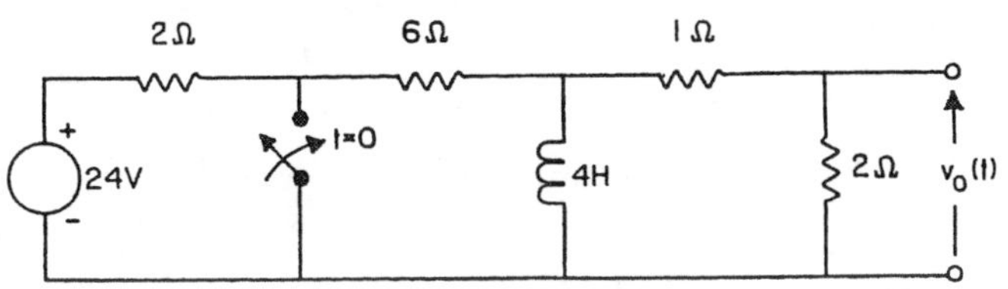

Solution: Note that $i_L(0-) = \frac{24}{8} = 3\text{A}$. Hence the network for $t>0$ is

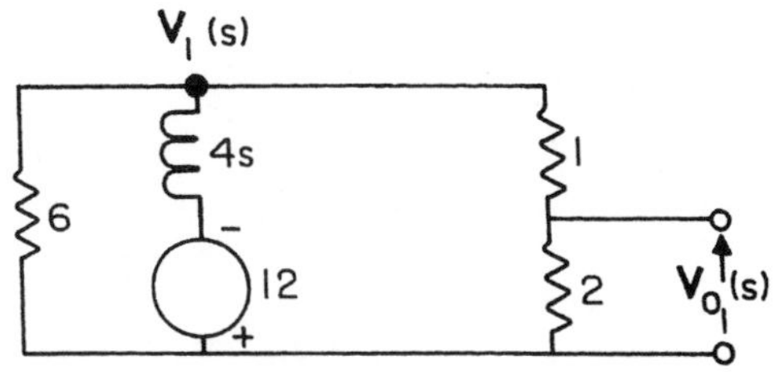

$$\frac{\mathbf{V}_1(s)}{6} + \frac{\mathbf{V}_1(s)+12}{4s} + \frac{\mathbf{V}_1(s)}{3} = 0$$

$$\mathbf{V}_1(s) = \frac{-6}{s+\frac{1}{2}}$$

Then $\mathbf{V}_o(s) = \frac{2}{3}\mathbf{V}_1(s) = \frac{-4}{s+\frac{1}{2}}$ and hence

$$v_o(t) = -4e^{-t/2}u(t)\text{V}$$

22. Find $v_o(t)$ for $t>0$ in the network below.

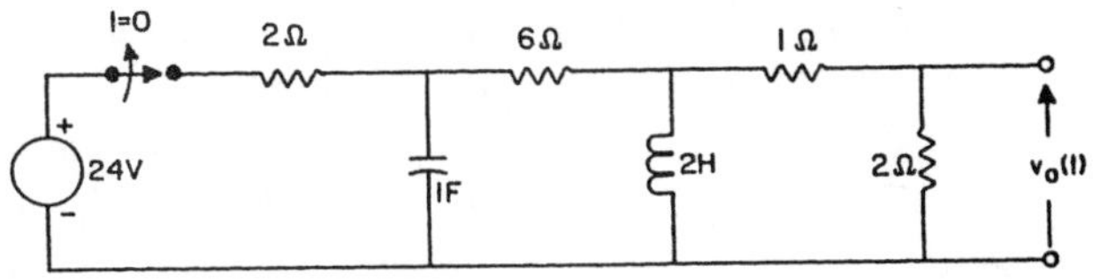

Solution: The initial conditions are found from the following network

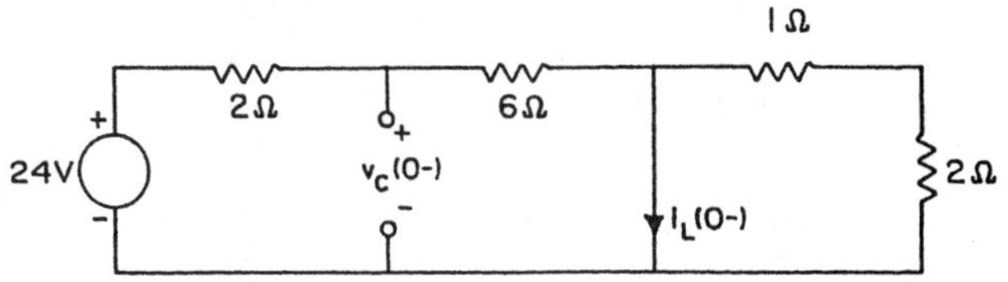

$$i_L(0-) = \frac{24}{8} = 3\text{A and } v_C(0-) = \frac{24}{8}(6) = 18\text{V}$$

The network for $t>0$ is

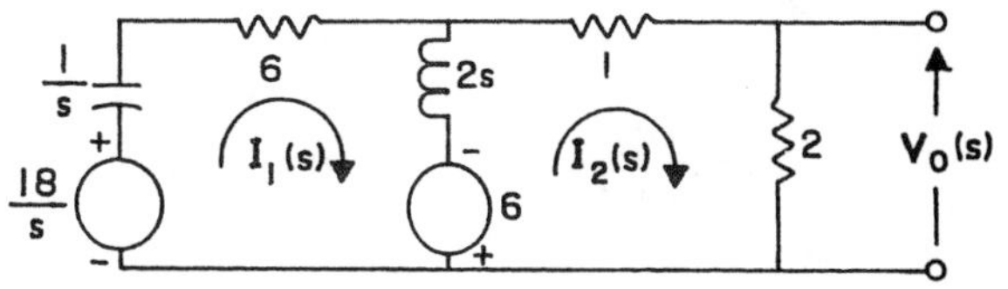

$$\mathbf{I}_1(s)\left(\frac{1}{s} + 6 + 2s\right) - \mathbf{I}_2(s)(2s) = \frac{18}{s} + 6$$

$$-\mathbf{I}_1(s)(2s) + \mathbf{I}_2(s)(2s+3) = -6$$

These equations yield

$$I_2(s) = \frac{-6}{18s^2+20s+3}$$

$$V_o(s) = 2I_2(s) = \frac{-0.66}{(s+0.94)(s+0.181)}$$

$$= \frac{k_1}{s+0.94} + \frac{k_2}{s+0.181}$$

$k_1 = 0.87$ and $k_2 = -0.87$. Therefore

$$v_o(t) = 0.87(e^{-0.94t}-e^{-0.181t})u(t)V$$

23. Find $v_o(t)$ in the network below.

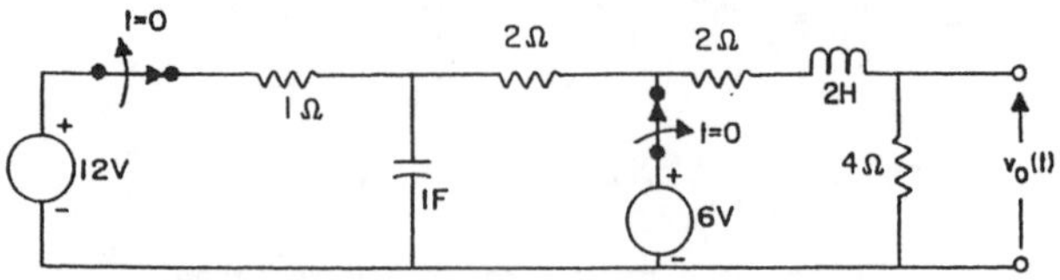

Solution: The initial conditions are found from the following network

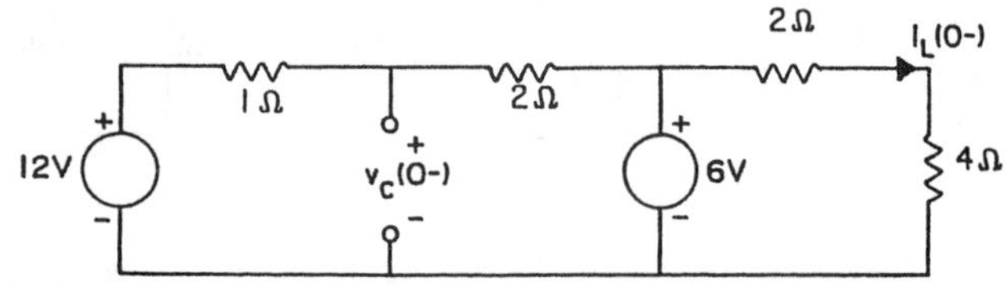

$$v_C(0-) = 6+\frac{12-6}{3}(2) = 10V \text{ and } i_L(0-) = \frac{6}{2+4} = 1A.$$

The network for t>0 is

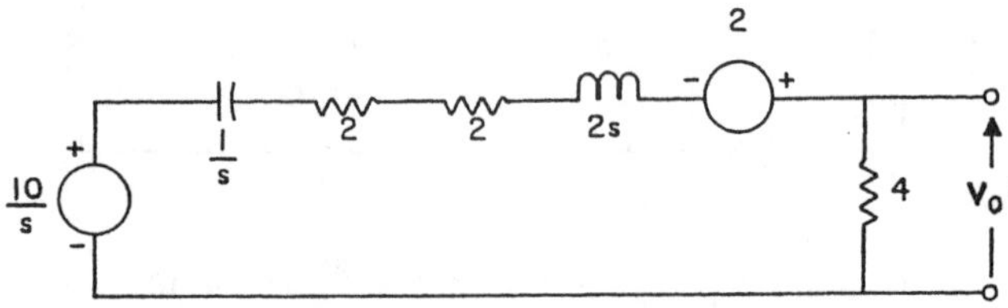

Therefore

$$V_o(s) = \frac{\frac{10}{s}+2}{\frac{1}{s}+2s+8}(4)$$

$$= \frac{4(s+5)}{s^2+4s+\frac{1}{2}} = \frac{4(s+5)}{(s+3.87)(s+0.13)}$$

$$= \frac{k_1}{s+3.87} + \frac{k_2}{s+0.13}$$

$k_1 = -1.21$, $k_2 = 5.21$ and hence

$$v_o(t) = [-1.21e^{-3.87t}+5.21e^{-0.13t}]u(t)V$$

24. Determine the initial and final values of the voltage $v_o(t)$ in the network below.

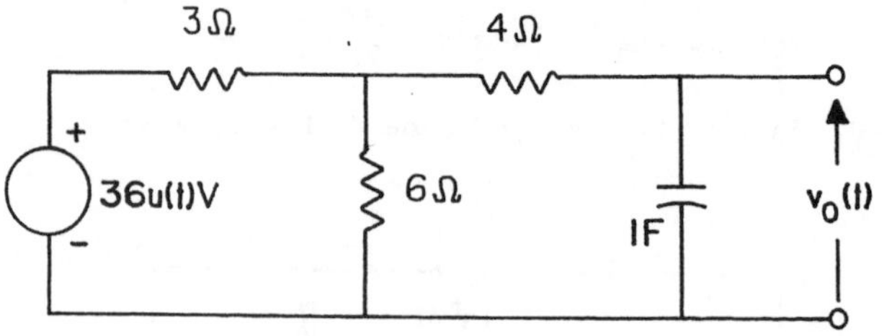

Solution: Using a Thevenin equivalent for the left side of the network reduces the transformed circuit to

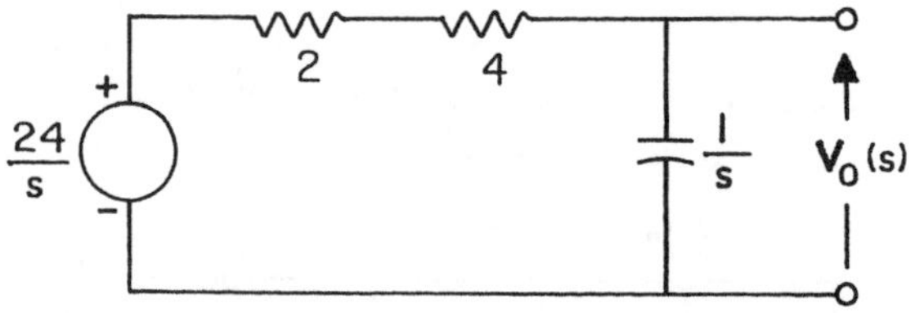

$$V_o(s) = \frac{(\frac{24}{s})(\frac{1}{s})}{6+\frac{1}{s}} = \frac{24}{s(6s+1)}$$

Then

$$v_o(0) = \lim_{s\to\infty} sV_o(s) = 0$$

$$v_o(\infty) = \lim_{s\to 0} sV_o(s) = 24V$$

25. Find the output voltage $v_o(t)$ in the following network if the input is given by the source shown below.

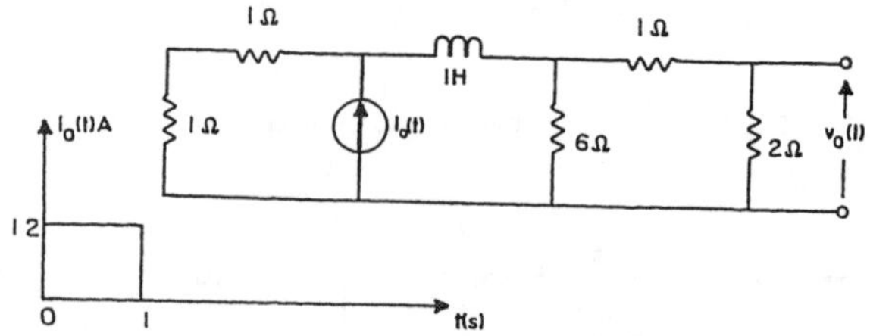

Solution: The transformed network is

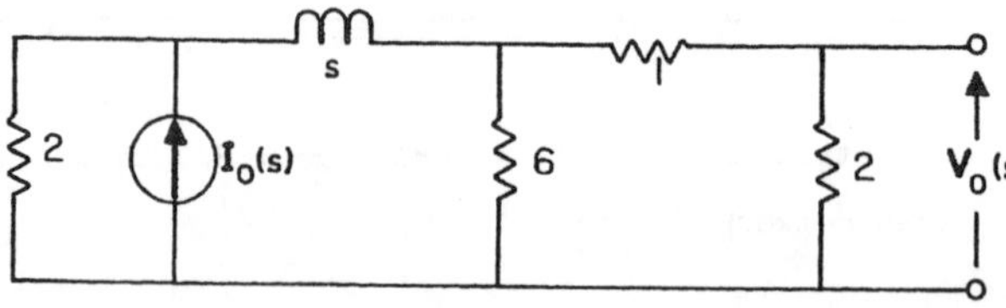

$$V_o(s) = \frac{I_o(s)(2)}{2+s+2}\left(\frac{6}{9}\right) = \frac{4I_o(s)}{3(s+4)}$$

$$I_o(s) = 12\left(\frac{1-e^{-s}}{s}\right)$$

Therefore

$$V_o(s) = \frac{16}{s+4}\left(\frac{1}{s} - \frac{e^{-s}}{s}\right)$$

Since $\frac{16}{s(s+4)} = \frac{4}{s} + \frac{-4}{s+4}$ then

$$v_o(t) = [4-4e^{-4t}]u(t)-[4-4e^{-4(t-1)}]u(t-1)V$$

26. Find $v_o(t)$ in Problem 25 if the input is as shown below.

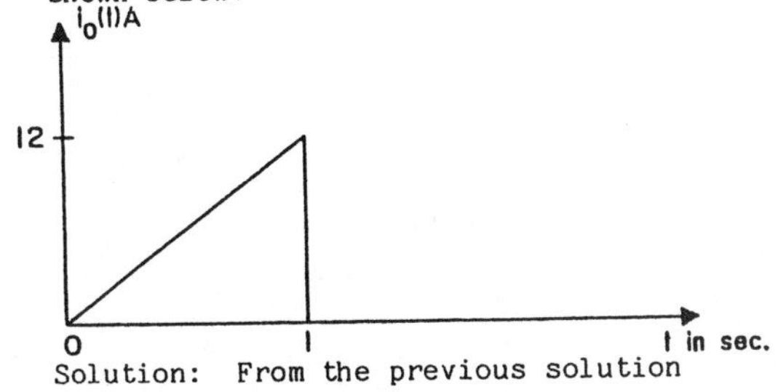

Solution: From the previous solution

$$V_o(s) = \frac{4I_o(s)}{3(s+4)} \text{ and } I_o(s) = \frac{12}{s^2}[1-(1+s)e^{-s}]$$

Therefore

$$V_o(s) = \frac{16}{s^2(s+4)}[1-(1+s)e^{-s}]$$

Since

$$\frac{16}{s^2(s+4)} = \frac{4}{s^2} - \frac{1}{s} + \frac{1}{s+4}$$

and

$$\frac{16(s+1)}{s^2(s+4)} = \frac{4}{s^2} + \frac{3}{s} - \frac{3}{s+4}$$

Then

$$v_o(t) = [4t-1te^{-4t}]u(t)-[4(t-1)+3-3e^{-4(t-1)}]u(t-1)V$$

27. If the input to a network is $x(t) = te^{-t}u(t)$ and the transfer function is $h(t) = 10e^{-4t}$, find the output $y(t)$.

Solution:

$$y(t) = \int_0^t 10e^{-4(t-\lambda)}\lambda e^{-\lambda}u(\lambda)d\lambda$$

$$= 10e^{-4t}\int_0^t \lambda e^{3\lambda}d\lambda, \quad t>0$$

Using integration by parts

$u = \lambda$ $\qquad dv = e^{3\lambda}d\lambda$

$du = d\lambda$ $\qquad v = \frac{1}{3}e^{3\lambda}$

Hence

$$y(t) = 10e^{-4t}\left[\frac{\lambda}{3}e^{3\lambda}\Big|_0^t - \int_0^t \frac{1}{3}e^{3\lambda}d\lambda\right], \quad t>0$$

$$= \left[\frac{10t}{3}e^{-t} - \frac{10}{9}e^{-t} + \frac{10}{9}e^{-4t}\right]u(t)$$

28. The input $u(t)$ is applied to a network $h(t)$ to produce an output $y(t) = [5-5e^{-4t}]u(t)$. Determine the network transfer function $h(t)$.

Solution: Transforming the output we obtain

$$Y(s) = \frac{5}{s} - \frac{5}{s+4} = \frac{20}{s(s+4)}$$

and

$$Y(s) = H(s)X(s)$$

$$\frac{20}{s(s+4)} = H(s)\left(\frac{1}{s}\right)$$

Therefore

$$H(s) = \frac{20}{s+4}$$

and

$$h(t) = 20e^{-4t}$$

29. Find the transfer function $\frac{V_o(s)}{V_i(s)}$ for the network below.

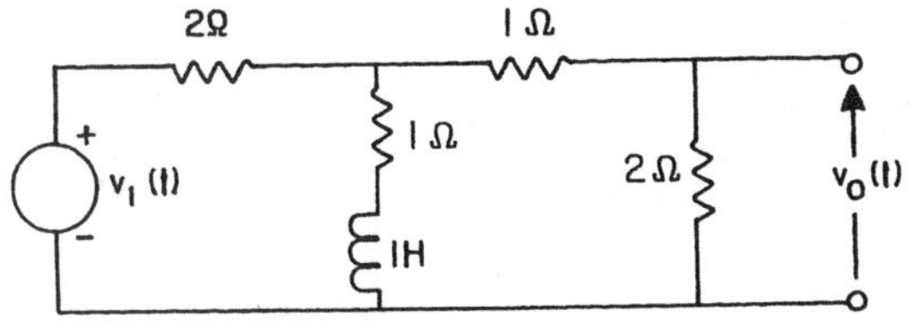

Solution: The transformed network is

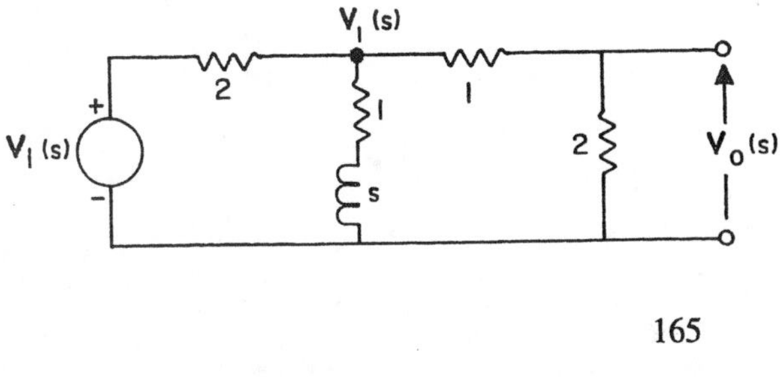

$$\frac{V_1(s)-V_i(s)}{2} + \frac{V_1(s)}{s+1} + \frac{V_1(s)}{3} = 0$$

Hence

$$V_1(s) = \frac{3(s+1)}{5s+11} V_i(s)$$

and since $V_o(s) = \frac{2}{3} V_1(s)$

$$\frac{V_o(s)}{V_1(s)} = \frac{2(s+1)}{5s+11}$$

30. The voltage response of a network to a unit step input is $V_o(s) = \dfrac{2(s+1)}{s(s^2+12s+37)}$. Is the response underdamped?

Solution: The characteristic equation is

$$s^2+12s+37 = (s+6)^2+1^2$$

The roots of the equation are s_1, s_2 = $-6\pm j1$ and therefore the network is underdamped.

31. The transfer function of a network is given by the expression $G(s) = \dfrac{2(s+10)}{s^2+6s+9}$. Determine the damping ratio, the undamped natural frequency, and the type of response that will be exhibited by the network.

Solution: The characteristic equation is

$$s^2+6s+9 = s^2+2\xi\omega_o s+\omega_o^2$$

Therefore $\omega_o = 3$ and since $2\xi\omega_o = 6$, $\xi = 1$ and the damping is critical.

Chapter 15 FOURIER ANALYSIS TECHNIQUES

THE FOURIER SERIES AND PSPICE

PSpice can determine the Fourier series of any node voltage or branch current. A transient simulation must be performed since the Fourier series is calculated from time-domain data points. The series data, consisting of the dc component, the fundamental and harmonic magnitudes and their phases, is tabulated in the output file. Note that PSpice uses a trigonometric series of sinewaves, not cosines.

PSPICE SIMULATIONS

Simulation One

To demonstrate this feature, we will find the Fourier series for the simple circuit in Figure 15.1 where

$$v_{1k}(t) = 2\sin(2\pi \cdot 1000t)\ \text{V} \qquad v_{3k}(t) = 0.5\sin(2\pi \cdot 3000t)\ \text{V} \qquad V_{DC} = 5\ \text{V} \tag{15.1}$$

To request a Fourier series, go to the **Transient** dialog box and select **Enable Fourier**, as shown in Figure 15.2. As for the editable fields, **Center Frequency** is the same as fundamental frequency. It should be the same as the lowest frequency in the waveform of interest, 1 kHz in this example. **Number of harmonics** is exactly that - how many harmonics are calculated and listed in the output file. **Output Vars** is a listing of the node voltages and branch currents you wish to include in the Fourier analysis. Examples of suitable node voltage variables are V(3), V(Vo) are V1(R1). Entries such as Vo will not work. Currents should be entered as I(R1), I(V1), etc.

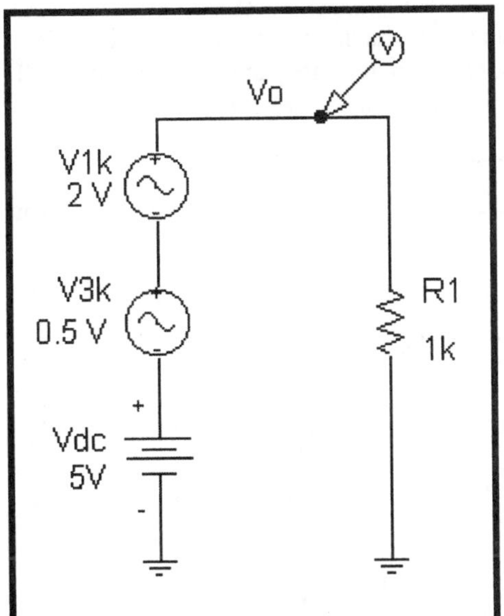

Figure 15.1. Schematics diagram for Simulation One.

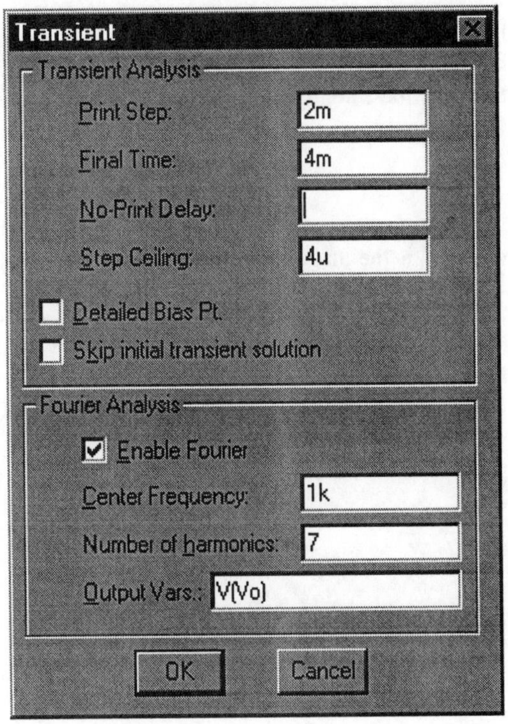

Figure 15.2. The Transient dialog box edited to find the first 7 Fourier components.

How does PSpice determine the series? Your **Center Frequency** entry is inverted to produce the fundamental period, $T_{FUND} = 1/f_C$. The data points from the last T_{FUND} time span are extracted from the PROBE data file. Assuming the extracted data to be periodic, PSpice constructs the series. This means that the duration of your transient simulations should extend at least one fundamental period beyond the point where transient behavior is died out. In general, increasing the number of data points by reducing the **Step Ceiling** in Figure 15.2 will improve the accuracy of the calculations.

Back to our simulation. The PROBE plot for Vo is shown in Figure 15.3. From the output file, the Fourier series data is listed below.

```
FOURIER COMPONENTS OF TRANSIENT RESPONSE V(Vo)

 DC COMPONENT =    5.000000E+00
```

HARMONIC NO	FREQUENCY (HZ)	FOURIER COMPONENT	NORMALIZED COMPONENT	PHASE (DEG)	NORMALIZED PHASE (DEG)
1	1.000E+03	2.000E+00	1.000E+00	1.167E-05	0.000E+00
2	2.000E+03	1.797E-05	8.987E-06	3.711E+01	3.711E+01
3	3.000E+03	4.998E-01	2.499E-01	4.214E-05	3.047E-05
4	4.000E+03	8.018E-06	4.009E-06	3.945E+01	3.945E+01
5	5.000E+03	6.420E-07	3.210E-07	1.445E+02	1.445E+02
6	6.000E+03	8.032E-06	4.016E-06	-1.475E+02	-1.475E+02
7	7.000E+03	4.242E-06	2.121E-06	-1.124E+02	-1.124E+02

```
TOTAL HARMONIC DISTORTION =   2.498946E+01 PERCENT
```

From the series listing, we see that the dc component, the first harmonic (same as the fundamental) and the third harmonic match the equations in (15.1). Although the other harmonics are non-zero, they are so extremely small that they have negligible effect on the series. The harmonic distortion is calculated as follows

$$harmonic\ distortion = \frac{\sqrt{V_2^2 + V_3^2 + V_4^2 + \cdots + V_N^2}}{V_1} \tag{15.2}$$

where V_i is the magnitude of the ith harmonic.

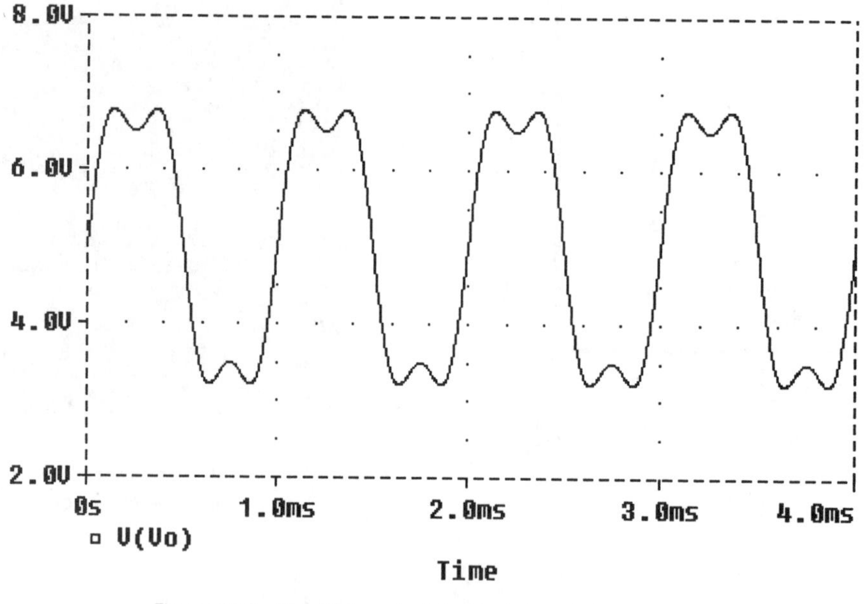

Figure 15.3. PROBE transient for Vo in Simulation One.

Simulation Two - A Fourier Faux Pas

As mentioned in Simulation One, PSpice uses only the last fundamental period of data when constructing the Fourier series. Let's investigate what can happen when the last period of data is corrupted. Figure 15.4 shows the network from Simulation One with a switch added. The switch opens at 5.5 ms. Setting the Final Time in the Transient dialog box (Figure 15.2) at 6.0 ms, produces the PROBE plot in Figure 15.5. Note that while the first 5 periods look fine, the last period is affected by the switch. From the output file, the Fourier calculations are listed at the bottom of the page.

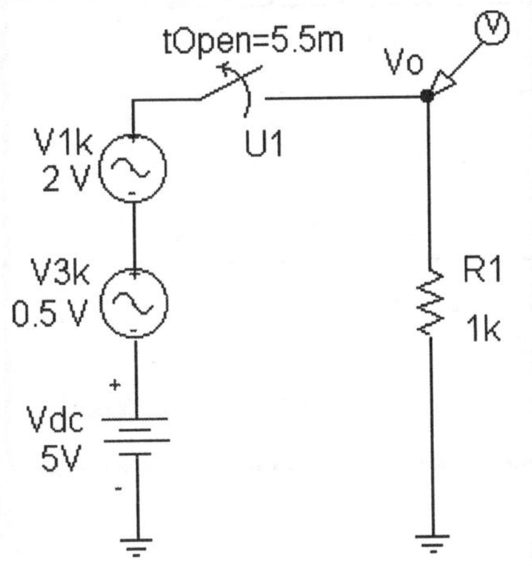

Figure 15.4. *Schematics* diagram for Simulation Two.

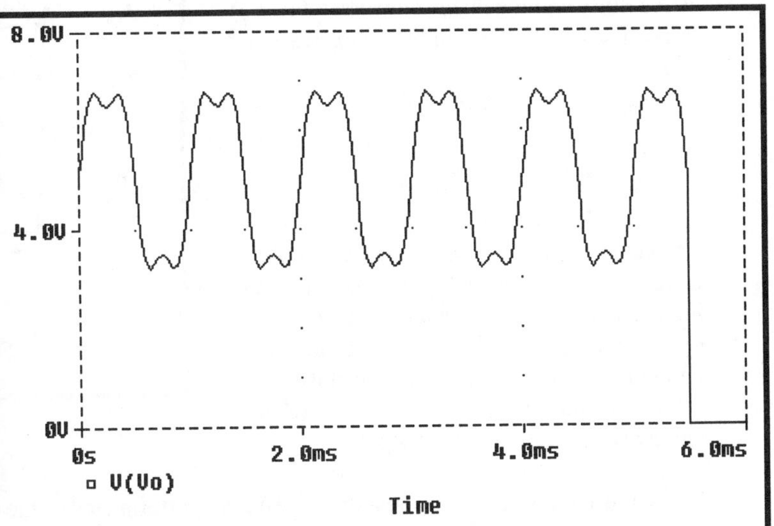

Figure 15.5. PROBE plot for Vo in Simulation Two. Notice how the last period of data is corrupted by the switch.

Comparing this data to that in Simulation One, notice how the harmonic components are all wrong and the harmonic distortion is much higher. Of course, this can be corrected by changing the Final Time to a value less than 5.5 ms. As a result, the switch will never move during the simulation and last period of data is preserved.

FOURIER COMPONENTS OF TRANSIENT RESPONSE V(Vo)

DC COMPONENT = 3.208369E+00

HARMONIC NO	FREQUENCY (HZ)	FOURIER COMPONENT	NORMALIZED COMPONENT	PHASE (DEG)	NORMALIZED PHASE (DEG)
1	1.000E+03	4.177E+00	1.000E+00	6.763E-01	0.000E+00
2	2.000E+03	1.908E-01	4.567E-02	-9.051E+01	-9.118E+01
3	3.000E+03	1.299E+00	3.108E-01	2.183E+00	1.507E+00
4	4.000E+03	1.640E-01	3.926E-02	-9.112E+01	-9.179E+01
5	5.000E+03	6.371E-01	1.525E-01	4.460E+00	3.783E+00
6	6.000E+03	2.019E-02	4.833E-03	-1.032E+02	-1.039E+02
7	7.000E+03	4.556E-01	1.091E-01	6.230E+00	5.554E+00

TOTAL HARMONIC DISTORTION = 3.680024E+01 PERCENT

Simulation Three

A European phone line is picking up 50 Hz noise from the power grid. The notch filter in Figure 15.6 is designed to extract the noise without altering frequencies above 250 Hz. The voltage sources Vnoise and Vin model the 50 Hz noise and the phone conversation respectively and are given by the expressions

$$v_{noise} = 1\sin(2\pi \cdot 50t)\ \text{V}$$

$$v_{in} = 0.2\sin(2\pi \cdot 250t)\ \text{V} \tag{15.3}$$

The key to the notch filter is the LC parallel combination. At the resonance frequency, their impedances are equal in magnitude but opposite in sign. Performing the standard parallel combination calculation yields an infinite impedance at resonance. As a result, no energy is transferred to the output.

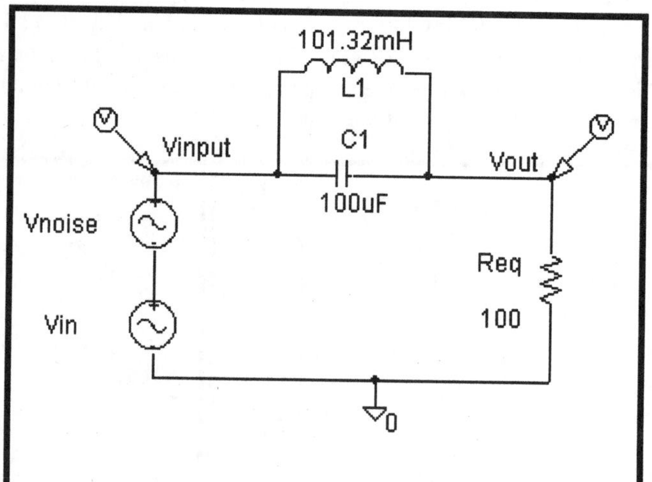

Figure 15.6. *Schematics* diagram for Simulation Three

We will use three simulation results to verify the performance of the filter: a Bode plot of the filter transfer function, a time-domain (transient) plot of the input and output, and an inspection of the Fourier series for the input and output voltages (center frequency = 50 Hz). Figure 15.7 shows the Bode plot of the filter demonstrating that the notch indeed occurs at 50 Hz and signals at 250 Hz are essentially unaltered. Transient simulation results are shown in Figure 15.8. After about 70 ms, all transients have died out, and the output is a 250 Hz sinewave - no 50 Hz noise is present. From the Fourier series data, listed at the bottom of the next page, we see that the 50 Hz component of the output voltage is roughly 1% of the input voltage component. Also, the 250 Hz component (5th harmonic) is within 1% of the input value.. Looks like our filter is a winner.

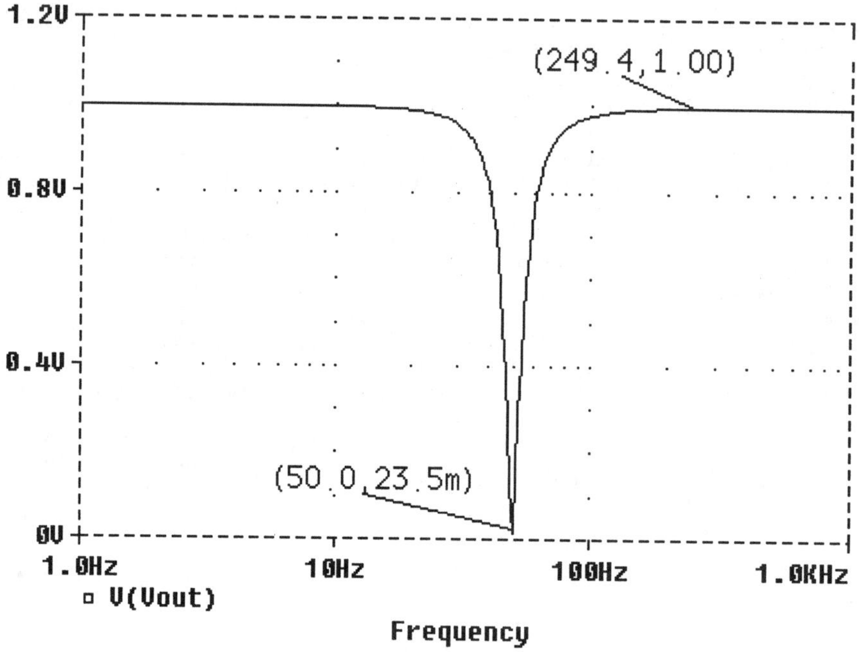

Figure 15.7. Bode plot for the filter in Simulation Three.

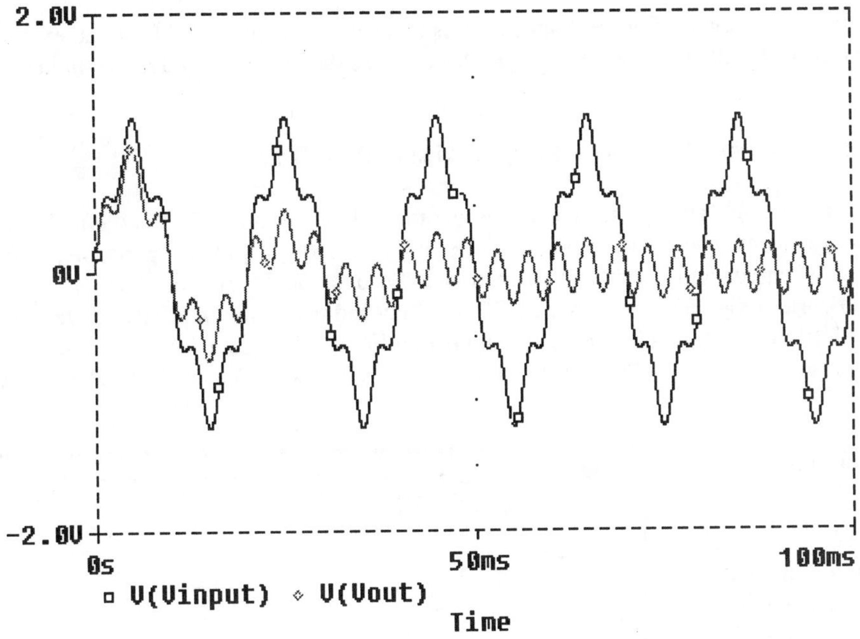

Figure 15.8. Transient results for the filter in Simulation Three.

```
FOURIER COMPONENTS OF TRANSIENT RESPONSE V(Vout)

 DC COMPONENT =   1.692078E-03

 HARMONIC    FREQUENCY     FOURIER      NORMALIZED      PHASE        NORMALIZED
    NO         (HZ)       COMPONENT     COMPONENT       (DEG)        PHASE (DEG)

     1      5.000E+01     1.113E-02     1.000E+00    -1.721E+01     0.000E+00
     2      1.000E+02     1.303E-03     1.171E-01    -1.101E+02    -9.293E+01
     3      1.500E+02     5.779E-04     5.191E-02    -1.265E+02    -1.093E+02
     4      2.000E+02     3.613E-04     3.246E-02    -1.366E+02    -1.194E+02
     5      2.500E+02     1.993E-01     1.790E+01     3.751E+00     2.096E+01

     TOTAL HARMONIC DISTORTION =    1.790308E+03 PERCENT

FOURIER COMPONENTS OF TRANSIENT RESPONSE V(Vinput)

 DC COMPONENT =  -2.916756E-07

 HARMONIC    FREQUENCY     FOURIER      NORMALIZED      PHASE        NORMALIZED
    NO         (HZ)       COMPONENT     COMPONENT       (DEG)        PHASE (DEG)

     1      5.000E+01     1.000E+00     1.000E+00    -2.014E-04     0.000E+00
     2      1.000E+02     5.430E-07     5.430E-07     3.242E+01     3.242E+01
     3      1.500E+02     6.536E-07     6.536E-07    -7.740E+01    -7.740E+01
     4      2.000E+02     5.071E-07     5.071E-07     1.741E+02     1.741E+02
     5      2.500E+02     1.999E-01     1.999E-01    -8.944E-04    -6.930E-04

     TOTAL HARMONIC DISTORTION =    1.999369E+01 PERCENT
```

THE FOURIER TRANSFORM AND PSPICE

Sometimes we prefer to view the Fourier characteristics of a network in graphical form rather than tabular. We want a Fourier transform rather than a series. In such cases, use the Fast Fourier Transform in the PROBE utility.

The Fast Fourier Transform (FFT) is activated by clicking on the FFT hotbutton. 

Before we look at some FFT examples, we must discuss the differences in the FFT and Fourier Series algorithms. First, as mentioned earlier, a Fourier Series is calculated using the last fundamental period of data. In the FFT, all data is used and is assumed to be periodic. Second, the duration of the transient simulation has no effect in the Fourier series accuracy. In the FFT, accuracy and resolution in particular are directly related to the simulation Final Time. Also, choosing a the simulation Final Time that is an integer number of waveform periods will improve the accuracy of the results. Finally, the FFT feature can be used on an expression such as V(Vo)*I(R1).

MORE SIMULATIONS

Simulation Four

Let's use the circuit in Figure 15.1 to investigate the effect of simulation duration on FFT resolution. From Simulation One, we know that the output voltage has components at dc, 1 kHz and 3 kHz with magnitudes of 5V, 1V and 0.5V respectively. We will perform transient simulations of 2 ms (2 fundamental periods) and 200 ms duration and compare the FFT results. In both simulations, the Step Ceiling was set equal to the Final Time/100. This ensures that both simulations have the same number of data points. Figure 15.9 shows the resulting FFTs. Obviously, the longer simulation has the better resolution. Of course, as the Final Time increases, simulations takes longer to run. The 2ms simulation process can be viewed in the Visual Tutor, FFT.EXE.

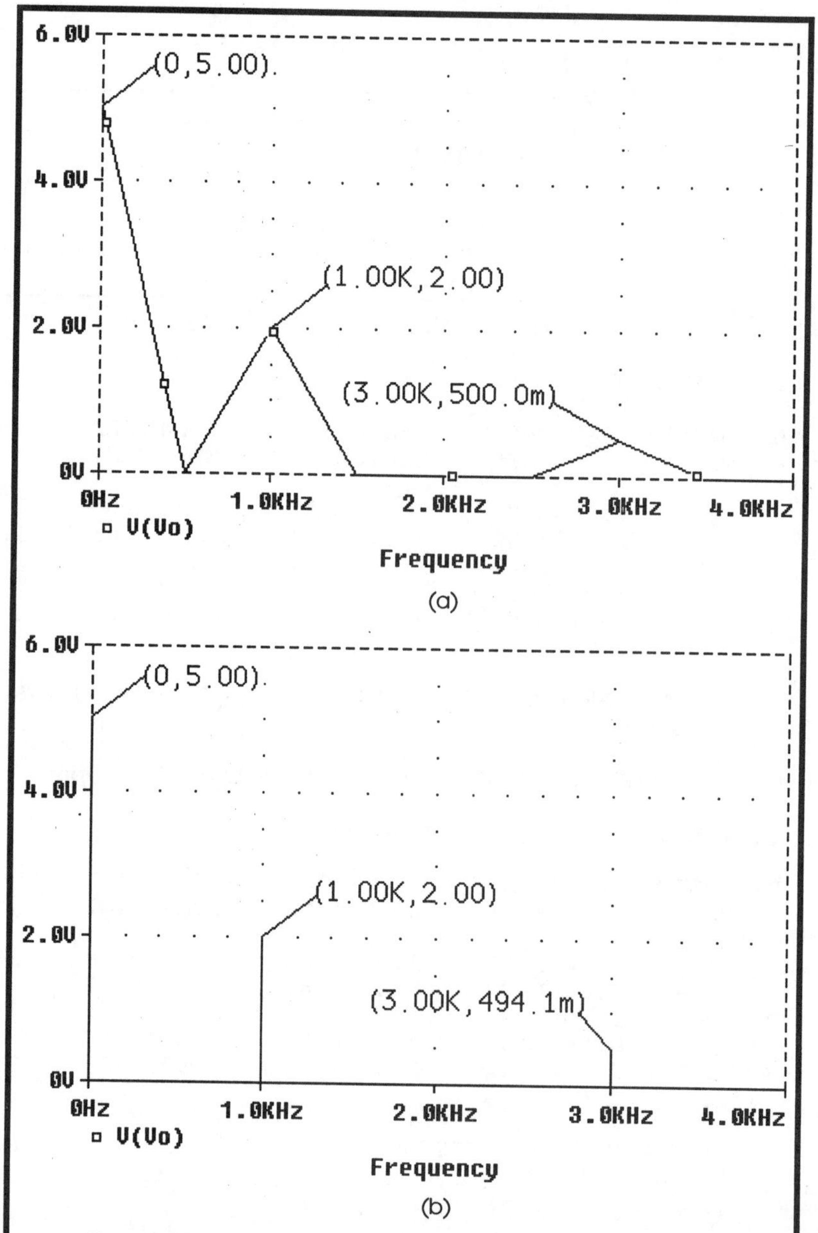

Figure 15.9. FFT results for Final Times of (a) 2 ms and (b) 200 ms.

Simulation Five

For our last PSpice simulation, we will return to the AM radio tuner filter discussed in Simulation Four of Chapter 14. In that example, we modeled a filter that could pass signals at 1.59 MHz while rejecting 1.61 MHz signals. Here, we will use the FFT feature in PROBE to examine the pre- and post- filtered frequency content. For convenience, the filter is redrawn in Figure 15.10. In the original circuit, station KOOL (1590 kHz) was modeled by the sinusoidal voltage source V_{KOOL}, which was fine fore ac simulations. Since we will now be performing transient simulations, we will model KOOL with a true amplitude modulated signal.

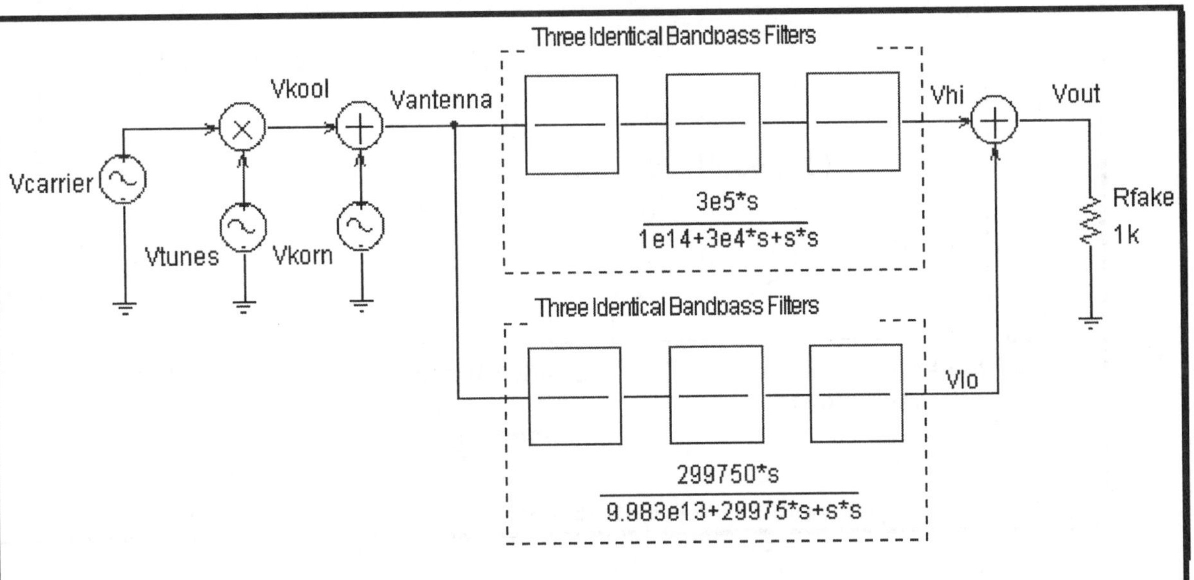

Figure 15.10. Schematics Diagram for the AM tuner filter discussed in Simulation Five.

The parts $V_{CARRIER}$, V_{TUNES} and the multiplier part (MULT in the ABM library) produce the voltage V_{KOOL} shown in Figure 15.11. To see how the AM effect is created, we simply multiply $V_{CARRIER}$ times V_{KOOL}.

$$V_{CARRIER} = 1\sin(2\pi \cdot 1.59\text{x}10^6 t)\ \mu\text{V} \qquad V_{TUNES} = 1 + 0.5\sin(2\pi \cdot 10^4 t)\ \text{V}$$

$$V_{KOOL} = V_{TUNES} * V_{CARRIER} = [1 + 0.5\sin(2\pi \cdot 10^4 t)]\sin(2\pi \cdot 1.59\text{x}10^6 t)\ \mu\text{V}$$

Obviously, as V_{TUNES} oscillates, the amplitude of the 1.59 MHz signal varies from 0.5 to 1.5 microvolts - that's amplitude modulation. As before, the source V_{KORN} models the nearby station KORN - 1610. Next, V_{KOOL} and V_{KORN} are added together to create $V_{ANTENNA}$, the voltage on the radio's antenna.

We want to see the frequency content at $V_{ANTENNA}$ and V_{OUT}.

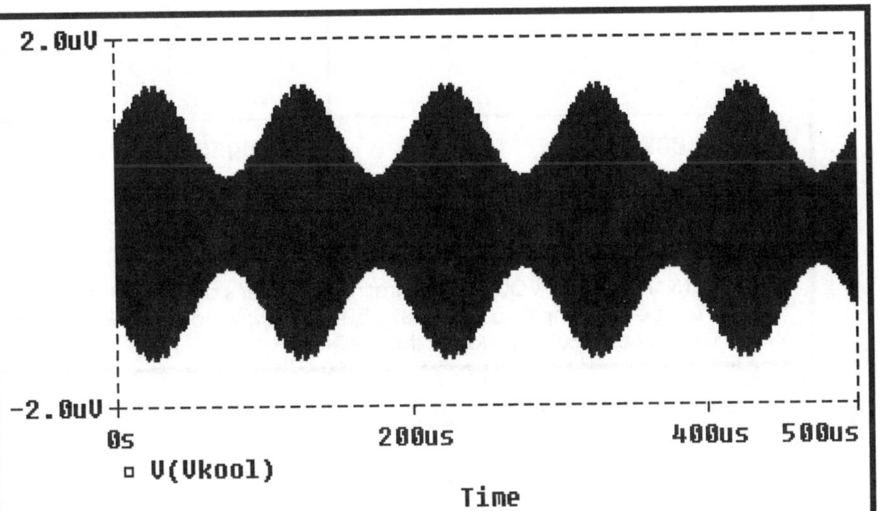

Figure 15.11. A typical AM signal with carrier and audio frequencies of 1.59 MHz and 10kHz respectively.

This simulation is very demanding. We hope to resolve 1.59 MHz from 1.6 MHz. This will require a transient simulation Final Time of many periods. In particular, we need many periods of the slowest signal we must resolve. That would be V_{TUNES} at a frequency of 10 kHz (period of 100 μs). Therefore, we have chosen a Final Time of 10 ms (100 periods). That's about 16,000 periods of V_{KOOL}! After several minutes of simulation, we obtain the FFT plots for $V_{ANTENNA}$ and V_{OUT} shown in Figure 15.12 and Figure 15.13. While both the KOOL and KORN signals are amplified, in the end, KOOL is bigger that KORN. This is exactly what we want.

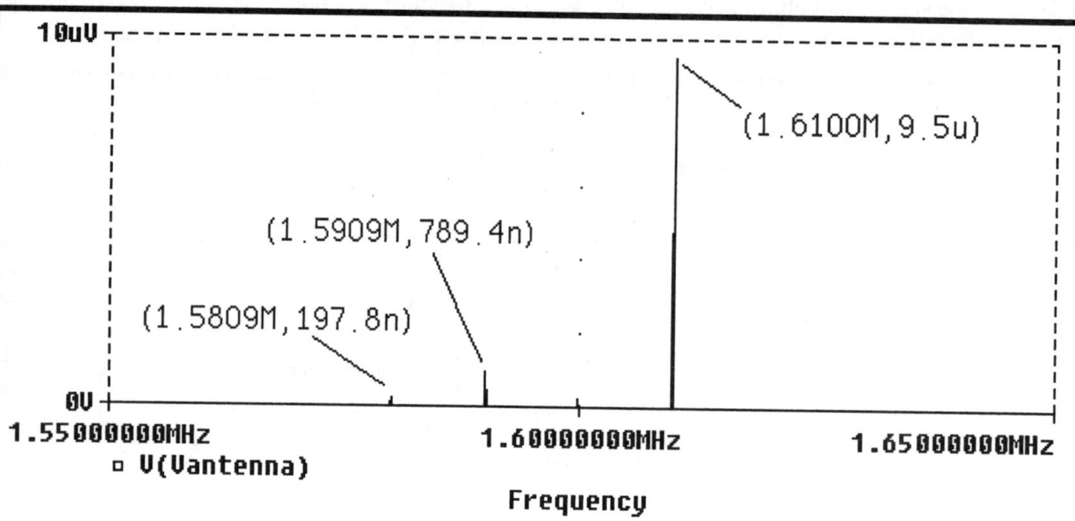

Figure 15.12. FFT plot for VANTENNA. Note the audio components symmetric around the KOOL carrier.

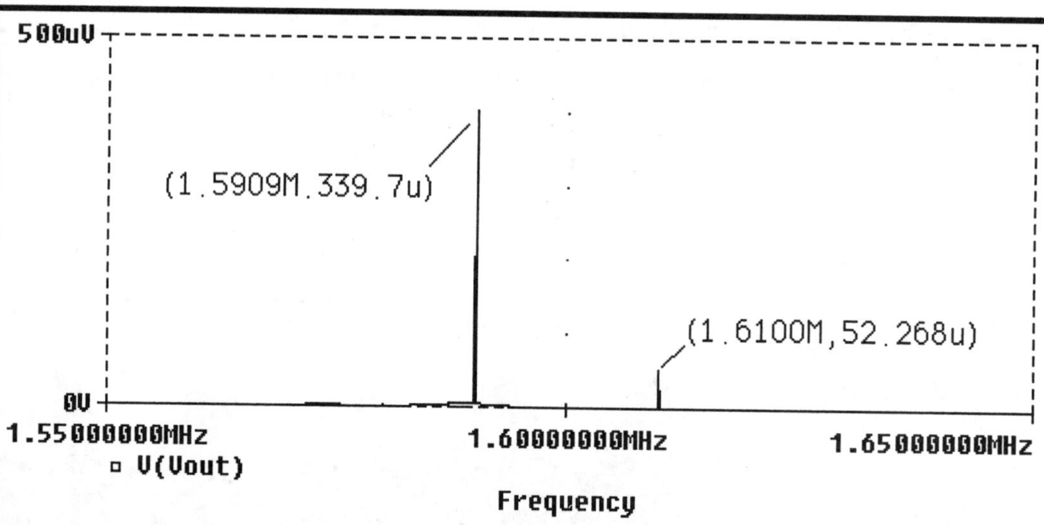

Figure 15.13. FFT plot for VOUT demonstrating the tuner's performance. Although the KORN signal has been amplified by about 5.5, the KOOL signal has been amplified by 430. Now VKOOL is about 6.5 times larger that VKORN.

EWB SIMULATION

EWB can calculate the Fourier series (magnitude and phase) of a time domain voltage. To request a Fourier series, simply select Fourier from the Analysis menu. As in PSpice, you must specify a center frequency and the number of harmonics you desire. Unlike PSpice, the results are plotted rather than listed. To demonstrate, we will use EWB to find the Fourier series of the waveform in Figure 15.14. The simple EWB circuit in Figure 15.16 will do nicely. The voltage source is a CLOCK part set to a magnitude of 10-V, a frequency of 1 Hz and a duty cycle of 50%. Duty cycle is the amount of time the signal is high per cycle divided by the period. When the duty cycle is 50%, the voltage is high half the time and low the other half.

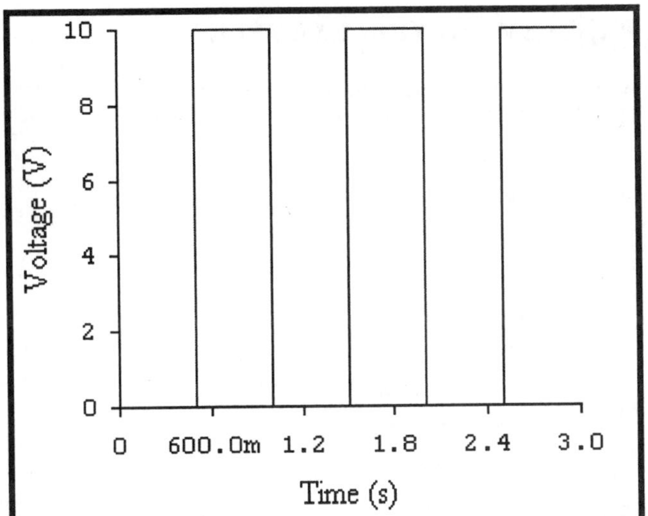

Figure 15.14. The EWB simulation voltage waveform.

The Fourier series for a request of 10 harmonics is shown in Figure 15.15. Comparing these results to exact answers reveals errors in the simulations harmonic of about 4%. However, if more harmonics are requested, the error decreases. Table 15.1 lists the exact value of the first 9 harmonics, the simulation results for request of 10 and 100 harmonics and the resulting simulation errors.

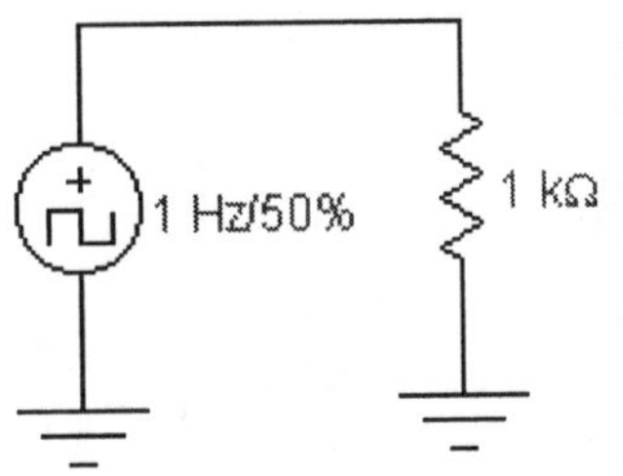

Figure 15.16. A simple EWB circuit that produces the desired waveform.

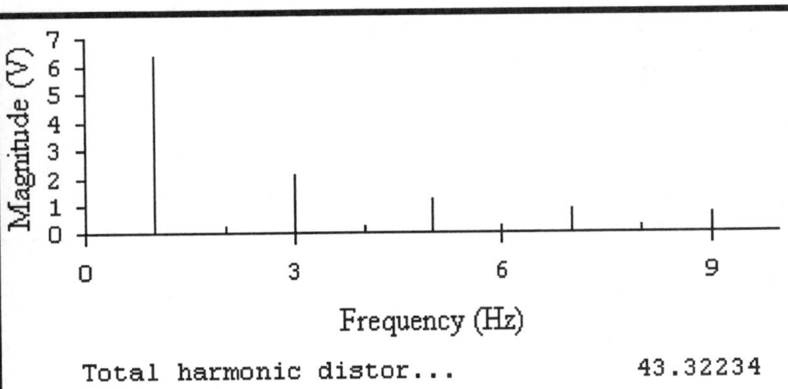

Figure 15.15. EWB Fourier analysis results (10 harmonics were requested).

TABLE 15.1

The Effect of the Requested Number of Harmonics on the Accuracy of EWB Fourier Analysis Results

Harmonic	Exact	10 Harmonic Sim.	Error	100 Harmonic Sim.	Error
dc	5.000	4.828	-3.44 %	4.974	-1.06 %
1	6.366	6.328	-0.6	6.361	-0.08
2	0	0.276	NA	0.099	NA
3	2.122	2.095	-1.27	2.118	-0.19
4	0	0.276	NA	0.099	NA
5	1.273	1.244	-1.28	1.267	-0.47
6	0	0.276	NA	0.099	NA
7	0.909	0.875	-3.74	0.902	-0.77
8	0	0.276	NA	0.099	NA
9	0.707	0.665	-5.94	0.698	-1.27

PROBLEM SOLVING EXAMPLES

1. Find the exponential Fourier series for the function f(t) with period 1 where f(t) is defined as

$$f(t) = e^t \qquad 0 \le t < 1$$

Solution:

$$c_n = \int_0^1 e^t e^{-jn\omega_o t} dt, \; T=1 \text{ and } \omega_o = 2\pi$$

$$c_n = \int_0^1 e^{(1-jn\omega_o)t} dt = \frac{e^{(1-jn\omega_o)}-1}{1-jn\omega_o}$$

$$= \frac{e-1}{1-jn\omega_o}$$

$$\text{Therefore } f(t) = \sum_{n=-\infty}^{\infty} \frac{e-1}{1-j2\pi n} e^{j2\pi nt}$$

2. Show that the exponential Fourier series for the periodic function $f(t) = e^{-t}$, $-1 < t < 1$ with period 2, can be expressed as

$$f(t) = \sum_{n=-\infty}^{\infty} \left[\frac{(-1)^n}{1+j\pi n}\right]\left[\frac{e^1-e^{-1}}{2}\right] e^{jn\omega_o t}$$

Solution:

$$c_n = \frac{1}{2}\int_{-1}^{1} e^{-t} e^{-jn\omega_o t} dt$$

$$= \frac{1}{2}\left[\frac{e^{-(1+jn\omega_o)t}}{-(1+jn\omega_o)}\right]_{-1}^{1}$$

$$= \frac{1}{2}\left[\frac{e^1 e^{jn\pi} - e^{-1} e^{-jn\pi}}{1+jn\pi}\right]$$

Since $e^{jn\pi} = \cos n\pi + j\sin n\pi = (-1)^n$ and $e^{-jn\pi} = (-1)^n$

$$c_n = \frac{(-1)^n}{2(1+jn\pi)}(e^1 - e^{-1})$$

3. Find the exponential Fourier series for the voltage waveform below

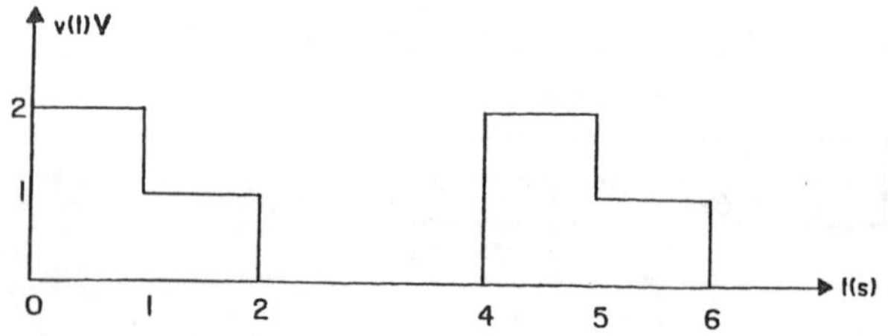

Solution:

$$c_o = \frac{3}{4} \text{ and } \omega_o = \frac{\pi}{2}$$

$$c_n = \frac{1}{4}\left[\int_0^1 2e^{-jn\omega_o t} dt + \int_1^2 e^{-jn\omega_o t} dt\right]$$

$$= \frac{1}{4}\left[\frac{2}{-jn\omega_o} e^{-jn\omega_o t}\Big|_0^1 - \frac{1}{jn\omega_o} e^{-jn\omega_o t}\Big|_1^2\right]$$

$$= \frac{1}{j\pi n}\left[1 - \frac{e^{-j\frac{n\pi}{2}}}{2} - \frac{e^{-jn\pi}}{2}\right]$$

4. If the waveform in Problem 3 is time delayed 1 second, we obtain the following waveform. Compute the exponential Fourier coefficients for this latter waveform and show that they differ from those for the waveform in problem 3 by n(90°).

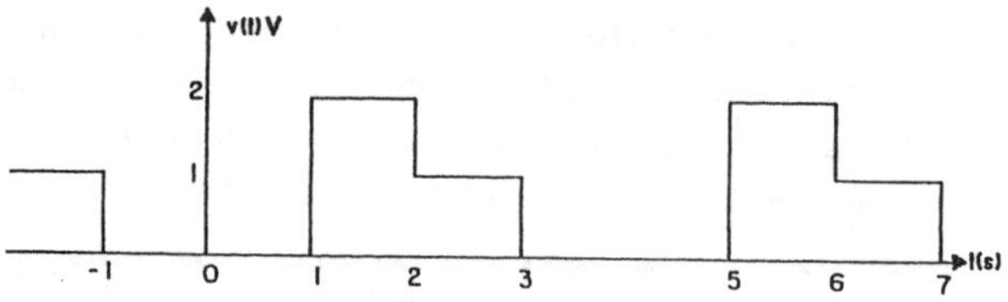

Solution: $c_o = \frac{3}{4}$, $\omega_o = \frac{\pi}{2}$

$$c_n = \frac{1}{4}\left[\int_1^2 2e^{-j\frac{n\pi}{2}t} dt + \int_2^3 e^{-j\frac{n\pi}{2}t} dt\right]$$

$$= \frac{1}{4}\left[\frac{2}{-jn\frac{\pi}{2}} e^{-j\frac{n\pi}{2}t}\Big|_1^2 + \frac{2}{-jn\frac{\pi}{2}} e^{-j\frac{n\pi}{2}t}\Big|_2^3\right]$$

$$= -\frac{1}{jn\pi}\left[\frac{e^{-jn\pi}}{2} - e^{-j\frac{n\pi}{2}} + e^{-j3\frac{n\pi}{2}}\right]$$

$$= \frac{1}{jn\pi}\left[1 - \frac{e^{-j\frac{n\pi}{2}}}{2} - \frac{e^{-jn\pi}}{2}\right] e^{-j\frac{n\pi}{2}}$$

Note that this term differs from the one in the previous problem by $e^{-j\frac{n\pi}{2}}$ or n90°.

5. Use the trigonometric Fourier series to show that the Fourier series for the following signal is given by the expression

$$v(t) = 1 - \frac{2}{\pi}\sum_{n=1}^{\infty} \frac{1}{n}\sin(2\pi nt)V$$

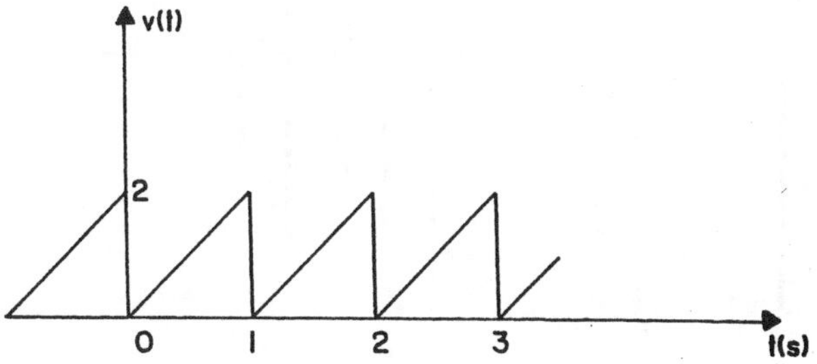

Solution: $v(t) = 2t$

$$a_o = \frac{1}{1} \int_0^1 2tdt = 1$$

$$a_n = \frac{2}{1} \int_0^1 2t\cos 2\pi ntdt$$

$$= 4\left[\frac{t}{2\pi n} \sin 2\pi nt \Big|_0^1 - \int_0^1 \frac{1}{2\pi n} \sin 2\pi ntdt \right]$$

$$= 4\left[\frac{1}{(2\pi n)^2} (\cos 2\pi n - 1) \right] = 0$$

$$b_n = \frac{2}{1} \int_0^1 2t\sin 2\pi ntdt$$

$$= 4\left[-\frac{t}{2\pi n} \cos 2\pi nt \Big|_0^1 - \int_0^1 -\frac{1}{2\pi n} \cos 2\pi ntdt \right]$$

$$= 4\left[-\frac{1}{2\pi n} \cos 2\pi n \right] = -\frac{2}{n\pi}$$

Therefore

$$v(t) = 1 - \sum_{n=1}^{\infty} \frac{2}{n\pi} \sin 2\pi nt V$$

6. Derive the trigonometric Fourier series for the function $v(t) = |\sin t|$ shown below.

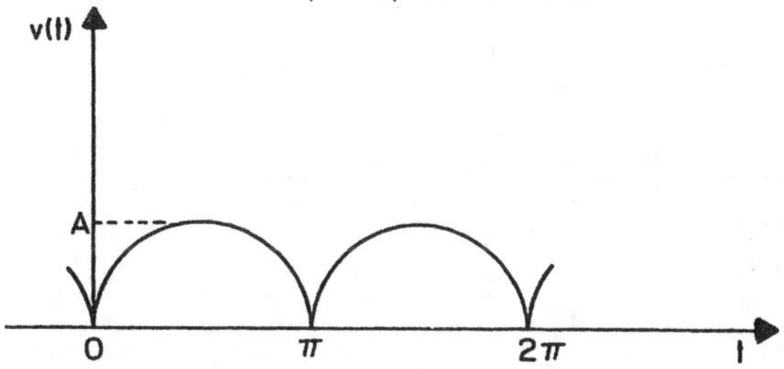

Solution: This is an even function, $T = \pi$ and $\omega_o = 2$.

$$a_o = \frac{2}{\pi} \int_0^{\frac{\pi}{2}} A\sin tdt = \frac{2A}{\pi} \left[-\cos t \Big|_0^{\frac{\pi}{2}} \right] = \frac{2A}{\pi}$$

$$a_n = \frac{4}{\pi} \int_0^{\frac{\pi}{2}} A\sin t\cos 2\pi ntdt$$

$$= \frac{4A}{\pi} \left[-\frac{\cos(1-2n)t}{2(1-2n)} \Big|_0^{\frac{\pi}{2}} - \frac{\cos(1+2n)t}{2(1+2n)} \Big|_0^{\frac{\pi}{2}} \right]$$

$$= \frac{2A}{\pi} \left[\frac{1}{2n+1} - \frac{1}{2n-1} \right] = \frac{-4A}{\pi(4n^2-1)}$$

Therefore

$$v(t) = \frac{2A}{\pi} + \sum_{n=1}^{\infty} \frac{-4A}{\pi(4n^2-1)} \cos 2nt$$

7. Sketch the missing section in the following periodic function between T/2 and T which will make (a) $a_o = 0$, $a_n = 0$ (b) $b_n = 0$.

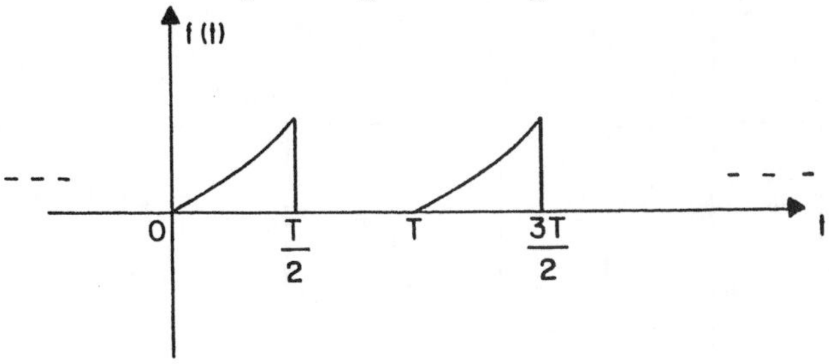

Solution: (a) If $a_o = 0$ and $a_n = 0$, the function must be odd. Therefore f(t) looks like

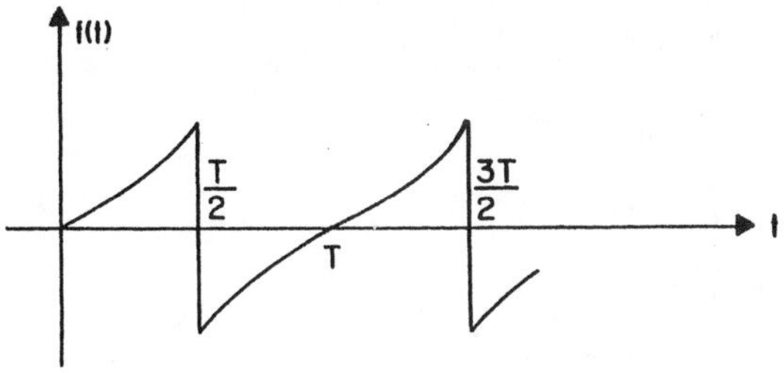

(b) If $b_n = 0$, the function must be even

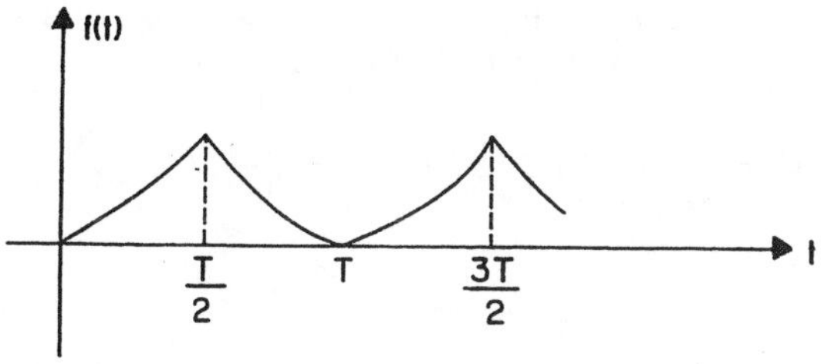

8. Determine which of the Fourier coefficients are zero for f(t) shown below.

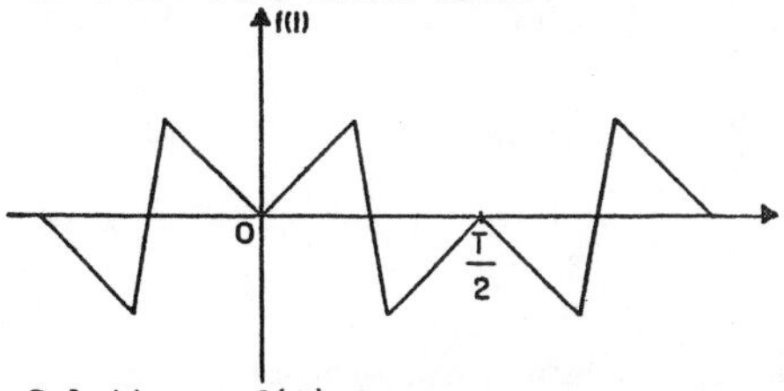

Solution: f(t) is even and has half wave symmetry therefore $b_n = 0$, $a_o = 0$ and $a_n = 0$ for n even.

9. Find the trigonometric Fourier series for the voltage waveform shown below.

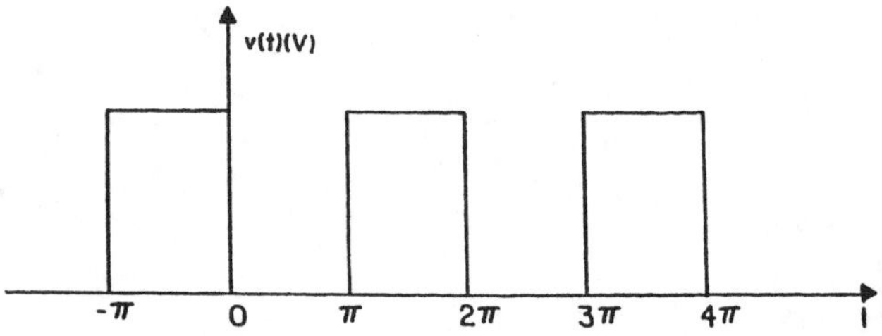

Solution: Note that this waveform can be analyzed by finding the series for the waveform

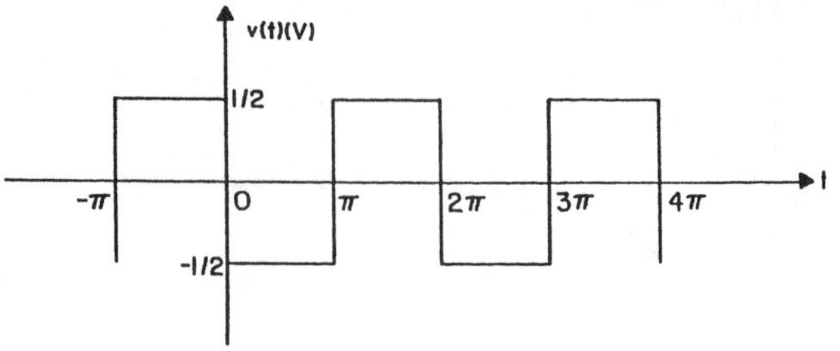

$\omega_o = 1$ and the waveform has odd symmetry.

$$b_n = \frac{4}{2\pi}\int_0^{\pi} -\frac{1}{2}\sin nt\,dt = \frac{1}{n\pi}\cos nt\Big|_0^{\pi}$$

$$= \frac{1}{n\pi}(\cos n\pi - 1)$$

Therefore

$$v(t) = \frac{1}{2} + \sum_{n=1}^{\infty} \frac{1}{n\pi}(\cos n\pi - 1)\sin nt\,\text{V}$$

10. Plot the first four terms of the amplitudes and phase spectra for the following signal.

$$f(t) = \sum_{\substack{n=1\\ n\ \text{odd}}}^{\infty} \frac{-2}{n\pi}\sin\frac{n\pi}{2}\cos n\omega_o t + \frac{6}{n\pi}\sin n\omega_o t$$

Solution: The terms are

$$D_1\angle\theta_1 = -\frac{2}{\pi} - j\frac{6}{\pi} = 2\angle{-108^\circ}$$

$$D_3\angle\theta_3 = \frac{2}{3\pi} - j\frac{6}{3\pi} = 0.66\angle{-71^\circ}$$

$$D_5\angle\theta_5 = \frac{-2}{5\pi} - j\frac{6}{5\pi} = 0.40\angle{-108^\circ}$$

$$D_7\angle\theta_7 = \frac{2}{7\pi} - j\frac{6}{7\pi} = 0.28\angle{-71^\circ}$$

Therefore the amplitude and phase spectra are

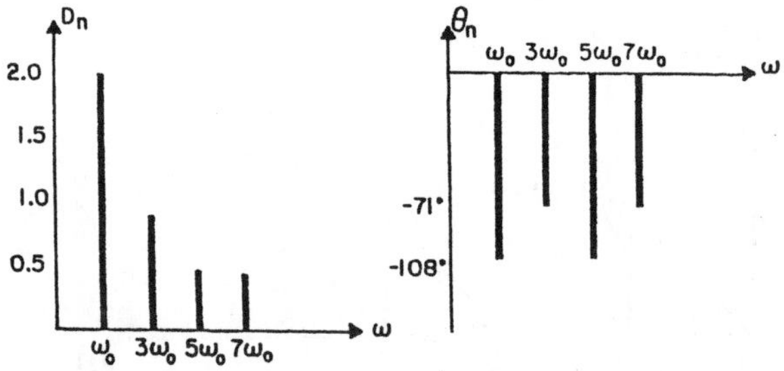

11. Compute the first four terms of the amplitude and phase spectra for the periodic signal defined in Problem 9.

Solution: See the solution to problem 9.

$$v(t) = \frac{1}{2} + \sum_{n=1}^{\infty} \frac{1}{n\pi}(\cos n\pi - 1)\sin nt$$

$$a_o = \frac{1}{2}$$

$$D_n\angle\theta_n = a_n - jb_n = -\frac{j}{2\pi}(\cos n\pi - 1)$$

$$D_1\angle\theta_1 = 0.637\angle{90^\circ}$$

$$D_2\angle\theta_2 = 0$$

$$D_3\angle\theta_3 = 0.212\angle{90^\circ}$$

12. For the amplitude and phase spectra shown below, express the function as a sum of sinusoidal functions using the Fourier series coefficients a_n and b_n.

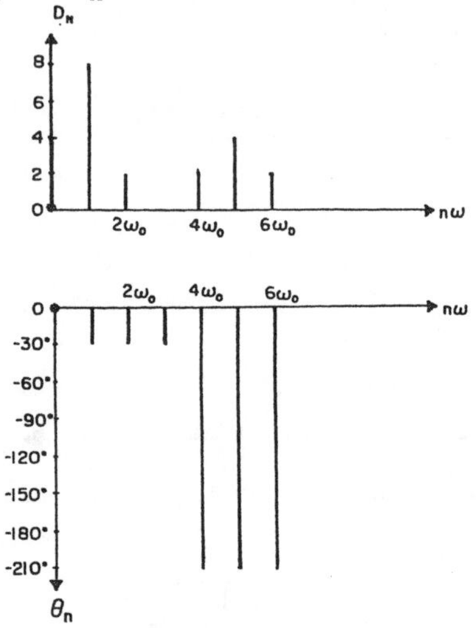

Solution: $D_n\angle\theta_n = a_n - jb_n$

$$D_1\angle\theta_1 = 8\angle{-30^\circ} = 6.93 - j4$$

$$D_2\angle\theta_2 = 2\angle{-30^\circ} = 1.73 - j1$$

$$D_3\angle\theta_3 = 0\angle{30^\circ} = 0$$

$$D_4\angle\theta_4 = 2\angle{-210^\circ} = -1.73 + j1$$

$D_5\angle\theta_5 = 4\angle -210° = -3.46+j2$

$D_6\angle\theta_6 = 2\angle -210° = -1.73+j1$

Therefore

$$f(t) = 4+6.93\cos\omega_o t-4\sin\omega_o t+1.73\cos 2\omega_o t$$
$$-\sin 2\omega_o t-1.73\cos 4\omega_o t+\sin 4\omega_o t+\ldots$$

13. Write a function f(t) in cosine form from the amplitude and phase spectra shown in problem 12.

Solution: See the solution to problem 12.

$$f(t) = 4+8\cos(\omega_o t-30°)+2\cos(2\omega_o t-30°)$$
$$+2\cos(4\omega_o t-210°)+4\cos(5\omega_o t-210°)$$
$$+2\cos(6\omega_o t-210°)$$

14. The rectified ac signal below is the input to the following low-pass filter. Determine the equation for the output signal $v_o(t)$.

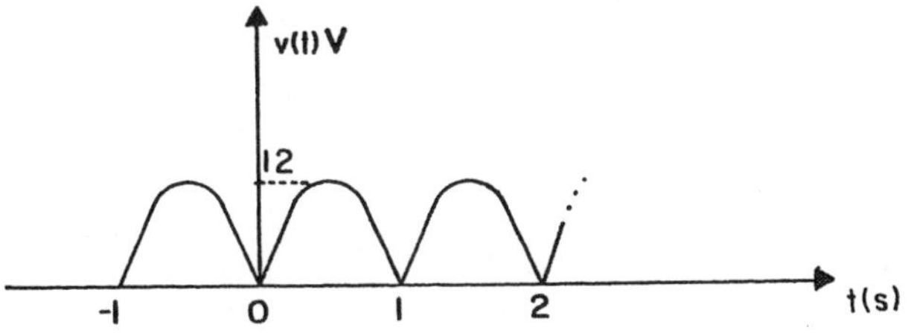

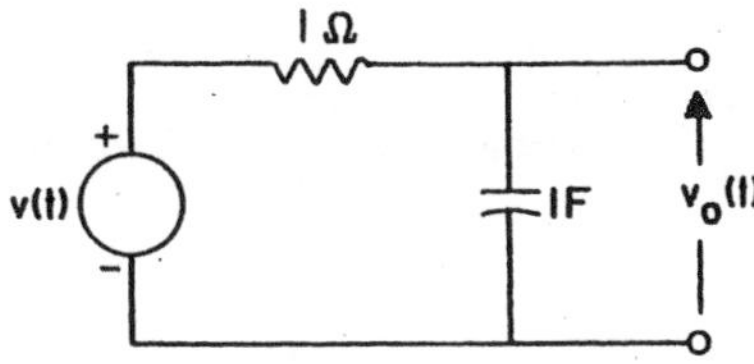

Solution: From the table of Fourier series for specific waveforms in Chapter 16 we find that.

$$v(t) = \frac{24}{\pi} + \sum_{n=1}^{\infty} \frac{48}{\pi(1-4n^2)} \cos 2\pi nt \text{ V}$$

for the network

$$\mathbf{V}(j\omega) = \frac{\mathbf{V}}{1+j\omega} \quad \text{or} \quad \mathbf{V}_o(n) = \frac{\mathbf{V}(n)}{1+j2\pi n}$$

Therefore

$$v_o(t) = \frac{24}{\pi} + \sum_{n=1}^{\infty} \frac{48}{\pi(1-4n^2)} \frac{1}{(1+4\pi^2n^2)^{1/2}}$$
$$\cos(2\pi nt-\tan^{-1}\frac{2\pi n}{1})\text{V}$$

15. The voltage $v_s(t)$ shown below is applied to the following circuit. Determine the expression for the steady-state voltage $v_o(t)$ using the first eight harmonics.

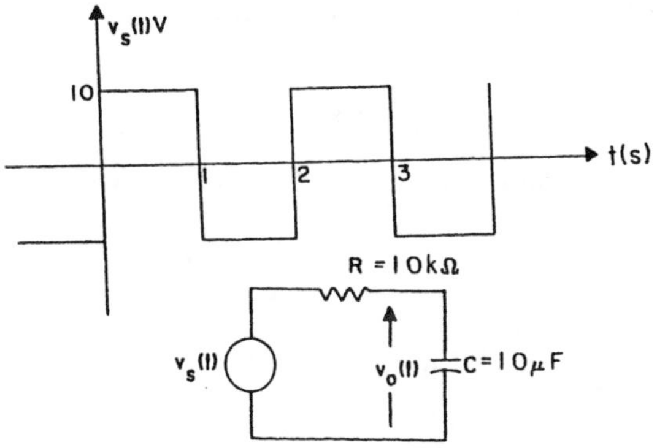

Solution: From the table of Fourier series

$$v_s(t) = \sum_{\substack{n=1\\ \text{odd}}}^{\infty} \frac{40}{n\pi} \sin n\pi t \quad \text{and} \quad \mathbf{V}_o(n) = \frac{\mathbf{V}_s(n)}{1+jn\omega_o RC} =$$

$(A_n\angle\theta_n)\mathbf{V}_s(n)$

$A_1\angle\theta_1 = 0.954\angle -17.4°$ $\quad A_5\angle\theta_5 = 0.537\angle -57.5°$

$A_2\angle\theta_2 = 0.728\angle -43.3°$ $\quad A_7\angle\theta_7 = 0.413\angle -65.5°$

$$v_o(t) = 12.15\sin(\pi t-17.4°)+3.08\sin(3\pi t-43.3°)$$
$$+1.37\sin(5\pi t-57.6°)+0.751\sin(7\pi t-65.5°)\text{V}$$

16. Find the Fourier transform of the function $f(t) = \sin\omega_o t$.

Solution:

$$F(\omega) = \int_{-\infty}^{\infty} \frac{e^{j\omega_o t}-e^{-j\omega_o t}}{2j} e^{j\omega t}dt$$

$$= -\frac{j}{2}\int_{-\infty}^{\infty} (e^{j(\omega-\omega_o)t}-e^{-j(\omega+\omega_o)t})dt$$

$$= -j\pi\delta(\omega-\omega_o)+j\pi\delta(\omega+\omega_o)$$

17. Derive the following properties of the Fourier transform.

(a) $\mathcal{F}[e^{j\omega_o t}f(t)] = \mathbf{F}(\omega-\omega_o)$

(b) $\mathcal{F}[\cos\omega_o t f(t)] = \frac{1}{2}[\mathbf{F}(\omega-\omega_o)+\mathbf{F}(\omega+\omega_o)]$

Solution:

(a) $$\mathcal{F}[e^{j\omega_o t}f(t)] = \int_{-\infty}^{\infty} f(t)e^{j\omega_o t}e^{-j\omega t}dt$$

$$= \int_{-\infty}^{\infty} f(t)e^{-j(\omega-\omega_o)t}dt$$

$$= \mathbf{F}(\omega-\omega_o)$$

(b) $\mathcal{F}[\cos\omega_o t f(t)] = \int_{-\infty}^{\infty} f(t)\left[\frac{e^{j\omega_o t}+e^{-j\omega_o t}}{2}\right]e^{-j\omega t}dt$

$$= \tfrac{1}{2}[\mathbf{F}(\omega-\omega_o)+\mathbf{F}(\omega+\omega_o)]$$

18. Find the Fourier transform of the function $f(t) = te^{-at}u(t)$

Solution:

$$\mathbf{F}(\omega) = \int_{-\infty}^{\infty} te^{-at}u(t)e^{-j\omega t}dt$$

$$= \int_0^{\infty} te^{-at}e^{-j\omega t}dt$$

integrating by parts

$$\mathbf{F}(\omega) = \frac{-t}{a+j\omega}e^{-(a+j\omega)t}\Big|_0^{\infty} + \int_0^{\infty}\frac{e^{-(a+j\omega)t}}{a+j\omega}dt$$

$$= \frac{1}{(a+j\omega)^2}$$

19. Show that

(a) $\mathcal{F}[f(at)] = (1/a)\mathbf{F}(\omega/a)$ for $a>0$

(b) $\mathcal{F}[f(t-a)] = e^{-j\omega a}\mathbf{F}(\omega)$

Solution:

(a) $u = at$, $du = adt$, and the limits of integration are unchanged

$$\mathcal{F}[f(at)] = \int_{-\infty}^{\infty} f(u)e^{-j\omega(\frac{u}{a})}\frac{du}{a}$$

$$= \frac{1}{a}\int_{-\infty}^{\infty} f(u)e^{-j\frac{\omega u}{a}}du$$

$$= \frac{1}{a}\mathbf{F}(\tfrac{\omega}{a}),\ a>0$$

(b) $u = t-a$, $du = dt$

$$\mathcal{F}[f(t-a)] = \int_{-\infty}^{\infty} f(u)e^{-j\omega(u+a)}du$$

$$= e^{-j\omega a}\int_{-\infty}^{\infty} f(u)e^{-j\omega u}du$$

$$= e^{-j\omega a}\mathbf{F}(\omega)$$

20. Given that $\mathbf{F}(\omega) = \dfrac{1+j\omega}{\omega^2+j8\omega+12}$ find the Fourier transform of

(a) $f(2t)$ (c) $8f(t/2)$
(b) $f(t-2)$ (d) $f(4t-1)$

Solution:

(a) $[f(2t)] = \frac{1}{2}\mathbf{F}(\frac{\omega}{2}) = \dfrac{2+j\omega}{\omega^2+16\omega j+48}$

(b) $[f(t-2)] = e^{-j\omega 2}\mathbf{F}(\omega) = e^{-j2\omega}\left[\dfrac{1+j\omega}{\omega^2+8\omega j+12}\right]$

(c) $[8f(\frac{t}{2})] = (8)(2)\mathbf{F}(2\omega) = 16\left[\dfrac{1+j2\omega}{4\omega^2+16\omega j+12}\right]$

(d) $[f(4t-1)] = \dfrac{e^{-j\frac{\omega}{4}}}{4}\mathbf{F}(\frac{\omega}{4}) = \dfrac{e^{-j\frac{\omega}{4}}}{4}\left[\dfrac{1+j\frac{\omega}{4}}{\frac{\omega^2}{16}+j2\omega+12}\right]$

21. Determine the total 1-Ω energy content of the signal $v(t) = (e^{-t}-e^{-2t})u(t)$.

Solution:

$$W_i = \int_0^{\infty}(e^{-t}-e^{-2t})^2dt$$

$$= \int_0^{\infty}(e^{-2t}-2e^{-3t}+e^{-4t})dt$$

$$= -\frac{1}{2}e^{-2t}+\frac{2}{3}e^{-3t}-\frac{1}{4}e^{-4t}\Big|_0^{\infty}$$

$$= \frac{1}{12}\ \text{J}$$

22. Determine the 1-Ω energy content of the signal in Problem 21 in the frequency band from 0 to 1 rad/sec.

Solution:

$$v(t) = (e^{-t}-e^{-2t})u(t)$$

$$\mathbf{V}(j\omega) = \frac{1}{j\omega+1}-\frac{1}{j\omega+2} = \frac{1}{(j\omega+1)(j\omega+2)}$$

$$|\mathbf{V}(j\omega)|^2 = \frac{1}{(\omega^2+1)(\omega^2+4)}$$

$$W_o = 2(\tfrac{1}{2\pi})\int_0^1\frac{d\omega}{(\omega^2+1)(\omega^2+4)}$$

$$= \frac{1}{3\pi}\left[\int_0^1\frac{d\omega}{\omega^2+1}-\int_0^1\frac{d\omega}{\omega^2+4}\right]$$

$$= \frac{1}{3\pi}[\tan^{-1}\omega-\tfrac{1}{2}\tan^{-1}\tfrac{\omega}{2}]$$

$$= \frac{1}{3\pi}[\tfrac{\pi}{4}-\tfrac{1}{2}(0.46)]$$

$$= 0.06\ \text{J}$$